Electrophysiology

Enzyme Assays

Essential Developmental Biology

Essential Molecular Biology I and II

Experimental Neuroanatomy

Extracellular Matrix

Fermentation

Flow Cytometry (2nd edition)

Gas Chromatography

Gel Electrophoresis of Nucleic Acids (2nd edition)

Gel Electrophoresis of Proteins (2nd edition)

Gene Targeting

Gene Transcription

Genome Analysis

Glycobiology

Growth Factors

Haemopoiesis

Histocompatibility Testing

HPLC of Macromolecules

HPLC of Small Molecules

Human Cytogenetics I and II (2nd edition)

Human Genetic Disease Analysis

Immobilised Cells and Enzymes

Immunocytochemistry

In Situ Hybridization

Iodinated Density Gradient Media

Light Microscopy in Biology

Lipid Analysis

Lipid Modification of Proteins

Lipoprotein Analysis

Liposomes

Lymphocytes

Mammalian Cell Biotechnology

Mammalian Development

Medical Bacteriology

Medical Mycology

Microcomputers in Biochemistry

Microcomputers in Biology

Microcomputers in Physiology

Mitochondria

Molecular Genetic Analysis of Populations

Molecular Genetics of Yeast

Molecular Imaging in
    Neuroscience

Molecular Neurobiology

Molecular Plant Pathology
    I and II

Molecular Virology

Monitoring Neuronal
    Activity

Mutagenicity Testing

Neural Transplantation

Neurochemistry

Neuronal Cell Lines

NMR of Biological
    Macromolecules

Nucleic Acid Hybridisation

Nucleic Acid and Protein
    Sequence Analysis

Nucleic Acids Sequencing

Oligonucleotides and
    Analogues

Oligonucleotide Synthesis

PCR

Peptide Hormone Action

Peptide Hormone
    Secretion

Photosynthesis: Energy
    Transduction

Plant Cell Biology

Plant Cell Culture

Plant Molecular Biology

Plasmids (2nd edition)

Pollination Ecology

Postimplantation
    Mammalian Embryos

Preparative Centrifugation

Prostaglandins and Related
    Substances

Protein Architecture

Protein Blotting

Protein Engineering

Protein Function

Protein Phosphorylation

Protein Purification
    Applications

Protein Purification
    Methods

Protein Sequencing

Protein Structure

Protein Targeting

Proteolytic Enzymes

Radioisotopes in Biology

Receptor Biochemistry

Receptor–Effector Coupling

Receptor–Ligand Interactions

Ribosomes and Protein
    Synthesis

RNA Processing I and II

# The Practical Approach Series

SERIES EDITORS

**D. RICKWOOD**
*Department of Biology, University of Essex*
*Wivenhoe Park, Colchester, Essex CO4 3SQ, UK*

**B. D. HAMES**
*Department of Biochemistry and Molecular Biology*
*University of Leeds, Leeds LS2 9JT, UK*

Affinity Chromatography

Anaerobic Microbiology

Animal Cell Culture (2nd edition)

Animal Virus Pathogenesis

Antibodies I and II

Behavioural Neuroscience

Biochemical Toxicology

Biological Data Analysis

Biological Membranes

Biomechanics—Materials

Biomechanics—Structures and Systems

Biosensors

Carbohydrate Analysis (2nd edition)

Cell–Cell Interactions

The Cell Cycle

Cell Growth and Division

Cellular Calcium

Cellular Interactions in Development

Cellular Neurobiology

Centrifugation (2nd edition)

Clinical Immunology

Computers in Microbiology

Crystallization of Nucleic Acids and Proteins

Cytokines

The Cytoskeleton

Diagnostic Molecular Pathology I and II

Directed Mutagenesis

DNA Cloning I, II, and III

Drosophila

Electron Microscopy in Biology

Electron Microscopy in Molecular Biology

Signal Transduction

Solid Phase Peptide
   Synthesis

Spectrophotometry and
   Spectrofluorimetry

Steroid Hormones

Teratocarcinomas and
   Embryonic Stem Cells

Transcription Factors

Transcription and
   Translation

Tumour Immunobiology

Virology

Yeast

# Carbohydrate Analysis

## A Practical Approach

### Second Edition

Edited by

**M. F. CHAPLIN**

*School of Applied Science, South Bank University*

and

**J. F. KENNEDY**

*Department of Chemistry, University of Birmingham*

**at**

OXFORD UNIVERSITY PRESS

Oxford  New York  Tokyo

Oxford University Press, Walton Street, Oxford OX2 6DP

Oxford   New York   Toronto
Delhi   Bombay   Calcutta   Madras   Karachi
Kuala Lumpur   Singapore   Hong Kong   Tokyo
Nairobi   Dar es Salaam   Cape Town
Melbourne   Auckland   Madrid

and associated companies in
Berlin   Ibadan

Oxford is a trade mark of Oxford University Press

A Practical Approach 🔵 is a registered trade mark
of the Chancellor, Masters, and Scholars of the University of Oxford
trading as Oxford University Press

Published in the United States
by Oxford University Press Inc., New York

© Oxford University Press, 1994

A catalogue record for this book is available from the British Library

Library of Congress Cataloging in Publication Data
Carbohydrate analysis: a practical approach/edited by M.F. Chaplin,
J.F. Kennedy.—2nd ed.
(The Practical approach series)
Includes bibliographical references and index.
1. Carbohydrates—Analysis. I. Chaplin, M.F. (Martin F.)
II. Kennedy, John F., 1942–. III. Series.
QD321.C24    1994    547.7'8046—dc20    94–11640
ISBN 0–19–963450–5 (hbk.)
ISBN 0–19–963449–1 (pbk.)

Typeset by Cambrian Typesetters, Frimley, Surrey
Printed in Great Britain by
Information Press Ltd, Eynsham, Oxford

# Preface

The involvement of carbohydrates in biological processes has greatly fuelled the current interest in this diverse range of molecules. This has resulted in a vast literature covering numerous analytical methods for these carbohydrates. The size of the literature arises in part from the large number of classes of carbohydrates—macromolecular and monomolecular—basic, neutral, and acidic—derivatized and underivatized—and the wide range of presentation of carbohydrates in everyday life—biosynthesis or chemicosynthesis—structural or gelatinous—edible or unmetabolizable. Superimpose upon just these selected aspects the notion of primary, secondary, tertiary, and quarternary structure, and the fact that the number of ways of covalently joining two carbohydrate monomers is at least one order of magnitude greater than the similar joining of two amino acids, then it is clear that analysis of the field of carbohydrates is, and must be, vast. For this reason, experts in particular areas of the field have been asked to contribute to this book—without them it would not have been possible.

It is not always obvious to the researcher or analyser which method, or even physical technique, is most appropriate to a particular investigation. This book has been produced in order to answer the need for a handbook of laboratory protocols in this field. It gives details of the approach needed to analyse a wide variety of carbohydrates and carbohydrate-containing molecules. We have attempted to show how particular analytical problems should be tackled, describing the most suitable, well tried and trusted methods in exact practical detail.

The chapters are arranged on the basis of carbohydrate moiety to facilitate choice of analytical method for specific applications. Thus the chapters act as a ready source of reference at the initial planning stage for the choice of approach to situations where carbohydrate analysis is required. Chapter 5 includes an extensive account of the way NMR spectroscopy is rapidly becoming a major weapon in the carbohydrate structural analysis armoury.

In this second edition we have endeavoured to ensure that new methods of analysis have been included, particularly those involving HPLC, mass spectrometry, supercritical fluid chromatography, capillary electrophoresis, and NMR spectroscopy, and that methods that have ceased to be well used have been removed. Overall this has resulted in a small but significant increase in size.

We are indebted to our co-authors, who have made this book a useful source of reference for carbohydrate analytical protocols which should find active and practical use within the laboratory.

*London*                                                                           M. F. C. and J. F. K.
May 1994

# Contents

*List of contributors*                                                      xvii

*List of abbreviations*                                                      xix

## 1. Monosaccharides                                                         1

*M. F. Chaplin*

  **1.** Introduction                                              1

  **2.** Extraction                                                 1

  **3.** Colorimetric assays                                        2
    Enzymatic methods                                     8
    Example calculations                                 10

  **4.** Thin-layer chromatography                                11
    Experimental approach                               11
    Detection methods                                    14
    General thin-layer chromatographic method            16

  **5.** Paper chromatography                                      17

  **6.** High performance liquid chromatography                    17
    Experimental approach                               18
    Detection methods                                    24

  **7.** Gas–liquid chromatography                                 27
    Experimental approach                               27
    Detection methods                                    28

  **8.** Mass spectrometry                                         34

  **9.** Infra-red spectroscopy                                    39

    References                                           40

## 2. Oligosaccharides                                                       43

*John F. Kennedy and Giampiero Pagliuca*

  **1.** Introduction                                             43

  **2.** Colorimetric methods                                      44
    Total sugar assays                                   44
    Reducing sugar assays                                44
    Automated assay system                               46

**3.** Thin-layer chromatography                                    48

**4.** Low pressure column chromatography                           50
    Ion-exchange chromatography                  50
    Gel permeation chromatography                53
    Affinity chromatography                      55

**5.** High performance liquid chromatography                       55
    Adsorption chromatography                    56
    Reversed-phase chromatography                56
    Bonded-phase chromatography                  57
    Ion-exchange chromatography                  59
    Gel permeation chromatography                61
    Affinity HPLC                                61
    Preparative HPLC                             62
    Detection                                    62

**6.** Gas chromatography                                           63

**7.** Supercritical fluid chromatography                           64

**8.** Capillary electrophoresis                                    65

**9.** Mass spectromtery                                            66

**10.** Nuclear magnetic resonance spectroscopy                     66

**11.** Infra-red spectroscopy                                      67

    References                                   68

# 3. Neutral polysaccharides                                        73

*John H. Pazur*

**1.** Introduction                                                 73

**2.** Extraction of polysaccharides                                74

**3.** Identification of monosaccharides                            76

**4.** Determination of D and L configuration                       76

**5.** Degree of polymerization                                     79

**6.** Position of glycosidic linkages                              81
    Methylation                                  81
    Periodate oxidation                          87
    Acetolysis                                   90
    Methanolysis                                 92
    Hydrolysis by acids                          93
    Hydrolysis with enzymes                      94
    Degradation reactions                        95

**7.** Ring structure of monosaccharides                            100

**8.** $\alpha$ and $\beta$ configuration of glycosidic linkages    101

Enzymic assays                                                    101
Nuclear magnetic resonance spectroscopy                          101
Oxidation with chromium trioxide                                 102

**9.** Immunological methods                                     103

**10.** Recent researches on complex polysaccharides             108
Anti-gum arabic antiobodies                                      108
Reductive cleavage of a polysaccharide from *Klebsiella*         111
Pneumococcal capsular $\beta$ (1 → 2)(1 → 3) glucan              112
Methylated peracetyl aldonitriles                                113
Gluco-galacto-glycan containing pyruvic acid                     114
Cleavage at the glucuronic moiety of a plant
arabinoglucuronomannan                                           116
Mycobacterial cell wall arabinogalactan                          117
Xanthan (manno-glucurono-glucan) from *Xanthomonas
campestris*                                                      118
Quantitative determination of acidic polysaccharides             119
Electron microscopic observation of waxy maize starch
($\alpha$(1 → 4)(1 → 6) glucan)                                  119
Monoclonal antibodies and the identification of *Brucella*
antigens                                                         120

Acknowledgements                                                 122

References                                                       122

**4.** Proteoglycans                                             125

*S. L. Carney*

**1.** Introduction                                              125

**2.** Proteoglycan component structure                          127
Connective tissue glycosaminoglycans                             127

**3.** Proteoglycan extraction                                   129

**4.** Proteoglycan purification                                 134
Purification by centrifugal techniques                           134
Proteoglycan preparation by chromatographic techniques           142

**5.** Proteoglycan analysis                                     150
Analysis of proteoglycans by gel electrophoresis                 150
Analysis of glycosaminoglycans by gel electrophoresis            157
Examination of chondroitin sulfate disaccharides by HPAE         166

**6.** Preparation of protein domains of aggregated
proteoglycans (aggrecan)                                         168

**7.** Functional assays for the degree of link stabilization
of proteoglycan aggregates                                       169
Determination of link stabilization by column chromatography     171

Determination of link stability by agarose–polyacrylamide
gel electrophesis 173
**8.** Proteoglycan constituent analytical techniques 173
Uronic acid assay using microtitre plate format 174
Dye binding assay for glycosaminoglycan using
1,9-dimethylmethylene blue 175
Competitive inhibition assay for hyaluronan using
proteoglycan G1 domain 176

Acknowledgements 179

References 179

# 5. Glycoproteins 181

*Jean Montreuil, Stéphane Bouquelet, Henri Debray,
Jérome Lemoine, Jean-Claude Michalski, Geneviève Spik,
and Gérard Strecker*

**1.** Introduction 181
Types of glycan–protein linkages 182
Primary structure of glycoprotein glycans 182
**2.** Chemical cleavage of *O*- and *N*-glycosidic linkages
of glycans 188
Alkaline cleavage of *O*-glycosidic linkages 188
Alkaline cleavage of *N*-glycosidic linkages 189
Hydrazinolysis of *N*-glycosidic linkages 190
Production of glycans from *O,N*-glycosylproteins 192
**3.** Glycopeptide and glycan isolation 193
Isolation of glycopeptides 193
Isolation of glycans by HPLC 196
**4.** Use of immobilized lectins 205
Introduction to the use of lectins 205
Immobilization of lectins 207
Fractionation of glycoproteins 213
Fractionation of glycopeptides and oligosaccharides 216
**5.** Colorimetric assays of carbohydrates in glycoproteins
and glycopeptides 222
Glycan composition 222
Colorimetric determination of neutral monosaccharides 223
Hexuronic acid 223
Hexosamines and *N*-acetylhexosamines 224
Sialic acids 226
**6.** Identification and determination of glycan
monosaccharides by gas–liquid chromatography 227
Glycoprotein hydrolysis 227

The use of alditol acetates                                        228
The use of methylglycoside trifluoroacetates                       228
The use of trimethylsilylated methylglycosides                     229

**7.** Methylation                                                 231
Analysis of monosaccharide methyl ethers                           234
Sequence and molecular weight determination of
(permethylated) oligosaccharides by mass spectrometry              239

**8.** Use of glycosidases                                         247
Experimental approach                                              247
Endoglycosidases                                                   256

**9.** ¹H-NMR spectra as fingerprints for glycan primary
structure determination                                            265
The structural-reporter groups                                     266
N-Glycosylprotein glycans                                          266
O-Glycosylprotein glycans                                          283

Acknowledgements                                                   289

References                                                         289

# 6. Glycolipids

295

*I. M. Morrison*

**1.** Introduction                                                295

**2.** Separation methods                                          296
Extraction of glycolipids                                          296
Purification of glycolipids                                        298
Separation of glycolipid mixtures                                  299

**3.** Chemical methods of analysis                                302
Composition of glycolipids                                         302
Methylation analysis of oligosaccharide chains                     306
Isolation of oligosaccharide chains                                308
Determination of anomeric configuration                            309

**4.** Physical methods of analysis                                309
Nuclear magnetic resonance spectroscopy                            309
Mass spectrometry                                                  312
Infra-red spectroscopy                                             314

**5.** Biochemical methods of analysis                             315
Enzymatic methods                                                  315
Immunological methods                                              315

References                                                         316

# Index

319

# Contributors

STEPHANE BOUQUELET
Laboratoire de Chimie Biologique, Universite des Sciences et Technologies de Lille, 59655 Villeneuve d'Ascq, France.

S. L. CARNEY
Lilly Research Centre Limited, Erlwood Manor, Windlesham, Surrey GU20 6PH, UK.

M. F. CHAPLIN
School of Applied Science, South Bank University, Borough Road, London SE1 0AA, UK.

HENRI DEBRAY
Laboratoire de Chimie Biologique, Universite des Sciences et Technologies de Lille, 59655 Villeneuve d'Ascq, France.

JOHN F. KENNEDY
Department of Chemistry, University of Birmingham, Birmingham B15 2TT, UK.

JEROME LEMOINE
Laboratoire de Chimie Biologique, Universite des Sciences et Technologies de Lille, 59655 Villeneuve d'Ascq, France.

JEAN-CLAUDE MICHALSKI
Laboratoire de Chimie Biologique, Universite des Sciences et Technologies de Lille, 59655 Villeneuve d'Ascq, France.

JEAN MONTREUIL
Laboratoire de Chimie Biologique, Universite des Sciences et Technologies de Lille, 59655 Villeneuve d'Ascq, France.

I. M. MORRISON
Scottish Crop Research Institute, Invergowrie, Dundee DD2 5DA, UK.

GIAMPIERO PAGLIUCA
Department of Chemistry, University of Birmingham, Birmingham B15 2TT, UK.

JOHN H. PAZUR
The Pennsylvania State University, 108 Althouse Laboratory, University Park, PA 16802, USA.

GENEVIEVE SPIK
Laboratoire de Chimie Biologique, Universite des Sciences et Technologies de Lille, 59655 Villeneuve d'Ascq, France.

# Contributors

GERARD STRECKER
Laboratoire de Chimie Biologique, Universite des Sciences et Technologies de Lille, 59655 Villeneuve d'Ascq, France.

# Abbreviations

| | |
|---|---|
| Ara | arabinose |
| Ara$_f$ | L-arabinofuranose |
| BP | boiling point |
| CE | capillary electrophoresis |
| Con A | concanavalin A |
| CZE | capillary zone electrophoresis |
| DCI | direct chemical ionization |
| DE | dextrose equivalent |
| DEAE | diethylaminoethyl |
| DMSO | dimethyl sulfoxide |
| DP | degree of polymerization |
| DSA | *Datura stramonium* agglutinin |
| EC | Enzyme Commission |
| EI-MS | electron impact mass spectrometry |
| FAB | fast atom bombardment |
| FID | flame ionization detector |
| FPLC | Pharmacia Fast protein liquid chromatography |
| FTIR | Fourier transform infra-red (spectroscopy) |
| Fru$_f$ | D-fructofuranose |
| Fuc | L-fucose |
| Gal | D-galactose |
| GalNAc | 2-acetamido-2-deoxy-D-galactose |
| galactosamine | 2-amino-2-deoxy-D-galactose |
| GC | gas chromatography |
| Glc | D-glucose |
| GlcA | D-glucuronic acid |
| GlcNAc | 2-acetamido-2-deoxy-D-glucose |
| glucosamine | 2-amino-2-deoxy-D-glucose |
| GPC | gel permeation chromatography |
| HexNAc | 2-acetamido-2-deoxyhexose |
| HPAE | high pH (performance) anion-exchange chromatography |
| HPCE | high performance capillary electrophoresis |
| HPTLC | high performance thin-layer chromatography |
| HRGC | high resolution gas chromatography |
| LCA | *Lens culinaris* agglutinin |
| LSI | liquid secondary ionization |
| MALDI | matrix-assisted laser desorption ionization (mass spectrometry) |
| Man | D-mannose |
| MS | mass spectrometry |
| *N*-acetylglucosamine | 2-acetamido-2-deoxy-D-glucose |
| NeuAc | neuraminic acid |
| NMR | nuclear magnetic resonance (spectroscopy) |

## Abbreviations

| | |
|---|---|
| PAAN | peracetyl aldonitrile |
| PAD | pulsed amperometric detection |
| PAGE | polyacrylamide gel electrophoresis |
| PBS | phosphate-buffered saline |
| PC | paper chromatography |
| PHA | *Phaseolus vulgaris* agglutinin |
| RCA | *Ricinus communis* agglutinin |
| Rha | L-rhamnose |
| RI | refractive index |
| SD | standard deviation |
| SDS | sodium dodecyl sulfate |
| SFC | supercritical fluid chromatography |
| SPE | solid phase extraction |
| TFA | (a) trifluoroacetic acid; (b) trifluoroacetyl |
| TLC | thin-layer chromatography |
| TMS | trimethylsilyl |
| TOF | time-of-flight (mass spectrometry) |
| U | standard unit of enzyme activity; catalyses 1 μmol of reaction in 1 min, under defined standard conditions |
| WCOT | wall coated open tubular |
| WGA | wheat germ agglutinin |
| Xyl | D-xylose |

# 1

# Monosaccharides

M. F. CHAPLIN

## 1. Introduction

Monosaccharides exhibit a great variety of structural types with vastly different chemical and physical properties. In contrast to the several methods available for the extensive analysis of the other major class of biological molecules, the amino acids, there is no single method which is suitable for the quantitative or qualitative analysis of all monosaccharides. The method of choice will depend on a number of factors including the accuracy required and resources available. There are two techniques available for the quantitative analysis of mixtures of monosaccharides, high performance liquid chromatography (HPLC) and gas–liquid chromatography (GLC). If only a qualitative analysis is needed, either paper chromatography (PC) or thin-layer chromatography (TLC) may be used. Single monosaccharides may be identified by means of mass spectrometry (MS), infra-red spectroscopy (IR), or proton or carbon-13 nuclear magnetic resonance ($^{1}$H-NMR, $^{13}$C-NMR), and quantified by use of specific colorimetric or enzymatic assays. An extensive and growing literature exists which describes methods within these techniques. In this chapter a number of these methods have been chosen and described. They encompass the most commonly occurring analytical circumstances but several volumes would be needed if a fully comprehensive text was to be presented.

## 2. Extraction

Often the sugars may be in a solution that can be analysed directly (e.g. after acid hydrolysis) or are easily extracted with water. Sometimes, however, the material is more intractable and an extraction and/or clean-up is necessary. Ethanol/water (80% v/v) has proved to be a good general purpose extractant for monosaccharides in which proteins, polysaccharides, and many oligosaccharides are insoluble. Such monosaccharide solutions are stable. Difficult extractions may be made by refluxing the finely divided material for about an hour, but care must be taken to avoid any unwanted hydrolysis. Where the samples are particularly complex (e.g. many solid foodstuffs), aqueous extracts can be clarified of proteins, other soluble macromolecules, and finely suspended or emulsified solids by use of Carrez reagent (*Protocol 1*).

---

**Protocol 1.** Carrez reagent

*Reagents*

A  85 mM Potassium ferrocyanide. Dissolve 3.6 g K$_4$[Fe(CN)$_6$].3H$_2$O in 100 ml solution.

B  0.25 M Zinc sulfate. Dissolve 7.2 g ZnSO$_4$.7H$_2$O in 100 ml solution.

*Method*

To the sample, add 5% (v/v) of each of reagents A and B sequentially followed by 10% (v/v) of 0.1 M NaOH (0.4 g/100 ml), shaking well between each addition. Note the final volume and filter.

---

# 3. Colorimetric assays

A number of colorimetric assays are presented for the main classes of monosaccharides, followed by an important specific enzymatic assay. A standard format has been chosen to describe these assays. The sensitivity describes the range over which the assay is fairly linear with a maximum absorbance of about 1.0 OD unit for a 1 cm path lengh cuvette. Distilled or deionized water is used throughout for aqueous solutions. The final volume, for spectrophotometric measurement, has normally been kept within the 1–2 ml range to allow microanalytical cuvettes to be used, but larger volumes can be arranged by increasing the sample and reagent volumes in proportion. In general, the protocols should be followed in as reproducible a manner as possible. Standards should always be run to check that the protocol is delivering the appropriate sensitivity and the spectrophotometer is functioning correctly.

The non-stoichiometric constant volume titration of Lane and Eynon (1) is still used as a standard procedure for reducing sugar in the food industry. It requires about 100 mg reducing sugar ($\sim$ 14 mM) and considerable skill and experience. It is not recommended except where required, by legislation or standard practice, in food analysis.

---

**Protocol 2.**  The general phenol–sulfuric acid assay for carbohydrate (2)

- Sensitivity: $\sim$ 1–60 µg glucose in 200 µl  • Final volume: 1.4 ml
  ($\sim$ 30 µM–2 mM)[a]

*Method*

1. Prepare the reagent by dissolving phenol in water (5% w/v).[b]

---

**2.** Mix samples, standards, and control solutions (200 µl containing up to 100 µg carbohydrate) with 200 µl of phenol reagent.

**3.** Add 1.0 ml of concentrated sulfuric acid rapidly and directly to the solution surface without allowing it to touch the sides of the tube.[c]

**4.** Leave the solutions undisturbed for 10 min before shaking vigorously.

**5.** Determine the absorbances at 490 nm after a further 30 min.

[a] Aldoses, ketoses, and alduronic acids respond to different degrees. Protein, cysteine, non-carbohydrate reducing agents, heavy metal ions, and azide interfere with this assay. However, it remains useful as a rapid non-specific method for the detection of neutral carbohydrate in column eluates, and is also applicable to solids containing carbohydrate, such as cereal flours, so long as all particles are milled to less than 50 µm diameter. The cysteine–sulfuric acid assay (3) may be used where a sixfold increase in sensitivity is required.
[b] This reagent is stable indefinitely.
[c] The reproducibility of this assay is strongly dependent on the manner of the addition of the sulfuric acid.

---

**Protocol 3.** The dinitrosalicyclic acid assay for reducing sugar (4)

● Sensitivity: ~ 5–500 µg glucose in 100 µl    ● Final volume: 1.1 ml
  (~ 0.3–30 mM)[a]

*Method*

**1.** Prepare the reagent by dissolving 0.25 g of 3,5-dinitrosalicyclic acid and 75 g sodium potassium tartrate (Rochelle salt) in 50 ml of 2 M NaOH (4 g NaOH in 50 ml water) and dilute to 250 ml with water.[b]

**2.** To samples, standards, and controls (100 µl) add 1.0 ml of the reagent. Mix well.

**3.** Heat the mixtures at 100 °C for 10 min.

**4.** After rapid cooling to room temperature, determine the absorbance at 570 nm.

[a] This assay does not cause detectable inversion of sucrose. Dissolved molecular oxygen interferes with this assay. This may be overcome either by purging the assay solutions with nitrogen or helium prior to the assay, or by the addition of a fixed known small amount of glucose (~ 20 µg) to all samples, in order to raise the total reducing sugar concentration above a critically low value. Non-carbohydrate reducing agents also interfere with this assay. Some metal ions (e.g. manganous, cobalt (II), and calcium) may increase the assay response. The neocuproine (5, 6) or Nelson–Somogyi assays (see *Protocol 4*) may be used where up to a 500-fold increase in sensitivity is required.
[b] The reagent is sensitive to $CO_2$. It is stable for several weeks if stored purged with helium or nitrogen, otherwise it should be freshly prepared.

**Protocol 4.** The Nelson–Somogyi method for reducing sugars (7)

- Sensitivity: ~ 10–100 μg glucose in 1 ml (~ 60–600 μM)[a]
- Final volume: 6.0 ml[b]

*Reagents*

A  Dissolve 15 g of sodium potassium tartrate and 30 g of anhydrous $Na_2CO_3$ in about 300 ml water. Add 20 g $NaHCO_3$. Dissolve 180 g of anhydrous $Na_2SO_4$ in 500 ml boiling water and cool. Mix the two solutions and make up to 1 litre with water.[a]

B  Dissolve 5 g $CuSO_4.5H_2O$ and 45 g anhydrous $Na_2SO_4$ in water and make up to 250 ml.[a]

C  Mix reagents A (4 vol.) and B (1 vol.) just before use.

D  Dissolve 25 g ammonium molybdate in 450 ml water. Carefully add 21 ml concentrated $H_2SO_4$ with stirring. Dissolve 3 g $Na_2HAsO_4.7H_2O$ in 25 ml water and add to the molybdate solution. Incubate for 24–28 h at 37 °C and store in a brown glass-stoppered bottle.[a]. Just before use this reagent should be diluted with 2 vol. of 0.75 M $H_2SO_4$ (4 ml concentrated $H_2SO_4$ in 100 ml solution).

*Method*

1. Mix the samples, standards, and control solutions (1.0 ml) containing up to 600 nmol reducing sugar with 1.0 ml of reagent C in small stoppered test-tubes.

2. Heat at 100 °C for 15 min. Cool the solution rapidly to room temperature.

3. Add reagent D (1.0 ml) and mix well.

4. Add 3.0 ml water, mix, and measure the absorbance at 520 nm.

[a] The reagents are stable in their concentrated unmixed forms.
[b] Care should be taken to minimize reoxidation by air (see *Protocol 3*).

---

**Protocol 5.** Ferric–orcinol assay for pentose (8)

- Sensitivity: ~ 0.2–20 μg xylose in 200 μl (~ 7–700 μM)[a]
- Final volume: 1.6 ml

*Reagents*

A  Trichloroacetic acid solution in water (10% w/v).[b]

B  Freshly prepared solution of ferric ammonium sulfate (1.15% w/v) and orcinol (0.2% w/v) in 9.6 M HCl (made by diluting five parts concentrated HCl with one part water).

*Method*

1. Mix the samples, standards, and control solutions (200 µl) containing up to 40 µg pentose with 200 µl of reagent A.

2. Heat at 100 °C for 15 min. Cool the solution rapidly to room temperature.

3. Add reagent B (1.2 ml) and mix well.

4. Re-heat the solution at 100 °C for a further 20 min.

5. Cool the solution to room temperature and determine the absorbance at 660 nm.

[a] Hexoses interfere in this assay but can be accounted for by additionally determining the absorbance at 520 nm at which wavelength they have a strong absorbance. It is recommended that both hexose and pentose standards are used where the hexose content of the samples might be considerable (see Section 3.2).
   [b] This is stable indefinitely.

---

**Protocol 6.** The phenol–boric acid–sulfuric acid assay for ketose (9)

- Sensitivity: ~ 0.1–9 µg fructose in 100 µl   • Final volume: 2.0 ml
  (~ 30–500 µM)[a]

*Method*

1. Prepare the reagent by dissolving 2.5 g phenol (recrystallized from methanol and ethanol) in 50 ml water. Add 1.0 ml of acetone drop-wise with constant stirring over a period of 10 min and stir the mixture for a further 10 min at room temperature. Dissolve 2.0 g of boric acid in the mixture.[b]

2. Mix the samples, standards, and controls (100 µl) with 0.5 ml of reagent and then rapidly add 1.4 ml of concentrated sulfuric acid directly to the surface, avoiding the sides of the tubes.

3. After thorough mixing, leave the solutions for 5 min at room temperature.

4. Incubate at 37 °C for 1 h.

5. Determine the absorbance at 568 nm.

[a] Different ketoses give differing absorbances in this assay. Interferences from non-ketose carbohydrate is slight (< 1%) to non-existent. The reproducibility of this assay is strongly dependent on the manner of the addition of the sulfuric acid.
   [b] The reagent is stable for at least two weeks at 4 °C.

---

**Protocol 7.** The Morgan–Elson assay for hexosamine (10)

- Sensitivity: ~ 0.06–6 µg 2-acetamido-2-deoxy-D-glucose in 250 µl (~ 1–110 µM)[a]
- Final volume: 1.8 ml

*Reagents*

A  Dissolve 6.1 g of dipotassium tetraborate tetrahydrate in 80 ml of water and make up to 100 ml with water.

B  Add 1.5 ml of water to 11 ml of concentrated HCl. Add a further 87.5 ml of glacial acetic acid and dissolve 10 g of 4-(*N,N*-dimethylamino)benzaldehyde in this mixture.[b] Dilute 10 ml to 100 ml with glacial acetic acid immediately prior to use.

*Method*

1. Add samples, standards, and controls (250 µl) to 50 µl of reagent A.

2. Heat each mixture at 100 °C for 3 min.

3. After cooling rapidly to room temperature, add 1.5 ml of reagent B, washing down any condensate formed.

4. Incubate the samples at 37 °C for 20 min.

5. After cooling to room temperature, determine the absorbance at 585 nm.

---

[a] 2-Acetamido-2-deoxy-D-galactose gives only one third the response of 2-acetamido-2-deoxy-D-glucose in this assay. Free amino hexoses may be *N*-acetylated prior to this assay by the addition of one part of freshly prepared 1.5% (v/v) acetic anhydride in acetone to eight parts of the aqueous solution and leaving for 5 min at room temperature.

[b] This solution may be stored for several weeks.

---

**Protocol 8.** The carbazole assay for uronic acids (11)

- Sensitivity: ~ 0.2–20 µg D-galacturonic acid in 250 µl (~ 4–400 µM)
- Final volume: 1.8 ml

*Reagents*

A  Dissolve 0.9 g of sodium tetraborate decahydrate in 10 ml of water and add 90 ml of ice-cold 98% concentrated sulfuric acid carefully to form a layer. Leave undisturbed overnight to mix without excessive heat production. Check it is thoroughly mixed and at room temperature before use.[b]

B  Dissolve 100 mg of carbazole (recrystallized from ethanol) in 100 ml of absolute ethanol.[b]

*Method*

1. Cool the samples, standards, and controls (250 μl) in an ice-bath.

2. Carefully add ice-cold reagent A (1.5 ml) with mixing and cooling in the ice-bath.

3. Heat the mixtures at 100 °C for 10 min.

4. Cool rapidly in the ice-bath.

5. Add 50 μl of reagent B and mix well.

6. Re-heat at 100 °C for 15 min.

7. Cool rapidly to room temperature and determine the absorbance at 525 nm.

[a] Neutral carbohydrates interfere with this assay to a greater (~ 10% on a molar basis for hexoses) or lesser extent (~ 2% on a molar basis for 6-deoxyhexoses). However interference can be reduced by use of appropriate controls as non-uronic acid carbohydrates give significantly different absorption spectra. Cysteine and other thiols increase the response of the assay but large amounts of protein may depress the colour development. Different uronic acids give different responses in this assay.

[b] These reagents are stable indefinitely if refrigerated.

---

**Protocol 9.** The Warren assay for sialic acid (12)

- Sensitivity: 0.08–8 μg *N*-acetylneuraminic acid in 80 μl (3–300 μM)[a]
- Final volume: 1.0 ml

*Reagents*

A Dissolve 4.278 g of sodium *meta*-periodate in 4.0 ml water. Add 58 ml of concentrated *ortho*-phosphoric acid and make up to 100 ml with water.[b]

B Dissolve 10 g of sodium arsenite, 7.1 g of sodium sulfate, and 10 mg potassium iodide in 0.1 M sulfuric acid (made by carefully diluting 5.7 ml concentrated sulfuric acid to one litre with water) to a total volume of 100 ml.[b]

C Dissolve 1.2 g of 2-thiobarbituric acid and 14.2 g of sodium sulfate in water to a total volume of 200 ml.[c]

D Redistilled cyclohexanone.[d]

*Method*

1. To the samples, standards, and controls (80 μl) add 40 μl reagent A and mix well. Leave at room temperature for 20 min.

2. Add 400 μl reagent B and then shake the tubes vigorously to expel the yellow coloured iodine. Leave for a further 5 min at room temperature.

**Protocol 9.** *Continued*

3. Add 1.2 ml reagent C, shake the tubes, stopper them, and heat at 100 °C for 15 min.

4. Cool rapidly to room temperature.

5. Extract the chromophore into 1.0 ml of reagent D by vigorous shaking.

6. Centrifuge the solutions using a bench centrifuge for a few minutes in order to properly separate the two layers.

7. Determine the absorbance of the upper cyclohexanone layer of 549 nm.

[a] DNA, 2-deoxy-D-ribose, and substances producing malondialdehyde on periodate oxidation interfere in this assay. This may be circumvented by additionally determining the absorbance at 532 nm and calculating from the resultant data (see Section 3.2). L-Fucose reduces the expected absorbance of this assay. Methoxyneuraminic acid and some acetylated neuraminic acids give no colour in this assay. The interfering O-acetyl groups may be removed by alkaline hydrolysis (0.1 M NaOH, 30 min, 37 °C) followed by neutralization with 0.2 M HCl. Alternatively the resorcinol–HCl assay (13) may be used if the presence of any of these is suspected. The assay is less specific than the Warren assay, however, and is not recommended for general use.
[b] This reagent is stable indefinitely.
[c] This reagent is stable for several weeks but eventually forms a yellow precipitate which indicates the need for its renewal.
[d] This is stable for several months until noticeably discoloured.

## 3.1 Enzymatic methods

Enzymatic methods utilize the specificity of enzymes to pick out their substrates from mixtures, and are ideal for the analysis of known carbohydrates in complex mixtures, such as clinical samples and foodstuffs. The methods used are generally variations of that used for glucose (*Protocol 10*).

**Protocol 10.** The hexokinase/dehydrogenase assay for glucose (14)

- Sensitivity: 0.4–40 µg glucose in 100 µl (20 µM–2 mM)[a]
- Final volume: 1.4 ml

*Reagents*[b]

A 0.33 M Triethanolamine, 4.3 mM $Mg^{2+}$. Dissolve 6.0 g triethanolamine hydrochloride and 0.11 g $MgSO_4 \cdot 7H_2O$ in 80 ml water. Adjust the pH to 7.6 with concentrated NaOH solution (~ 20% w/v) and make up the solution to 100 ml with water.[c]

B 5.5 mM NADP. Dissolve NADP (disodium salt, 4.3 mg/ml) in distilled water.[c]

C  35 mM ATP, 0.26 M NaHCO$_3$. Dissolve ATP (disodium salt hydrate, 22 mg/ml) and NaHCO$_3$ (22 mg/ml) in water.[c]

D  3.2 M Ammonium sulfate. Add 0.6 g ammonium sulfate to 1 ml of water and allow to dissolve.[d]

E  Dissolve hexokinase (ATP: D-hexose-6-phosphotransferase, EC 2.7.1.1 *ex.* yeast, 280 U/ml, ~ 2 mg/ml) and glucose-6-phosphate dehydrogenase (D-glucose-6-phosphate: NADP 1-oxidoreductase, EC 1.1.1.49, *ex.* yeast, 140 U/ml, ~ 1 mg/ml) in solution D.[c]

*Method*

1. Add each sample and standard solution (100 μl) to a mixture containing 1.0 ml buffer solution A, 100 μl reagent B, and 100 μl reagent C. Mix well.

2. Start the reaction with 100 μl of enzyme solution E. The control solutions lack enzyme and so should consist of sample solution (100 μl) plus reagents A (1.1 ml), B (100 μl), and C (100 μl).

3. After further mixing, incubate the solutions at 37 °C for 30 min.

4. Cool and determine the absorbance at 340 nm. The reaction should have stopped at this stage. However, check whether there is a significant change (> 5%) in absorbance at 340 nm after a further 30 min incubation at 37 °C. If so, check the reagents and/or extend the incubation time.

---

[a] The sensitivity of this spectrophotometric assay may be increased 10–100-fold (0.2–20 μM) by using the fluorescence change rather than the absorption change. The excitation wavelength is 340 nm and the fluorescence is emitted at about 465 nm. This improvement in sensitivity is achieved at an extra cost in the care needed for the assay. All solutions should be dust-free and all glassware scrupulously cleaned. The cuvettes should be of low fluorescence glass or quartz and temperature-equilibrated before the determinations are made. Because of the higher background variability, this method is best chosen only when the additional sensitivity over the spectrophotometric assay is essential.
[b] Some of these reagents are available in a kit form from manufacturers such as Boehringer Mannheim GmbH.
[c] This solution is stable for a month at 4 °C.
[d] This solution is stable.

*Protocol 10* can be adapted for the determination of fructose and/or mannose in the presence or absence of glucose. Determine fructose by the addition of 100 μl phosphoglucose isomerase (D-glucose-6-phosphate ketol-isomerase, EC 5.3.1.9, *ex.* yeast, 65 U/ml, ~ 0.2 mg/ml in buffer solution A) to the reaction mixture after the glucose content has been determined. Mix this solution, incubate at 37 °C for 30 minutes, cool, and re-measure the absorbance at 340 nm. Determine mannose in a similar manner subsequent to the addition of 100 μl phosphomannose isomerase (D-mannose-6-phosphate ketol-isomerase EC 5.3.1.8, *ex.* yeast, 60 U/ml, ~ 1.0 mg/ml in buffer solution A). The sensitivities for fructose and mannose are similar to those for glucose. Starch and sucrose may also be determined (14) after enzymatic conversion to monosaccharides by glucoamylase or invertase respectively. As the enzymes are optimally active at substantially lower pH than hexokinase it is recommended that a separate procedure be adopted whereby the hydrolysed and unhydrolysed samples are analysed in the above assay. Similar assay systems may be set-up for the determination of other carbohydrates where the appropriate enzyme is available, for example L-fucose using fucose dehydrogenase (15) and D-galactose using β-D-galactose dehydrogenase (16). In these cases the buffer solution A should be replaced by a buffer appropriate to the determination and reagents B, C, D, and E will all probably be different. The principle of the assay however will remain as the change in absorbance at 340 nm due to the formation of the reduced dinucleotide.

## 3.2 Example calculations

A mixture of two components (A, B) can simply be determined spectrophotometrically if each component absorbs maximally at different wavelengths (X nm and Y nm).

(a) Using a suitable range of standard concentrations for both components separately determine the absorbance produced by each at both wavelengths (i.e. component A at concentration $C_A$ gives absorbance $\Delta A_X$ at wavelength X nm and $\Delta A_Y$ at Y nm. Similarly component B at concentration $C_B$ gives absorbance $\Delta B_X$ at X nm and $\Delta B_Y$ at Y nm).

(b) Determine the absorbance difference between blank and unknown sample S at the same wavelengths X ($\Delta S_X$) and Y ($\Delta S_X$).

(c) The concentration of A in the sample is

$$C_A \times \frac{\Delta S_X \times \Delta B_Y - \Delta S_Y \times \Delta B_X}{\Delta A_X \times \Delta B_Y - \Delta A_Y \times \Delta B_X}$$

The concentration of B in the sample is

$$C_B \times \frac{\Delta S_X \times \Delta A_Y - \Delta S_Y \times \Delta A_X}{\Delta B_X \times \Delta A_Y - \Delta B_Y \times \Delta A_X}$$

For example, for the pentose assay (*Protocol 5*) the absorbance maxima X and Y are 660 nm and 520 nm, respectively. The absorbance of 20 μg xylose (A) at these maxima are $\Delta A_X = 1.0$ and $\Delta A_Y = 0.25$, whereas 180 μg of glucose (B) gives $\Delta B_X = 0.25$ and $\Delta B_Y = 0.60$.

# 4. Thin-layer chromatography (17, 18, 19)

## 4.1 Experimental approach

Thin-layer chromatography is a simple and rapid technique that is very useful for the preliminary examination of carbohydrate mixtures and as a screening technique for multiple samples. A number of solid supports have been used with a vast number of solvents and detection methods for separating a wide range of carbohydrates. Adding to the difficulty of choosing a system for recommendation is the fact that workers in this field never seem to choose the same range of standard sugars to calibrate and compare their systems. It is clear, however, that no single system is available that will separate all possible combinations of carbohydrates. The best plan is to try likely systems until a suitable one is found.

There are two solid supports which have proved themselves to be particularly useful, microcrystalline cellulose and silica gel. Cellulose separates essentially by liquid–liquid partition. The sugar is distributed between the mobile phase and the cellulose-bound water complex, dependent upon the solubility of the sugar in the eluent and the ease with which it can enter the structures of the complex and/or solid support. This latter ability is determined by its size and steric configuration. Generally cellulose TLC has the same chromatographic characteristics as paper with the advantages that elution times are shorter and the sensitivity enhanced.

Silica gel separates in a similar manner but with an additional adsorption component. Often inorganic salts (e.g. phosphate, bisulfite, or citrate) are impregnated into the gel by wetting with the salt solution followed by thorough drying after the plates have been coated, or by inclusion in the slurry solvent. In these cases, the selectivity of the inorganic salt greatly influences the carbohydrate separation and is, in turn, determined by its concentration and ionic form. In use, a gradient of salt is formed up the plate according to the composition of the eluent.

Thin-layer plates can either be prepared in the laboratory or purchased ready-coated with cellulose, silica gel, or impregnated silica gel as the solid support. Plates prepared in the laboratory should be thoroughly dried in an oven at 100 °C and stored in a desiccator until used. In general, precoated plates are to be preferred to home-made plates as they give excellent reproducibility, a higher sensitivity to detection reagents, and because of their bonded strength, they allow multiple elutions and reagent applications without their surface breaking up. Simple, inexpensive, and rapid preliminary investigation of a system is possible by use of coated microscope slides. Clean dry slides are coated by dipping in a slurry of the chromatographic material, dried, and run in a covered beaker.

The choice of a suitable solvent for TLC is not easy unless a very simple sugar mixture is anticipated. The initial choice should lie between the solvent

**Table 1.** Retention data[a] and sample visualization[b]

| Monosaccharide | TLC | | | | | PC | Visualization | |
|---|---|---|---|---|---|---|---|---|
| Solvent[c] | (a) | (b) | (c) | (d) | (e) | (f) | 11A | 11B |
| D-Xylose | 1.32 | —[d] | 1.34 | 1.70 | 1.54 | 1.79 | Faint grey blue | Brown |
| D-Ribose | 1.56 | — | 1.49 | 1.95 | 1.46 | 2.14 | Faint grey blue | Light blue |
| L-Arabinose | 1.24 | — | 1.30 | 1.45 | 1.15 | 1.36 | Faint grey blue | Light blue |
| D-Glucose | 1.00 | 1.00 | 1.00 | 1.00 | 1.00 | 1.00 | Blue grey | Violet |
| D-Galactose | 1.24 | 0.89 | 0.93 | 0.87 | 0.73 | 0.80 | Blue grey | Blue |
| D-Fructose | — | — | 1.21 | 1.26 | 1.12 | 1.38 | Light red | Purple/red |
| D-Mannose | 1.20 | 1.13 | 1.17 | 1.15 | 1.22 | 1.43 | Blue grey | Blue |
| L-Fucose | — | 1.34 | 1.44 | 1.83 | 1.59 | 1.96 | Olive intense blue | Pink |
| D-Glucosamine | — | 0.72 | — | 0.0 | 0.0 | — | Pale grey | Grey |
| D-Galactosamine | — | 0.55 | — | 0.0 | 0.0 | — | Pale grey | Grey |
| Reference | (17) | (20) | (18, 19) | (21)[e] | (22)[e] | (23) | (17)[e] | (17)[e] |

[a] Values given are $R_{glucose}$.
[b] Colours using silica gel TLC after incubation for the times indicated in *Protocol 11*.
[c] See Section 4.1 for details.
[d] Not reported.
[e] Chaplin, M. F. (1993). Unpublished results.

systems suggested below (see *Table 1*) but if all of these prove unsatisfactory there is an abundance of choice in the literature (17, 18, 19). It should be noted, however, that even under optimal conditions a maximum of about ten carbohydrates may be separated in a one-dimensional run, increasing to about 20 if a suitable two-dimensional system is appropriate. Carbohydrates with relative $R_f$s closer than about 5% cannot normally be resolved unless a discriminating detection system is used. Solvents are generally binary, tertiary, or quaternary and always include an aqueous solution, usually between 10% and 20% by volume. Small changes in the composition of such mixtures may have large and possibly unpredictable effects on the relative movement of the carbohydrate ($R_f$) and the efficiency of the separations; for example, the elution order of glucose, mannose, and galactose may change. Therefore, $R_f$ or $R_{glucose}$ values should not be regarded as constant parameters, especially between laboratories. They may vary with temperature, humidity, coating batch, coating method, and thickness, any pretreatment, and chromatographic tank size. They should be established in the system under scrutiny and literature values should be used for guidance purposes only. As a general rule, a carbohydrate has a higher $R_f$ if it is more hydrophobic or of lower molecular weight. There are many apparent exceptions to this rule however.

Given below is a list of solvent systems for use in TLC and PC.

(a) Ethyl acetate/pyridine/water: mix 100 ml of ethyl acetate with 35 ml of pyridine and 25 ml of water. This solvent is suitable for the TLC and PC analysis of hexoses, deoxyhexoses, and some disaccharides on cellulose using three successive developments.

(b) Butanol/pyridine/0.1 M HCl: mix 50 ml of *n*-butanol with 30 ml of pyridine and 20 ml 0.1 M HCl (made up by adding 1.0 ml of concentrated HCl to 114 ml of water). This solvent is suitable for TLC and PC use on cellulose in order to separate monosaccharides derived from the acid hydrolysis of glycoproteins, i.e. galactose, mannose, fucose, glucosamine, and galactosamine.

(c) Formic acid/ethyl methyl ketone/*tert*-butanol/water: mix 30 ml of formic acid, 60 ml of ethyl methyl ketone (2-butanone), 80 ml of *tert*-butanol, and 30 ml of water. This solvent is suitable for TLC and PC use on cellulose for the analysis of carbohydrates derived from plant extracts, including uronic acids. D-Arabinose may be distinguished from its L-isomer.

(d) Acetonitrile/water: mix 85 ml acetonitrile (HPLC grade) with 15 ml water. This solvent is suitable for use, employing three successive developments, on silica plates that have been pretreated by spraying with 0.1 M sodium *meta*-bisulfite (1.9 g/100 ml), followed by drying and spraying with 0.009 M sodium citrate/citric acid buffer pH 4.8 (159 mg trisodium citrate dihydrate + 76 mg citric acid monohydrate in 100 ml), finally drying at 100 °C for 1 h and keeping desiccated until use.

(e) 2-propanol/acetone/0.1 M lactic acid: dissolve 148 mg of lactic acid in 20 ml of water. Add 40 ml of 2-propanol and 40 ml of acetone. This solvent system is recommended for use on phosphate impregnated silica gel plates. The system separates many of the more common constituents found in the clinical analysis of urine and plasma (e.g. glucose, galactose, and mannose). The impregnated plates are obtained by either:

    *i.* buying phosphate-activated precoated silica gel plates

    *ii.* preparing the coating silica gel slurry in 0.5 M $NaH_2PO_4$ or $KH_2PO_4$ (dissolve 19.5 g of $NaH_2PO_4.2H_2O$ in 250 ml of water)

    *iii.* spray dampening precoated plates in 0.5 M $NaH_2PO_4$ followed by drying.

(f) *n*-Butanol/pyridine/water: mix 100 ml of *n*-butanol with 30 ml of pyridine and 30 ml of water. This solvent is suitable for the PC analysis of carbohydrates derived from glycoproteins. A run takes about 25 hours at room temperature. The chromatogram should be dried in a fume cupboard for about 4 hours (until only a faint odour of *n*-butanol remains) before use of detection reagents.

(g) *n*-Butanol/*n*-propanol/0.1 M HCl: mix 25 ml of *n*-butanol with 50 ml of *n*-propanol and 25 ml of 0.1 M HCl. This solvent is suitable for the PC and TLC analysis, on cellulose, of sialic acids derived from glycoproteins.

## 4.2 Detection methods

There has been a large number of spray reagents described in the literature for the detection of carbohydrates. Two are suggested here (*Protocol 11*A and B, see *Table 1*) which cover most analyses. They give different colours with different carbohydrates and are useful in identifying components in mixtures and overlapping unresolved carbohydrates. The third reagent (*Protocol 11*C) may be used for dipping silica HPTLC plates prior to densitometric analysis. The colour of background and samples produced by these detection systems is dependent on both the temperature and the duration of the heating period. This might affect the selectivity of the reaction. Additionally trace amounts of eluent persisting in the chromatographic layer may affect both the colour and sensitivity of the reactions. Care should be taken with precoated plates to avoid destroying the coating by overheating during colour development.

It is possible to quantify material separated by TLC. Dipping of the plates give a much more uniform colour development than spraying. Direct densitometric assessment of the sprayed plates is possible but, due to difficulties arising from spot irregularity, perhaps should only be used with samples applied as bands in one-dimensional TLC. In addition, the spray detection techniques give poor reproducibility of quantitative results even on the same plate. It is best to locate the bands by use of co-chromatographed standards, identified by spray detecting parts of the TLC plates, while protecting other parts by shielding with a glass plate. The samples may then be scraped off from the unsprayed part of the plate, eluted, and analysed by a specific colorimetric assay. Care must be taken that different impurities in standard or sample do not cause a difference in $R_f$. The dipping method largely avoids this problem. Plates are dipped at uniform pace into and out of the reagent (2–4 sec), avoiding pauses which may cause tide-marks. Dipping must be carried out in such a way that the sugars are not lost from the plates; e.g. by leaving in a wet state for too long allowing spots to diffuse away. Instrumentation is being developed for the mass spectrometric scanning of TLC plates (24).

---

**Protocol 11.** Detection reagents for use in TLC and PC

A. *Diphenylamine/aniline/phosphoric acid*

1. Prepare the spray reagent solutions:

   • dissolve 4 g of diphenylamine in 80 ml of acetone and make up to 100 ml with more acetone

---

- add 4 ml of aniline to 96 ml of acetone and mix well
- 20 ml of 85% *ortho*-phosphoric acid

Mix the three solutions just prior to use.

2. Spray the plate, air dry, and then heat at 100 °C for 10 min. The colour appears after 2–4 min. This spray reagent can be used for aldoses, ketoses, deoxysugars, oligosaccharides, and uronic acids and may be used on both cellulose and silica thin-layer plates. It gives a wide variation of colours with different carbohydrates, aldoses producing blue grey spots whereas ketoses give light red spots. The sensitivity is about 1 µg.

3. For the PC dip reagent, dissolve 0.15 g of diphenylamine in 25 ml of ethyl acetate. Add 0.8 ml of aniline and 75 ml of ethyl acetate. Finally add 1.0 ml of water and 10 ml of concentrated *ortho*-phosphoric acid. The reagent should be made up immediately before use.

4. Dip the dry chromatograms through the reagent, dry, and heat at 95–100 °C until the background is faintly grey. This is a good general purpose reagent giving similar results to the spray reagent.

B. *Naphthoresorcinol/ethanol/sulfuric acid*

1. Prepare solution A by dissolving 0.2 g of naphthoresorcinol (naphthalene-1,3-diol) in 100 ml of 95% ethanol. Diphenylamine (0.4 g) may be added in order to reduce the background coloration.

2. For the spray reagent, carefully add 4 ml of concentrated sulfuric acid to 96 ml of solution A just prior to use.

3. Spray the plate and then heat at 100–150 °C for 5 min.

4. This spray reagent can be used for aldoses, ketoses, uronic acids, deoxysugars, glycosides, and oligosaccharides. It is recommended for use only on silica thin-layer plates. It gives a wide variation of distinctive colours with different carbohydrates; aldoses producing blue or violet spots, ketoses producing pink or red spots, and uronic acids producing characteristic blue spots. The sensitivity is between 0.1 µg (L-sorbose) and 4 µg (D-glucose).

C. *Ceric sulfate/sulfuric acid*

1. Mix 5 ml of 0.1 M ceric sulfate with 100 ml of 15% (v/v) sulfuric acid. This can be used as a dip reagent for silica HPTLC plates.

2. Heat rapidly to 120 °C for 15 min to develop the colour.

3. Hexoses, deoxyhexoses, and pentoses appear as blue grey, brown, or pinkish-grey spots, respectively, on a white background. The spots can be quantified by densitometric scanning at 440 nm. Gentle brushing of the plates with a camel hair brush immediately prior to dipping

**Protocol 11.** *Continued*

provides a cleaner background by removing small dust particles that otherwise char. The sensitivity is between about 1 μg (hexoses and pentoses) and 10 μg (hexosamines).

## 4.3 General thin-layer chromatographic method

For one-dimensional development, samples and standard mixtures (made by dissolving 0.05–2 mg dry component in 0.5 ml water plus 0.5 ml isopropanol) are applied, via a drawn-out capillary tube, as short streaks (1–2 cm long) 1.0 cm from the edge to be dipped in the eluent and parallel to that edge. Specialist spray applicators may be used. The volume applied should depend on the sensitivity of the detection system and the estimated concentration of the carbohydrate mixture. If more than one application is necessary, the previous application should first be dried well by a gentle draught of warm (not hot) air from a hand-held hair dryer. After sample application, the dry plates are placed in chromatographic tanks to which the elution solution has already been placed sufficient in advance to saturate its atmosphere. Filter paper placed along one side of the tank will help this process. Given a choice, the coated face of the plate should be pointed towards the eluent rather than away from it. The top should be placed on the tank and the chromatogram developed at room temperature. This usually takes about 2–6 min/cm. If a multiple run is necessary (e.g. using a cellulose plate in ethyl acetate/pyridine/water (solvent system (a)) for hexoses), the plate should be removed and dried thoroughly in a stream of warm air before running again, in order to avoid streaking and diffuse spots. Each run should be slightly longer than the previous one.

If a two-dimensional development is to be attempted, the sample is applied as a spot to a corner position on the plate 1.0 cm from each side. The second run will be at right angles to the first in a significantly different solvent. The plate should be thoroughly dried in a stream of warm air before the second run is commenced. A difficulty with this method concerns the standards since they have to be included with the unknown sample and co-chromatographed. Urea is reported to be a useful internal standard as it is generally well separated from carbohydrates but can be identified with the naphthoresorcinol reagent (*Protocol 11*B). The plates should be thoroughly dried in a stream of warm air before using the spray reagents for detecting the sugars. It is suggested that a commercial aerosol spray gun be used, spraying into a fume cupboard.

HPTLC precoated plates are now available from a number of suppliers. These have a much narrower particle size and particle size range enabling an improved resolution at higher efficiency. In addition more samples may be applied to each plate due to the small degree of spreading and, hence, spot size. The smoother, more homogeneous, media gives improved reproducibility and samples may be quantified by densitometry in a far more satisfactory manner than standard TLC plates. Some of these plates also come with a concentration zone for improved sample application. The major and only drawback to the use of these plates is their relatively high cost (> £3 per 10 × 10 cm plate, in 1993).

## 5. Paper chromatography (25)

Paper chromatography has similar characteristics to TLC on cellulose. It is mainly a partition process with the possibility of some adsorption. It is a cheap and simple method and easier to use, on a preparative scale than TLC. The use of dip reagents for visualizing the spots (*Protocol 11*A, step 4, see *Table 1*) is also an easier and more uniform process than spraying. Although the solvents, developed for cellulose-coated TLC may be used in paper chromatography, some systems have been developed particularly for the paper medium. One such solvent system is *n*-butanol/pyridine/water (see Section 4.1, solvent system (f)). In general, the preferred conditions are descending chromatography on Whatman No. 1 paper. Multiple or continuous development may be used. For this latter technique pinking shears are used on the bottom end of the paper to ensure a uniform flow of the solvent. Thicker paper (e.g. Whatman No. 17) can be used for preparative separations of up to about 0.5 grams, bands being visualized by using a dip reagent on a copy of the chromatogram 'blotted' onto thin chromatographic paper (e.g. Whatman No. 1). Chromatography paper is most often used straight from the pack but pre-washing (e.g. with 0.25% (w/v) hydroxyquinoline in 8% (w/v) acetic acid or sequentially with 0.1 M HCl, water, and chloroform/methanol (2:1 v/v)) may be beneficial. In either case the paper should be well equilibrated in the solvent vapours for several hours before use. For reproducible $R_f$s the chromatographic runs should be performed under constant temperature conditions in cold (slower running) or warm (faster running) rooms. Many a run has been spoilt by running overnight in a laboratory with no overnight heating.

## 6. High performance liquid chromatography

The separation and quantification of the components in mixtures of monosaccharides forms an important part of carbohydrate analysis. There are two main methods available, HPLC and GLC, which both deliver quantitative data where standards are available. The choice between these two depends on a number of factors in addition to personal preferences and prejudice. One clearly important factor is the available resources and expertise. Both methods need fairly expensive equipment and are facilitated by helpful and experienced operators. Experience in running these systems can only be built up over a fairly long period through a number of mishaps and, therefore, it is best if it is gained second-hand. As a broad generalization, HPLC is preferred for the analysis of simple monosaccharide mixtures, oligosaccharide analysis and purification, whereas GLC can be used on very complex monosaccharide mixtures. There are, however, protocols available for separating and analysing most mixtures of carbohydrates by either method. Advances in both methods, but particularly HPLC, are constantly being made and reported in the literature (e.g. in *Analytical Biochemistry* and the *Journal of Chromatography*) and by the column manufacturers and suppliers. This section describes the practical use of HPLC for carbohydrate analysis and Section 7 describes GLC.

## 6.1 Experimental approach

There are a number of different HPLC columns and processes for separating carbohydrates which depend on different chemical and physical properties for resolution (see (26, 27) and *Table 2*). This is a somewhat confusing state of affairs but no single method is useful over the entire range of possible separations. Advice and literature from prospective column suppliers, concerning the separation required, should be sought before any column is bought.

Sulfonated polymeric columns containing metal-loaded cation exchangers, at moderately high temperatures ($\sim$ 85 °C), or amino-bonded silica columns operating at around ambient temperature are popular. The cation exchangers act by ion-moderated partition, forming linkages to the sugar hydroxyl groups, and size exclusion. The variations in resolution between columns containing the different metal counterions ($Ag^+$, $Ca^{2+}$, and $Pb^{2+}$) are due to the different modes of complex formation. Elution is roughly in order of increasing affinity but decreasing molecular weight; the resolution normally increasing with the temperature as long as the ion-exchange resin is stable. Of these cation exchangers, the $H^+$ form is suited for carbohydrate mixtures including acids and the separation of amino sugars, the $Ca^{2+}$ form is the most popular general purpose column which is used for corn sugars and syrups and at low temperatures (1.5 °C) separates $\alpha$- and $\beta$-anomers, the $Pb^{2+}$ form is used for mixed hexoses and pentoses such as those released from cellulosic materials, and the $Ag^+$ form is most useful where oligosaccharides also need to be separated from monosaccharides. It is essential to remove $Na^+$ and $K^+$ ions (e.g. by prior de-ashing ion-exchange or use of a suitable guard column) from samples when using these metal-loaded columns or they will exchange so reducing the efficiency of the separations. Where particularly high levels of $Na^+$ might be expected (e.g. molasses) a $Na^+$-loaded column may be useful, so avoiding the de-ashing.

Amino-bonded silica columns use an acetonitrile/water mobile phase and separate by hydrophobic and polar interactions and partition between the acetonitrile-rich mobile phase and the water-enriched stationary phase. Separation processes involving partition may make use of a number of different types of stationary phase: amino-derivatized anion-exchange resins, cation-exchange resins (H[+] form, see ref. 28 and *Figure 1*), and derivatized silica. Monosaccharides may be separated in all these processes using aqueous acetonitrile solvents. The amino-derivatized columns tend to lose their

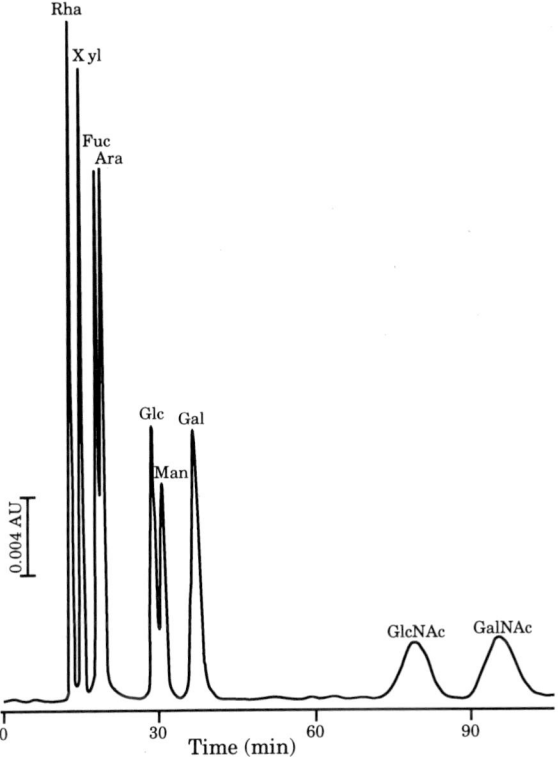

**Figure 1**. Analysis of an equimolar mixture of aldoses and *N*-acetylhexosamines. 5 nmol each of a mixture of L-rhamnose, D-xylose, L-fucose, L-arabinose, D-glucose, D-mannose, D-galactose, *N*-acetyl-D-glucosamine, and *N*-acetyl-D-galactosamine were dissolved in 20 μl water/acetonitrile (8:92 v/v) and applied to a Shodex DC-613 column (4 mm i.d. × 25 cm). The column was a highly cross-linked sulfonated polystyrene resin in the H[+] form which separates by partition chromatography. It was run isocratically at 30 °C, 0.6 ml/min, using aqueous 92% (v/v) acetonitrile with post-column detection by UV 280 nm absorbance after reaction with 2-cyanoacetamide/borate (*Protocol 12*). This system can be used for the analysis of carbohydrates derived from glycoproteins. The amino sugars produced by hydrolysis must be reacetylated (see *Protocol 13, step 6*) before analysis. *N*-Acetyl- and *N*-glycollylneuraminic acids may be analysed (as *N*-acylmannosamines) after hydrolysis and conversion using *N*-acetylneuraminidase (0.5 U) and *N*-acetylneuraminate pyruvate-lysate (0.3 U, 1 h 37 °C, 200 μg glycoprotein in 800 μl 0.06 M phosphate buffer pH 7.0). From ref. 28 with permission.

**Table 2.** Typical conditions for the HPLC of monosaccharides

| Column type | Mobile phase | Separation mechanism | Typical analysis | Commercial analytical columns |
|---|---|---|---|---|
| Anion exchange (quaternary ammonium)[a] | (a) Sodium hydroxide | Anion exchange | Alditols, mono-saccharides | Dionex CarboPac MA1 (0.4 × 25), Dionex CarboPac PA1 (0.4 × 25)[b] |
| | (b) Acetate buffers | | Sialic acids, uronic acids | Bio-Rad Aminex A-29 (0.9 × 99) |
| Anion exchange (amino-propylsilane bonded silica, OH⁻ form) | Acetonitrile/water | H-bonding between hydroxyls and amines | Monosaccharides, oligosaccharides | Waters Carbohydrate Analysis (0.39 × 30), Supelco Supercosil LC-$NH_2$ (0.46 × 25), Whatman Partisil 10 PAC (0.4 × 25)[c], Varian MicroPak AX-5 (0.4 × 30), Merck Lichrosorb $NH_2$ (0.4 × 25), Alltech Carbohydrate 1[d] |
| Cation exchange (sulfonate, H⁺ form)[a] | (a) Citrate buffers (b) Acetonitrile/water | Cation exchange Ion-moderated partition | Hexosamines Glycoprotein-derived carbohydrates, uronic acids, lactones | Pierce PC-6A (0.9 × 30), Bio-Rad Aminex HPX-87H (0.78 × 30) |
| Cation exchange (sulfonate, $Ca^{2+}$ form)[a] | Water | Ion-moderated partition | Alditols, mono-saccharides, and oligosaccharides | Waters Sugar-Pak I (0.65 × 30), Bio-Rad Aminex HPX-87C (0.4 × 25) |

| | Mobile phase | Separation mechanism | Analytes | Columns (dimensions in cm) |
|---|---|---|---|---|
| Cation exchange (sulfonate, Pb$^{2+}$ form)[a] | Water | Ion-moderated partition | Pentoses and hexoses | Bio-Rad Aminex HPX-87P (0.78 × 30), Bio-Rad Fast Carbohydrate (0.78 × 10) |
| Cation exchange (sulfonate, Ag$^+$ form)[a] | Water | Ion-moderated partition | Monosaccharides, oligosaccharides | Bio-Rad Aminex HPX-42A (0.78 × 30), Bio-Rad Aminex HPX-65A (0.78 × 30) |
| Silica, straight phase | (a) Acetonitrile/water | Polar interactions | Prederivatized carbohydrates | Whatman Partisil O (0.46 × 25), Merck (LiChrosorb Si60 (0.4 × 25) |
| | (b) Acetonitrile/water + diaminoalkanes | H-bonding between hydroxyls and amines | Monosaccharides, oligosaccharides | Waters μPorosil (0.39 × 30) |
| Silica, reverse phase | Acetonitrile/water | Hydrophobic interactions | Prederivatized carbohydrates, oligosaccharides | Waters μBondakpak C$_{18}$ (0.39 × 30), Merck LiChrosorb RP-18 (0.4 × 25), Waters Radialpak C$_{18}$ (0.4 × 30), Supelco Supelcosil LC-18 (0.45 × 25), DuPont Zorbax ODS (0.46 × 25), Toso Haas TSK ODS-120-T (0.46 × 25) |

[a] Bonded to polymerized styrene/divinylbenzene (PS-DVB).
[b] Ethylvinylbenzene/divinylbenzene pellicular non-porous.
[c] Contains cyano and secondary amine groups (2:1); resistant to Schiff base formation.
[d] Cross-linked amino-propyl phase; resistant to hydrolysis.

resolving power quicker than average due to stripping of the aminopropyl ligand and Schiff base formation. A similar chromatographic separation is achievable using an underivatized silica solid phase plus a diaminoalkane co-eluent with the aqueous acetonitrile. Anion-exchange chromatography, at high pH using strongly basic anion-exchange resins, is a reproducible method which is rapidly becoming preferred (*Figure 2*). Carboxylic acids and polyhydroxyl compounds, such as the carbohydrates, are negatively charged under such conditions of high pH and are separated by anion exchange. Although base-catalysed rearrangements were originally thought to be a possible drawback to this method, they have been shown not to occur under the conditions used. Derivatized carbohydrates may be analysed by standard techniques using normal-phase or reverse-phase silica columns at ambient temperatures (*Table 2*).

Optimization of the conditions for the HPLC separation of a number of known carbohydrates is a straightforward if somewhat lengthy and tedious process. The variation of elution times of all the components with the separation parameters (e.g. temperature, acetonitrile concentration, buffer pH) must be determined. The optimum conditions may then be determined by computer analysis of the resultant curves or by visual inspection of the curves generated for the resolution of all pairs of components against the separation parameter chosen (see ref. 28 for further details).

The columns used in HPLC are usually bought ready-packed. They may be packed in-house, however, with a saving in cost and a possible increase in efficiency. The best method appears to be slurry-packing. The calculated amount of the packing material is dispersed in the slurry medium (often

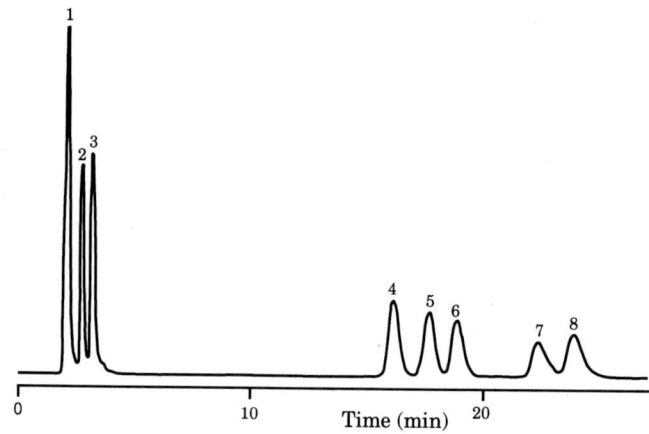

**Figure 2.** Separation of monosaccharides by anion-exchange HPLC on a Dionex CarboPac PA1 column eluted with 26 mM NaOH. Detection was by PAD after post-column addition of 0.5 M NaOH. The sugars (1 nmol each) are: 1, *myo*-inositol plus glycerol; 2, D-arabitol; 3, D-sorbitol; 4, D-galactose; 5, D-glucose; 6, D-mannose; 7, D-fructose; 8, D-ribose. From ref. 37 with permission.

water) by use of an ultrasonic bath and thoroughly degassed. The slurry is then pressed into the column by use of the expected elution solution at a high enough pressure to achieve at least twice the analytical flow rate. In use, the columns gradually deteriorate. This process of deterioration can be reduced by careful attention to procedure. All solvents to be passed through the column should be as pure as possible and have been filtered through a 0.5 μm filter, to avoid plugging of the frits by fine particles, and thoroughly degassed, preferably by use of helium bubbled through air stones. Where possible, microbial growth inhibitor (e.g. 0.01% (w/v) sodium azide) should be present in all aqueous phases. Samples applied to columns should also be free from particulate matter and other gross contaminants (e.g. proteins). The analytical column should, where feasible, be protected from contamination by use of a short 'guard' column containing the same type of resin (although possibly with a larger bead size). It should also be used only within the manufacturer's recommended range of pH and solvent. This is particularly important for silica-based stationary phases which dissolve readily at alkaline pH. At the end of each analytical session the column and all eluent delivery lines should be rinsed through with pure solvent (e.g. water) especially if salt solutions were being used. Where the column is constantly being removed from the HPLC apparatus it should be stored carefully and not dropped. Even with all these precautions some alkylated silica columns have a fairly short life of as little as three months due to particle dissolution and loss of the bonded phase. If, however, the analytical method fails suddenly, suspect the sample preparation before the column.

Amino sugars can be released by hydrolysis from glycoproteins (4 M HCl, 4 hours, 100 °C, under nitrogen) and determined by use of an amino acid analyser with ninhydrin detection or after derivatization (*Tables 2–4*).

**Table 3.** Pre-column derivatization of carbohydrates for use in $C_{18}$ reverse-phase HPLC

| Carbohydrate | Reagent | Sensitivity[a] | Reference |
|---|---|---|---|
| Reducing sugars | (a) Aminopyridine | 10 | (30) |
| | (b) Dansyl hydrazine | 10 | (31) |
| Monosaccharides | Benzoyl chloride | 10[b] | (32) |
| Sialic acids | 1,2-Diamino-4,5-methylenedioxybenzene | 3 | (33) |
| Glycoprotein hexoses and hexosamines | Reductive amination/phenylisothiocyanate | 20 | (34) |
| Amino sugars | Reduction/o-phthalaldehyde | 20 | (35) |

[a] pmol sample, as determined by fluorescence; about 10% applied to chromatographic column.
[b] By UV absorbance.

**Table 4.** Post-column detection of specific classes of carbohydrates

| Carbohydrate | Reagent | Sensitivity[a] | Reference |
|---|---|---|---|
| Carbohydrates, sugar alcohols | (a) Ammoniacal cupric sulfate | 3 | (36) |
| | (b) 0.5 M NaOH | 0.01[b] | (37) |
| Reducing sugars | Tetrazolium blue | 1 | (8) |
| Amino sugars | (a) o-Phthalaldehyde | 0.04[c] | (38) |
| | (b) Ninhydrin | 1 | (8, 39) |
| Sialic acids | 300 mM NaOH | 0.2[b] | (40) |

[a] nmol; by UV/colorimetric detection, except where indicated.
[b] Pulsed amperometric detection.
[c] Fluorimetric assay.

## 6.2 Detection methods

### 6.2.1 Direct detection

The direct detection of carbohydrates is not straightforward as they do not absorb light in the normal UV or visible range and have no fluorescence. Monosaccharides, however, do absorb at wavelengths in the far-UV. The absorption maxima are at about 188 nm but, due to noise in the detection signal below 190 nm, detection is normally performed at wavelengths between 192 nm and 200 nm. The response, depending largely on the freedom of the carbonyl group, differs between the monosaccharides, being six times higher for D-fructose ($\sim$ 2 µg in 20 µl, 0.6 mM) than D-glucose; the response of most other carbohydrates lying between these extremes. Analyses are restricted to solvents which do not absorb significantly. The response is linear over a wide range but the absolute response depends on other parameters such as the solvent purity, the reference cell, and the spectrophotometer's condition. Detection in this range requires an expensive instrument and a very pure solvent. Acetonitrile/water mixtures are often used. A reliable supplier of non-absorbing HPLC grade acetonitrile should be sought as this has often been found to be the critical component and it is not easy or practicable to purify in-house.

Another detection method of general applicability makes use of changes in refractivity. The refractive index detector is inherently as sensitive as absorbance measurement in the far-UV but, in practice, is at least ten times less sensitive due to the background noise caused by pump pulsations and temperature fluctuations. The method is generally applicable to carbohydrate detection but is not suitable for separations involving gradient elution. The sensitivity of detection varies between carbohydrates and with the solvent composition but is generally linear.

Pulsed amperometric detection (PAD) is an electrochemical method (29) for detecting carbohydrates that is rapidly proving most popular and has reduced the need for post-column detection, by other means. It is a very sensitive technique (~ 1 ng glucose) which makes use of the electrochemical oxidation, at a gold electrode poised at positive potential, of the carbohydrates made anionic at about pH 13. The subsequent pulsing of the electrode with high positive and negative potentials (which provides the name for the device) being necessary in order to clean the electrode surface. This detection method is usually used after anion-exchange chromatography on quaternary ammonium resins using NaOH as eluent, but may also be used after post-column reaction of column eluates produced by other methods; particularly if base-labile groups (e.g. *O*-acetyl) groups are present. It is very important to ensure that the NaOH contains minimal carbonate on preparation and by degassing with helium gas and storing under helium. A metal-free chromatography system must be used if strong alkali is employed.

### 6.2.2 Pre-column derivatization

In order to increase the sensitivity of detection of carbohydrates, and allow the use of highly efficient normal or reverse-phase silica columns and gradient elution profiles, they may be derivatized to give light-absorbing or fluorescent materials before separation on the HPLC column. Pre-column derivatization not only increases the carbohydrate sensitivity to detection but also substantially changes its chromatographic behaviour. Suitable procedures are given in *Table 3*.

### 6.2.3 Post-column derivatization

In general, post-column derivatization should be the method of choice if a greater sensitivity than that achieved by direct detection is required, since using this approach the separation process can make full use of the differences between the carbohydrates without the addition of a number of groups with intrinsically similar properties. Post-column derivatization is, in general, a straightforward technique requiring one or two additional pumps, a mixing coil of Teflon tubing, and a thermostatted bath. The derivatization process must be compatible with the mobile phase used. A generally applicable but simple and sensitive method is fully described below (*Protocol 12*). The assay uses non-corrosive reagents, shows good linearity, and is highly sensitive for aldoses, hexosamines, and alduronic acids. A number of other methods are given in *Table 4*. Quantification from the detector's response is made by direct comparison with standards (or standard curves). The amounts are usually proportional to the area of the peak in the detector output. This may be measured by use of an integrator or, more cheaply, by physically weighing on a balance the peak cut out from a photocopy of the trace. This method is particularly useful for overlapping peaks. If the system gives reproducible elution times, the peak heights are normally proportional to the amount of material for any given compound.

## 6.2.4 Sialic and uronic acids

Sialic (40) and uronic acids (41) can be separated using anion-exchange chromatography on strong anion-exchange resins using a pH gradient of acetate buffers and detected by PAD after post-column addition of NaOH.

## 6.2.5 Inositols

The naturally occurring inositols can be separated on $Ca^{2+}$ cation-exchange resins using deionized water at 50 °C and detected by PAD after post-column addition of NaOH (42).

---

**Protocol 12.** Post-column detection using 2-cyanoacetamide (28, 43)

- Sensitivity: photometric; 18 ng glucose in 20 μl (5 μM)
- Sensitivity: fluorometric; 2 ng glucose in 20 μl (0.5 μM)

*Reagents*

A  Dissolve 2-cyanoacetamide in methanol, decolorize with activated carbon, and recrystallize from methanol. Dissolve 1.0 g of the purified 2-cyanoacetamide in 100 ml of water. This solution may be stored in a refrigerator within a dark bottle for at least one month.

B  Dissolve 31 g of boric acid in 600 ml of water. Make the solution pH 9.5 by the addition of 10% (w/v) potassium hydroxide and dilute with water to a final volume of 1 litre (0.5 M borate).

*Method*

1.  Mix the column eluate (0.6 ml/min; if a different elution rate is used, scale all the rates accordingly) successively with reagent A (0.5 ml/min) and reagent B (0.5 ml/min) by use of Teflon Y-shaped joints.

2.  Pass the mixture through an open tubular Teflon reaction coil (10 m × 0.5 mm i.d.) held at 100 °C in a thermostatted bath containing glycerol (80 sec contact time) followed by an air-cooled Teflon cooling coil (1 m × 0.5 mm i.d.).

3.  Determine the absorbance at 280 nm (274 nm maximum) or the fluorescence at 331 nm (excitation)/383 nm (emission).

*Note*: The fluorescence is quenched by acetonitrile, if present in the eluate, but the absorbance is unaffected.

---

# 7. Gas–liquid chromatography

## 7.1 Experimental approach

GLC is preferred to HPLC in a number of instances. It is a sensitive technique, allowing the analysis of sub-nanomolar amounts of carbohydrates, and is generally less prone to interference (e.g. from salts and protein). Detection is usually by means of a flame ionization detector (FID) which responds to all carbohydrate related molecules over an extremely wide linear range. GLC separation is dependent upon the differential extractive distillation of the components in the mixture. It is fundamental to the technique, therefore, that volatile derivatives of the carbohydrates are prepared. There are two schools of thought concerning this derivatization. One is to produce a single derivative from each carbohydrate and the other allows several derivatives to be produced dependent upon the anomeric composition. The advantage of the former is that the chromatograms are simpler and the sensitivity may be marginally greater. However by producing more than one derivative in a well defined mass ratio, a recognizable 'fingerprint' can be obtained which is confirmatory for a particular component. This reduces the possibility of a chromatographic peak being incorrectly assigned to a carbohydrate when actually due to an unexpected impurity. It is not necessary to simplify analyses of complex mixtures to one peak/ component as highly efficient, fused-silica, wall coated open tubular (WCOT) columns are generally powerful enough ($\geq$ 100 000 theoretical plates) to resolve a large number of components ($>$ 30). The much better resolution ($>$ tenfold) and speed ($>$ fivefold) of WCOT columns, compared with packed bed columns, recommends them for routine analytical work; packed bed columns being rarely used these days. There is no substantial change in the relative elution positions between these two types of column if similar liquid phases are used (*Table 5*). WCOT columns are more fragile and deteriorate

**Table 5.** Alternative liquid phases for the GLC of carbohydrates

| Polarity | Typical composition | Alternative phases |
|---|---|---|
| Most nonpolar | Dimethyl silicone | SE-30, OV-1, BP-1, OV-101, CP-Sil 5CB |
| Nonpolar | 5% Phenyl, methyl silicone | SE-52, SE-54, BP-5, CP-Sil 8CB |
| Intermediate polarity | 50% Phenyl, methyl silicone Cyanopropyl, phenyl, methyl silicone | OV-17, BP-10 CP-Sil 19CB |
| Polar | Trifluoropropyl, methyl silicone | OV-202, OV-210 |
| Most polar | Cyanopropyl silicone, or similar | OV-275, CP-Sil 88, Silar 9CP, BP-75, SP-2330 |

more easily than packed bed columns. In general, this deterioration may be greatly reduced by the use of oxygen-free dry helium as the carrier gas. This helium should be passed through the appropriate gas purifiers before entering the column as oxygen has a particularly devastating effect on the column. If well treated, and run below the supplier's maximum operating temperature limit, the WCOT columns should last for over a year of continuous use, which is longer than most HPLC columns last.

Internal standards are generally necessary for GLC analyses since it is difficult to inject a reproducible proportion of a sample into the column. Standards may be any similar compound which is not already present in the mixture to be analysed and which is clearly separable from the other components. *Myo*-inositol (mesoinositol) may be used, except for plant or membrane carbohydrate analyses where it is usually already present. An alternative internal standard is pentaerythritol. Glucose should be avoided as an internal standard, even if it is known not be be expected in the samples, as it is a common, and annoying, contaminant.

## 7.2 Detection methods

Three GLC derivatization methods are described in *Protocols 13–15*. *Protocol 13*, involving trimethylsilylation, is a good, generally applicable,

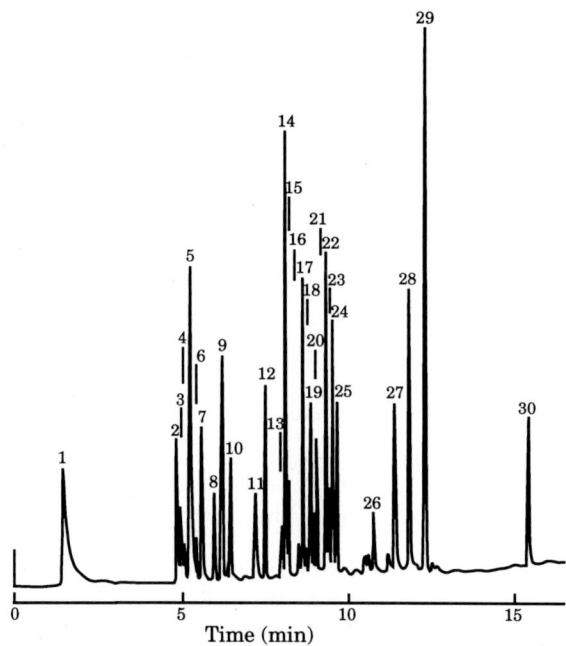

**Figure 3.** GLC of trimethylsilylated monosaccharides. Approximately 0.5 nmol of each monosaccharide was derivatized (see *Protocol 13*). The peaks are identified in *Table 6*. From ref. 45 with permission.

**Table 6.** GLC data of trimethylsilylated monosaccharides

| Parent carbohydrate | Peak[a] | Time (min)[b] | Peak height[c] |
|---|---|---|---|
| (Solvent front) | 1 | 1.00 | |
| L Arabinose | 2 | 4.82 | 0.271 |
| | 3 | 4.95 | 0.122 |
| | 6 | 5.36 | 0.024 |
| D-Ribose | 4 | 5.03 | 0.064 |
| | 5 | 5.26 | 0.317 |
| L-Rhamnose | 5 | 5.25 | 0.433 |
| | 6 | 5.36 | 0.046 |
| L-Fucose | 5 | 5.28 | 0.054 |
| | 7 | 5.52 | 0.266 |
| | 8 | 5.80 | 0.156 |
| D-Xylose | 9 | 6.10 | 0.377 |
| | 10 | 6.37 | 0.188 |
| D-Glucuronic acid | 11 | 7.14 | 0.069 |
| | 23 | 9.37 | 0.073 |
| | 24 | 9.49 | 0.240 |
| D-Galacturonic acid | 12 | 7.31 | 0.186 |
| | 13 | 7.90 | 0.056 |
| | 19 | 8.79 | 0.160 |
| | 20 | 8.90 | 0.051 |
| D-Mannose | 14 | 8.02 | 0.598 |
| | 16 | 8.42 | 0.042 |
| D-Galactose | 15 | 8.12 | 0.109 |
| | 17 | 8.54 | 0.382 |
| | 18 | 8.62 | 0.048 |
| | 21 | 8.97 | 0.153 |
| D-Glucose | 22 | 9.27 | 0.478 |
| | 25 | 9.59 | 0.201 |
| *N*-Acetyl-D-glucosamine | — | 10.57 | 0.031 |
| | — | 11.17 | 0.029 |
| | 28 | 11.70 | 0.359 |
| *N*-Acetyl-D-galactosamine | — | 10.45 | 0.031 |
| | 26 | 10.70 | 0.082 |
| | 27 | 11.35 | 0.257 |
| *Myo*-inositol | 29 | 12.24 | 1.000 |
| *N*-Acetylneuraminic acid | 30 | 15.35 | 0.200 |

[a] Refers to *Figure 3*.
[b] SD = 2 sec.
[c] SD = 3.5%; peak height is proportional to peak area under these conditions.

method whereas *Protocol 14*, involving acetylation, is recommended particularly for methoxy-derivatized monosaccharides derived from the methylation of more complex molecules (e.g. glycoproteins, polysaccharides, or glycolipids). *Protocol 15*, involving oxime formation gives only one peak per sugar, and is particularly useful where fructose is present. Quantification of the components in a mixture is by analysis of the peak areas (heights) as described for HPLC (Section 6.2.3).

Most monosaccharides give a mixture of isomers on trimethylsilyl derivatization (*Table 6, Figure 3*). The ratio of the peak areas is constant, however, and may be used to confirm small amounts of material having a low signal-to-noise ratio. Labile carbohydrates (e.g. *N*-acetylneuraminic acid) are not destroyed by this analytical process and it is therefore a good method for the analysis of all the carbohydrate moieties derived from glycoproteins. Fructose, however, is not analysable by this method and should be derivatized as the trimethylsilyl or oxime derivative of the free sugar, because the methylglycoside is not formed on methanolysis. In general, free sugars may be determined by GLC analysis, as above, after derivatization of the dry material (i.e. without any methanolysis or re-*N*-acetylation steps). The retention times, number of peaks, and relative peak heights will all be different to those obtained for the methylglycosides and should be determined separately. The D and L configurations of neutral monosaccharides may be determined in a similar manner to methanolysis above but by the use of $R(-)$-2-butanol in place of methanol throughout. The trimethylsilylated $(-)$-2-butylglycosides are separable on an SE30 capillary column (44).

---

**Protocol 13.** Preparation of trimethylsilyl derivatives (45)

*Reagents*

A  Dry methanol. It may be possible to purchase this. Otherwise methanol may be dried by refluxing 500 ml with 2.5 g of magnesium turnings and 0.1 g of iodine for 1 h, followed by distillation into a clean dry container.

B  Anhydrous pyridine, bought or prepared by the addition of 30 g dry KOH pellets to 500 ml of pyridine. The mixture should be well stirred and allowed to settle for 24 h before use.

C  Methanolic hydrogen chloride. This may be prepared by either of two methods:

   (a)  Bubble hydrogen chloride gas through a train of moisture traps, containing concentrated sulfuric acid, into dry methanol (reagent A) until the weight of the methanol solution increased by about 3%. Care should be taken over preventing the sulfuric acid traps sucking back when the process is finished. The methanolic HCl should be standardized to 0.625 M by the titration of an aliquot against standard 0.1 M NaOH using phenolphthalein as an indicator, and dilution with more dry methanol.

   (b)  Add 4.65 ml of good quality acetyl chloride carefully to 100 ml of dry methanol.

D  Trimethylsilylation reagent. This may be purchased, ready mixed and ampouled (e.g. Tri–Sil® Reagent from Pierce), but can be prepared by mixing dry pyridine (10 vol.), with hexamethyldisilazane (2 vol.), and trimethylchlorosilane (1 vol.). It is best stored in sealed or serum-capped ampoules under nitrogen. An alternative reagent for use on carbohydrate syrups can be prepared by mixing dry pyridine (10 vol.), with hexamethyldisilazane (9 vol.), and good quality trifluoroacetic acid (1 vol.). This may be used in the presence of up to 5.0 mg water/ml reagent if the carbohydrate content is fairly high ($\sim$ 1 mg/ml reagent).

*Method*

A dry cabinet (glove box) should be used throughout for all reagent manipulations.

1. Dry the samples containing glycoproteins ($\sim$ 0.5–100 µg), polysaccharide or monosaccharide mixtures, or standards ($\sim$ 0.1–100 nmol) together with *myo*-inositol ($\sim$ 100 ng; or other suitable internal standard) over $P_2O_5$, under vacuum, in 100 µl or 300 µl Reacti-Vials (screw topped, serum-capped vials, e.g. as supplied by Pierce. Note that crystalline standard *N*-acetylneuraminic acid contains acetic acid of crystallization which prevents the use of $P_2O_5$ for drying since it causes decomposition).

2. Add a mixture of dry methanolic HCl (4 vol.) and methyl acetate (1 vol., total volume of about 40–200 µl depending on the expected carbohydrate content). If reagent C(b) is used (see above), which already contains some methyl acetate, the ratio of methanolic HCl to additional methyl acetate should be 5:1 (v/v). Seal the tube by means of the Teflon-lined septa in the screw caps and mix thoroughly be use of a vortex mixer.

3. Incubate the tubes at 70 °C in an oven for about 16 h. Check the screw caps for tightness after the first 10 min of this incubation and vortex mix the tubes again before replacing in the oven. Only very occasionally will this system of sealing fail ($\sim$ 1 in 50), usually due to use of a cracked or chipped tube.

4. After cooling, add redistilled 2-methyl-2-propanol (*t*-butyl alcohol, 20% v/v).

5. Vortex mix the samples and evaporate to dryness using a stream of dry nitrogen (oxygen-free grade) at room temperature.

6. Add 50 µl of dry methanol, 5 µl of pyridine, and 5 µl of acetic anhydride successively with intermediate mixing in order to re-*N*-acetylate the amino sugars (twice these volumes should be used if more than 50 nmol of amino sugars is present; fatty acid methyl esters may be removed, if suspected, by extraction with hexane).

7. Leave the solutions at room temperature for 15 min before evaporating to near dryness by use of a stream of dry nitrogen (oxygen-free grade).

8. Completely dry the residue using gentle vacuum over $P_2O_5$.

**Protocol 13.** *Continued*

9. After this thorough drying stage, add the silylation reagent (reagent D, ~ 20–100 μl depending on the carbohydrate content), vortex mix the tubes for 30 sec, and leave for 1 h at room temperature. The derivatized samples should be analysed the same day.

10. Re-evaporate using the dry nitrogen stream and immediately dissolve each sample in 14–50 μl of redistilled hexane.

11. Apply some or all of this sample to the capillary column. The set-up recommended is a fused-silica WCOT column (25 m × 0.32 mm i.d.) coated with CP Sil 5 liquid phase (see *Table 5* for alternative liquid phases; any other reasonably equivalent fused-silica column is suitable), using an all-glass solid injector (moving needle, Chrompack, previously inactivated by reaction with the silylation reagent Rejuv 8; supplied by Supelco, but any similar product is suitable). Other injection techniques may be used but this method has the advantage that all the sample is injected onto the column without excessive solvent.

12. Perform the analyses using dry oxygen-free helium as carrier gas and a temperature program (140 °C for 2 min, then increasing at 8 °C/min up to 240 °C). A higher final temperature improves the analysis but tends to cause the more rapid deterioration of the shrinkable Teflon link between the injector and the column.

**Protocol 14.** Preparation of alditol acetate derivatives (46, 47)

*Reagents*

A Dry dimethyl sulfoxide, prepared by storage of material, from a previously unopened bottle, over molecular sieve type 4A. Care should be taken not to disturb any fine sediment that may be present when removing aliquots for use.

B Dissolve 2 g of sodium borohydride in 100 ml of anhydrous dimethyl sulfoxide (reagent A) at 100 °C. This reagent is stable if kept dry at 4 °C. If deuterium is to be inserted (see *Table 7*) then $NaBD_4$ should be used in place of the $NaBH_4$; no special precautions are needed.

*Method*

1. Dissolve the samples and standards containing up to about 500 μg of the monosaccharides in 0.1 ml of 1 M ammonia solution (prepared by diluting 3 ml of concentrated ammonia solution, of sp. gr. 0.88, with 50 ml of water).

2. Reduce the carbohydrates by the addition of 1.0 ml of reagent B followed by incubation at 40 °C for 90 min.[a]

3. After reduction, decompose the excess sodium borohydride by the addition of 0.1 ml of glacial acetic acid.

4. The sample can be cleaned somewhat of contaminants, such as surfactants and lipids, at this stage by selective extraction with *n*-heptane (2 ml) from an equal volume of sample dissolved in acetonitrile/water (9:1 v/v), keeping the bottom aqueous acetonitrile layer, followed by rotary evaporation.

5. Peracetylate the reduced monosaccharides by the addition of 0.2 ml of 1-methylimidazole (catalyst), followed by 1.0 ml of acetic anhydride, and thorough mixing.

6. After 10 min at room temperature, add 2 ml of water to decompose the excess acetic anhydride.

7. After cooling, extract the peracetylated carbohydrates into the lower layer formed after vortex mixing the aqueous solution with 1 ml of dichloromethane. This layer may be removed via a Pasteur pipette and stored in a septum-capped vial at −20 °C until analysis on a polar OV-275, or similar, column (*Table 7*, *Figure 4*, see ref. 48).

[a] Some carbohydrate pairs are reduced to the same alditol and cannot be resolved by this method, e.g. lyxose/arabinose, glucose/gulose, and altrose/talose. In addition, ketoses are not reduced stereospecifically and give rise to isomeric hexitols, e.g. D-fructose gives D-glucitol and D-mannitol, and D-sorbose gives L-glucitol and D-iditol.

---

**Protocol 15.** Preparation of oxime derivatives

*Reagent*

Dissolve 2.5 g hydroxylamine hydrochloride in 100 ml redistilled pyridine.

*Method*

1. Dry the samples and standards containing up to about 5 mg of the monosaccharides in a screw topped, Teflon-capped vial. Add 500 μl of the hydroxylamine reagent,[a] cap, and heat for 30 min at 70–75 °C.

2. Cool to room temperature, add 500 μl of hexamethyldisilazane, and mix.

3. Add 50 μl anhydrous trifluoroacetic acid, cap, vortex mix, and leave at room temperature for 30 min to allow the white precipitate to settle.

**Protocol 15.** *Continued*

4. Add 500 µl *N*-trimethylsilylimidazole, cap, vortex mix, and leave at room temperature for a further 30 min.

5. Apply some or all of this sample to an intermediate polarity capillary column, such as OV-17.

---

*ª* Where sucrose is thought to be present, add 27 µl dimethylaminoethanol to avoid its slight, but detectable, hydrolysis.

### 7.2.1 Inositol analysis

The naturally occurring inositols can be simply separated by GLC as hexakis-*O*-trimethylsilyl derivatives. The trimethylsilyl derivatives may be produced by the silylation reagent used for monosaccharide derivatization (*Protocol 13*) but allowing the reaction to proceed for 48 hours due to the slowness with which *neo*-inositol reacts (49).

## 8. Mass spectrometry

Electron impact mass spectrometry (EI–MS) is an extremely powerful analytical technique and is usually used in conjunction with GLC (as GLC–MS) (50). Together they have become the fastest technique for the analysis of complex mixtures using two sets of data, the retention times and the mass spectra. EI–MS is based on the positive ionization of molecules due to bombardment with a beam of electrons. As carbohydrates are normally converted into volatile derivatives before analysis, the linkage of MS to GLC, which also requires volatile derivatives, is clearly beneficial. The mass spectra from monosaccharide derivatives consist of primary fragments formed from the initial cleavage of the molecular ion plus many secondary fragments due to the elimination of neutral molecules from these primary fragments. The fragmentation pattern may be used for identification by comparison with spectra from known materials or by deduction from the known cleavage and elimination probabilities (51). In general, the molecular ion of a derivatized carbohydrate is not seen and large primary fragments tend to eliminate neutral molecules quite readily. Most of the usable spectra are, therefore, below about 300 m/z. Care should also be taken not to over-rely on the absolute or relative peak heights in the spectra as these may vary considerably from day to day, depending on the precise conditions used. However if care is taken, and suitable peaks are chosen, a reproducibility of better than ± 10% is possible. One further drawback to the use of mass spectrometry is that there is generally little difference between isomers. This is easily overcome by the additional use of the retention data. GLC–MS is a particularly powerful technique for the analysis of partially methylated alditol acetates (*Figure 4, Table 7*) especially as known standards are not always available. The mass

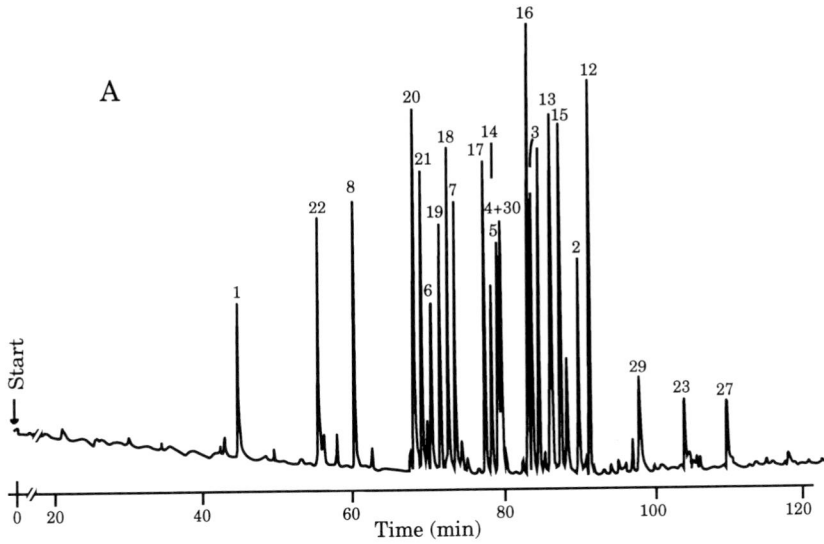

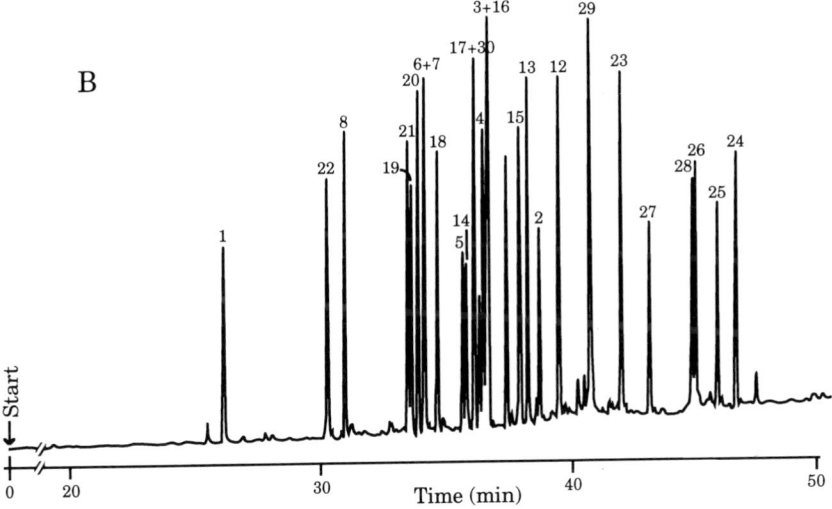

**Figure 4.** GLC of peracetylated hexitols and methyl hexitols. 25–100 pmol of each of 32 *O*-methylated hexitol and 2-deoxy-2-(*N*-methyl)acetamidohexitol acetates, most of which are commonly found during the methylation analysis of *N*-glycosidically linked glycoprotein oligosaccharides, were analysed by (A) capillary GLC using polar (22 m × 0.25 mm i.d. WCOT column containing Silar 9CP) and (B) nonpolar (85 m × 0.25 mm i.d. WCOT column containing OV 101) liquid phases. The peaks are identified in *Table 7*. The analytical conditions were (A) 20 min at 100 °C then raised to 230 °C at 1 °C/min, and (B) 100 °C to 240 °C at 3 °C/min. From ref. 48 with permission.

**Table 7.** Retention times and mass spectra primary of methyl hexitol and hexitol acetates

| Peak[a] | Positions of methylation[b] | Retention time[c] | | Primary fragments in the mass spectra | |
| --- | --- | --- | --- | --- | --- |
| | | Silar 9CP | OV 101 | Sodium borohydride | Sodium borodeuteride |
| **Fucitol** | | —[d] | 1.107 | 73, 145, 217, 231, 289 | 74, 146, 218, 231, 290 |
| 1. | 2,3,4 | 0.808 | 0.864 | 117, 131, 161, 175 | 118, 131, 162, 175 |
| **Galactitol** | | —[d] | 1.385 | 73, 145, 217, 289 | 73, 74, 145, 146, 217, 218, 289, 290 |
| 2. | 2,4 | 1.619 | 1.273 | 117, 189, 233, 305 | 118, 189, 234, 305 |
| 3. | 3,6 | 1.512 | 1.208 | 45, 189, 233 | 45, 190, 233 |
| 4. | 4,6 | 1.439 | 1.201 | 45, 161, 261 | 45, 161, 262 |
| 5. | 2,3,4 | 1.431 | 1.176 | 117, 161, 189, 233 | 118, 162, 189, 233 |
| 6. | 2,4,6 | 1.270 | 1.126 | 45, 117, 161, 233 | 45, 118, 161, 234 |
| 7. | 3,4,6 | 1.327 | 1.126 | 45, 161, 189 | 45, 161, 190 |
| 8. | 2,3,4,6 | 1.086 | 1.022 | 45, 117, 161, 205 | 45, 118, 161, 162, 205 |
| **Glucitol** | | —[d] | 1.376 | 73, 145, 217, 289 | 73, 74, 145, 146, 217, 218, 289, 290 |
| 32. | 3,4 | 1.593 | 1.197 | 189, 233 | 189, 190, 233, 234 |
| 9. | 2,4,6 | 1.220 | 1.104 | 45, 117, 161, 233 | 45, 118, 161, 234 |
| 10. | 3,4,6 | 1.236 | 1.099 | 45, 161, 189 | 45, 161, 190 |
| 11. | 2,3,4,6 | 0.991 | 0.997 | 45, 117, 161, 205 | 45, 118, 161, 162, 205 |
| **Mannitol** | | —[d] | 1.376 | 73, 145, 217, 289 | 73, 74, 145, 146, 217, 218, 289, 290 |
| 12. | 2 | 1.647 | 1.299 | 117, 333 | 118, 333 |
| 31. | 2,3 | 1.528 | 1.233 | 117, 161, 261, 305 | 118, 162, 261, 305 |
| 13. | 2,4 | 1.559 | 1.260 | 117, 189, 233, 305 | 118, 189, 234, 305 |
| 14. | 2,6 | 1.415 | 1.180 | 45, 117, 305 | 45, 118, 305 |

| No.[a] | Positions[b] | | | | |
|---|---|---|---|---|---|
| 15. | 3,4 | 1.578 | 1.249 | 189, 233 | 189, 190, 233, 234 |
| 16. | 3,6 | 1.501 | 1.208 | 45, 189, 233 | 45, 190, 233 |
| 17. | 4,6 | 1.399 | 1.191 | 45, 161, 261 | 45, 161, 262 |
| 18. | 2,3,4 | 1.309 | 1.143 | 117, 161, 189, 233 | 118, 162, 189, 233 |
| 19. | 2,3,6 | 1.290 | 1.109 | 45, 117, 161, 233 | 45, 118, 162, 233 |
| 20. | 2,4,6 | 1.230 | 1.118 | 45, 117, 161, 233 | 45, 118, 161, 234 |
| 21. | 3,4,6 | 1.247 | 1.104 | 45, 161, 189 | 45, 161, 190 |
| 22. | 2,3,4,6 | 1.000 | 1.000 | 45, 117, 161, 205 | 45, 118, 161, 162, 205 |
| *N*-Methyl-2-amino-2-deoxygalactitol | | —[d] | 1.553 | 158 | 159 |
| 23. | 3,4,6 | 1.880 | 1.379 | 45, 158, 161, 202, 205 | 45, 159, 161, 203, 205 |
| *N*-Methyl-2-amino-2-deoxyglucitol | | —[d] | 1.529 | 158 | 159 |
| 24. | 3 | —[e] | 1.520 | 158, 202, 261 | 159, 203, 261 |
| 25. | 6 | —[e] | 1.497 | 45, 158 | 45, 159 |
| 26. | 3,4 | —[e] | 1.466 | 158, 189, 202, 233 | 159, 189, 203, 233 |
| 27. | 3,6 | 1.984 | 1.417 | 45, 158, 202, 233 | 45, 159, 203, 233 |
| 28. | 4,6 | —[e] | 1.462 | 45, 158, 161, 274 | 45, 159, 161, 275 |
| 29. | 3,4,6 | 1.766 | 1.339 | 45, 158, 161, 202, 205 | 45, 159, 161, 203, 205 |
| 30. | 1,3,5,6 | 1.439 | 1.191 | 45, 89, 130, 205 | 45, 46, 89, 131, 205 |

[a] Refers to *Figure 4*.

[b] Numbers refer to positions of methylation. All the carbohydrates are otherwise peracetylated (e.g. mannitol 2,3,4 refers to 1,5,6-tri-*O*-acetyl-2,3,4-tri-*O*-methyl-D-mannitol, from which the 6-linked mannose can be deduced).

[c] Relative to 1,5-di-*O*-acetyl-2,3,4,6-tetra-*O*-methyl-D-mannitol (peak 22) = 1.000, on WCOT capillary columns ref. 48.

[d] Not determined.

[e] Most partially methylated *N*-methyl hexitol acetates are lost during chromatography over the polar Silar 9CP phase.

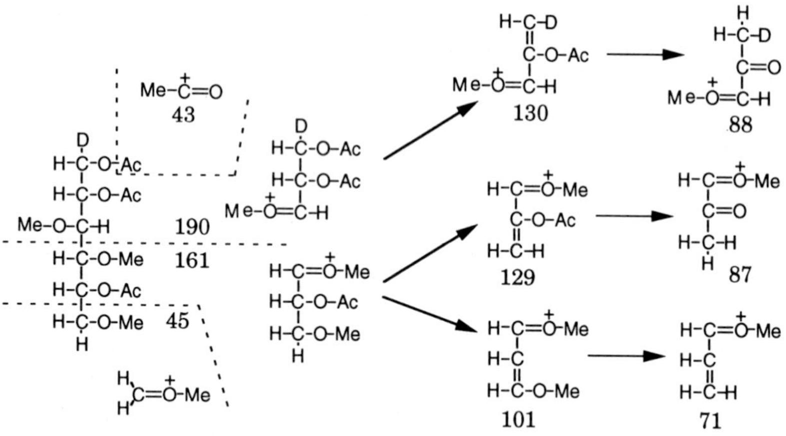

Me-Ċ=O
43

D
H-Ċ-O-Ac
H-Ċ-O-Ac
Me-O-Ċ-H        190
Me-O=Ċ-H
H-Ċ-O-Me        161
H-Ċ-O-Ac
H-Ċ-O-Me        45
H

H
  C=O-Me
H

D
H-Ċ-O-Ac
H-Ċ-O-Ac
Me-O=Ċ-H

H-Ċ=O-Me
H-Ċ-O-Ac
H-Ċ-O-Me
H

H-C-D
C-O-Ac
Me-O=Ċ-H
130

H-Ċ=O-Me
C-O-Ac
H-Ċ-H
129

H-Ċ=O-Me
H-C
H-Ċ-O-Me
101

H
H-Ċ-D
C=O
Me-O=Ċ-H
88

H-C=O-Me
C=O
H-Ċ-H
H        87

H-C=O-Me
H-C
H-Ċ-H
71

**Figure 5.** The fragmentation pattern in mass spectrometry of 1,2,5-triacetyl-3,4,6-tri-*O*-methyl-D-glucitol. The initial cleavage pattern giving about 43% m/z 43, 27% m/z 161 (including 12% m/z 129, 5% m/z 87, 3% m/z 101, and 2% m/z 71 secondary cleavage products), 18% m/z 190 (including 10% m/z 130 and 6% m/z 88 secondary cleavage products), and 12% m/z 45.

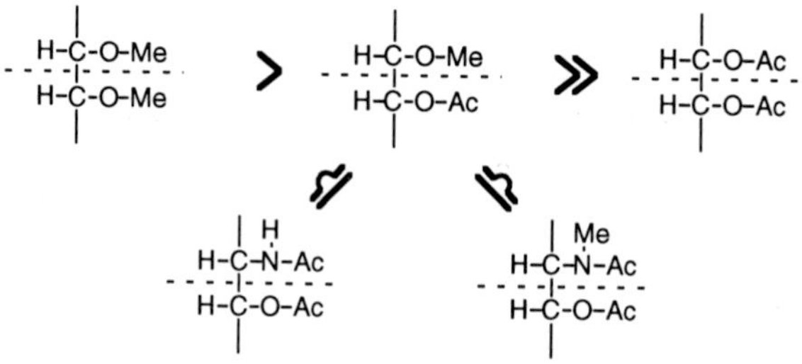

H-Ċ-O-Me          H-Ċ-O-Me          H-Ċ-O-Ac
H-Ċ-O-Me    >     H-Ċ-O-Ac    >>    H-Ċ-O-Ac

H
H-Ċ-N-Ac
H-Ċ-O-Ac

Me
H-Ċ-N-Ac
H-Ċ-O-Ac

**Figure 6.** The relative ease of cleavage, in mass spectrometry between carbon atoms in peracetylated methylated alditols and amino alditols. The unmethylated amino group often arises due to under-methylation.

spectra may be analysed with reference to a few simple rules (see also *Figure 5*).

(a) Fission prefers to take place next to the methoxyl or *N*-acetyl groups rather than the *O*-acetyl groups, with the positive ion stabilized by the methoxyl or *N*-acetyl grouping (*Figure 6*).

(b) Secondary fragments are derived from the primary fragment by β-elimination of methanol (mol. wt. 32) or acetic acid (mol. wt. 60) from the carbon atom two-removed from the formal charge on the carbonium ion. Some fission of acetic acid from the α-carbon atom is detected where the β-carbon is methoxylated (see *Figure 5*). Additional fission of ketene (mol. wt. 42) and, more rarely, formaldehyde (mol. wt. 30) also occurs, (e.g. fragments such as $R-CH=\overset{+}{N}(CH_3)Ac$ immediately eliminate ketene to form $R-CH=\overset{+}{N}HCH_3$).

(c) Large primary fragments eliminate neutral molecules more readily than small primary fragments. A base fragment at m/z 43 ($CH_3\overset{+}{C}=O$) is normally found.

There is no significant difference between isomeric alditols having the same substitution pattern, alditols from 2,3- and 3,4-di-*O*-methyl pentoses or from 3- and 4-*O*-methyl hexoses would be indistinguishable unless sodium borodeuteride was used in the reduction step.

The trimethylsilyl derivatives of alditols, monosaccharides, or methyl-glycosides do not give molecular ions but weak $(M - 15)^+$ ions may be seen in aldoses. They cleave between carbon atoms carrying trimethylsilyl ether or methoxyl groups to form the primary fragments. The secondary fragments are formed from these by the consecutive elimination or trimethylsilylhydroxide ($Me_3SiOH$, mol. wt. 90) molecules to give ions such as $(M - 15 - 90)^+$. A base fragment at m/z 73 ($Me_3\overset{+}{Si}$) is normally found. The mass spectra of the trimethylsilyl derivatives of cyclic carbohydrates are complicated because the initial cleavage does not reduce the m/z of the resulting fragment; secondary cleavages must occur and these produce fragments at m/z 147 ($Me_3Si-\overset{+}{O}=SiMe_2$) from pairs of derivatized hydroxyl groups, 204 [ $(Me_3\overset{+}{Si}O-CH)_2$] from vicinal hydroxyl groups, and 217 ($Me_3SiO-CH=CH-\overset{+}{CH}-OSiMe_3$) and 305 [$Me_3SiO-CH=C(-OSiMe_3)-\overset{+}{CH}-OSiMe_3$] from three neigh-bouring hydroxyl groups, in addition to the base peak at m/z 73 and a peak at m/z 191 [ $(Me_3SiO)_2\overset{+}{CH}$] due to the straight chain form at C-1. The ratio of these five peaks may be used to distinguish between carbohydrates if reproducible analytical conditions are possible. In particular, for aldohexoses, aldopentoses, and 6-deoxyaldohexoses, the pyranoid forms give an m/z 204/217 ratio of greater than one ($\sim$ 1 to > 1 for aldopentoses), whereas the furanoid forms show a ratio of less than one. An intense ion at m/z 173 ($Me_3SiO-CH-\overset{+}{CH}-NHCOMe$) is seen in the spectra of the TMS ethers of the *N*-acetylhexosamines.

# 9. Infra-red spectroscopy

Infra-red spectroscopy is mainly of use in monosaccharide analysis for the comfirmation of the identity of a molecule by comparison with a published standard spectrum or the spectra of known material. About one milligram of pure dry carbohydrate is needed for an analysis although a smaller amount may give useful spectra. The preferred method makes use of salt discs made by the intimate mixture of about one milligram of carbohydrate with 300 milligrams of pure dry KBr followed by pressing into a disc. The whole IR spectrum (4000–650 cm$^{-1}$) should be used for comparison (52).

Fourier transform IR spectroscopy (FTIR) may be used for the on-line quantitative analysis of relatively simple sugar solutions, such as soft drinks by use of a matrix of absorbances within the mid-infrared range (4000–400 $cm^{-1}$). The sugar concentrations may be determined from matrix constants previously deduced from standard mixtures (53).

# References

* Indicates a modification of the cited method.

1. Kirk, R. S. and Sawyer, R. (1991). *Pearson's composition and analysis of food*, 9th edn, pp. 195–9. Longman Scientific and Technical, Essex, UK.
2.* Dubois, M., Gilles, K. A., Hamilton, J. K., Rebers, P. A., and Smith, F. (1956). *Anal. Chem.*, **28**, 350.
3. Dische, Z., Shettles, L. B., and Osnos, M. (1949). *Arch. Biochem.*, **22**, 169.
4.* Bernfeld, P. (1955). In *Methods in enzymology*, Vol. 1 (ed. S. P. Colowick and N. O. Kaplan), p. 149. Academic Press, New York.
5. Dygert, S., Li, L. H., Florida, D., and Thoma, J. A. (1965). *Anal Biochem.*, **13**, 367.
6. Peris-Tortajada, M., Puchades, R., and Maquiera, A. (1992). *Food Chem.*, **43**, 65.
7.* Somogyi, M. (1952). *J. Biol. Chem.*, **195**, 19.
8.* White, C. A. and Kennedy, J. F. (1981). *Tech. Life Sci.*, **B3**, B312/1.
9. Boratynski, J. (1984). *Anal. Biochem.*, **137**, 528.
10.* Reissig, J. L., Strominger, J. L., and Leloir, L. F. (1955). *J. Biol. Chem.*, **217**, 959.
11.* Taylor, K. A. and Buchanan-Smith, J. G. (1992). *Anal. Biochem.*, **201**, 190.
12.* Warren, L. (1959). *J. Biol. Chem.*, **234**, 1971.
13. Svennerholm, L. (1957). *Biochim. Biophys. Acta*, **24**, 604.
14.* Anon. (1989). In *Methods of enzymatic food analysis*, Boehringer Mannheim, GmbH, Germany.
15. Morris, J. B. (1982). *Anal. Biochem.*, **121**, 129.
16. Fujimura, Y., Ishii, S., Kawamura, M., and Naruse, H. (1981). *Anal. Biochem.*, **117**, 187.
17. Ghebregzabher, M., Rufini, S., Monaldi, B., and Lato, M. (1976). *J. Chromatogr.*, **127**, 133.
18. Zweig, G. and Sherma, J. (ed.) (1972). *CRC handbook of chromatography*, Vol. 1. CRC Press Inc., Boca Raton, Florida.
19. Churms, S. C. (ed.) (1982) *Carbohydrates*, Vol. 1. CRC Press Inc., Boca Raton, Florida.
20. Got, R., Cheftel, R.-I., Font, J., and Moretti, J. (1967). *Biochim. Biophys. Acta*, **136**, 320.
21. Fell, R. D. (1990). *Comp. Biochem. Physiol.*, **95A**, 539.
22. Hansen, S. A. (1975). *J. Chromatogr.*, **107**, 224.
23. Hough, L. and Jones, J. K. N. (1962) In *Methods in carbohydrate chemistry* (ed.

R. L. Whistler and M. L. Wolfrom), Vol. 1, p. 21. Academic Press, New York and London.

24. Busch, K. L. (1992). *Trends Anal. Chem.*, **11**, 314.
25. Sherma, J. and Zweig, G. (1971). *Paper chromatography and electrophoresis* Vol. 2, Paper Chromatography. Academic Press, New York and London.
26. Ball, G. F. M. (1990). *Food Chem.*, **35**, 117.
27. Macrae, R. (1991). *Adv. Carbohydr. Anal.*, **1**, 293.
28.* Honda, S. and Suzuki, S. (1984). *Anal. Biochem.*, **142**, 167.
29. Melzi d'Eril, G. V. and Achilli, G. (1992). *Chromatogr. Anal.*, **24**, 5.
30. Takemoto, H., Hase, S., and Ikenaka, T. (1985). *Anal. Biochem.*, **145**, 245.
31. Alpenfels, W. F. (1981). *Anal. Biochem.*, **114**, 153.
32. Kang, E. Y. J., Coleman, R. D., Pownall, H. J., Gotto, A. M., Jr., and Yang, C.-Y. (1990). *J. Protein Chem.*, **9**, 31.
33. Hara, S., Yamaguchi, M., Takemor, I. Y., Furuhata, K., Ogura, H., and Nakamura, M. (1989). *Anal. Biochem.*, **179**, 162.
34. Spiro, M. J. and Spiro, R. G. (1992). *Anal. Biochem.*, **204**, 152.
35. Altmann, F. (1992). *Anal. Biochem.*, **204**, 215.
36. Grimble, G. K., Barker, H. M., and Taylor, R. H. (1983). *Anal. Biochem.*, **128**, 422.
37. Tomiya, N., Suzuki, T., Awaya, J., Mizuno, K., Matsubara, K., Nakano, K., and Kurono, M. (1992). *Anal. Biochem.*, **206**, 98.
38. Perini, F. and Peters, B. P. (1982). *Anal. Biochem.*, **123**, 357.
39. James, L. B. (1984). *J. Chromatogr.*, **284**, 97.
40. Manzi, A. E., Diaz, S., and Varki, A. (1990). *Anal. Biochem.*, **188**, 20.
41. De Ruiter, G.A., Schols, H. A., Voragen, A. G. J., and Rombouts, F. M. (1992). *Anal. Biochem.*, **207**, 176.
42. Wang, W. T., Safar, J., and Zopf, D. (1990). *Anal. Biochem.*, **188**, 432.
43. Bach, E. and Schollmeyer, E. (1992). *Anal. Biochem.*, **203**, 335.
44. Gerwig, G. J., Kamerling, J. P., and Vleigenthart, F. G. (1978). *Carbohydr. Res.*, **62**, 349.
45. Chaplin, M. F. (1982). *Anal. Biochem.*, **123**, 336.
46. Blakeney, A. B., Harris, P. J., Henry, R. J., and Stone, B. A. (1983). *Carbohydr. Res.*, **113**, 291.
47. Anumula, K. R. and Taylor, P. B. (1992). *Anal. Biochem.*, **203**, 101.
48. Geyer, R., Geyer, H., Kühnhardt, S., Mink, W., and Stirm, S. (1982). *Anal. Biochem.*, **121**, 263.
49. Wells, W. W., Pittman, T. A., and Wells, H. J. (1965). *Anal. Biochem.*, **10**, 450.
50. Carpita, N. C. and Shea, E. M. (1989). In *Analysis of carbohydrates by GLC and MS* (ed. C. J. Biermann and G. D. McGinnis), pp. 157–216. CRC Press Inc., Boca Raton, Florida.
51. Wait, R. (1991). *Adv. Carbohydr. Anal.*, **1**, 335.
52. Tipson, R. S. and Parker, F. S. (1980). In *The carbohydrates* (2nd edn), (ed. W. W. Pigman and D. Horton), Vol. 1B, p. 1394. Academic Press, New York.
53. Kemsley, E. K., Zhuo, L., Hammouri, M. K., and Wilson, R. H. (1992). *Food Chem.*, **44**, 299.

# 2

# Oligosaccharides

JOHN F. KENNEDY and GIAMPIERO PAGLIUCA

## 1. Introduction

Oligosaccharides are traditionally defined as polymers of monosaccharides containing from two to ten residues. However, since the naturally occurring polysaccharides rarely contain less than 25–30 residues, it is possible to consider the range of polymers having between two and about 20–25 residues as oligosaccharides, particularly when considering the analytical methods involved. As oligosaccharides fall into the classification between monosaccharides and polysaccharides it is not surprising that the methods used for oligosaccharide analysis are extensions of those used for either monosaccharide or polysaccharide analysis. This chapter uses a slightly different format to that of Chapter 1 in order to give the maximum information, within the limitations of available space, on the modifications which have to be made in adapting the methods described in detail for the analysis of monosaccharides and polysaccharides. Rather than repeat descriptions of practical details etc. this chapter concentrates on the types of methods which can be applied and their relative merits and shortcomings.

As with monosaccharide analysis, the analysis of oligosaccharides can be divided into qualitative identification and quantitative determination but, in addition to determination of the number and type of component residues, the actual linkage between successive residues is an area not encountered in monosaccharide analysis. Due to the complexities of linkage analysis, discussion of the methodology is not described in this chapter but can be found in subsequent chapters concerning the analysis of polysaccharides and carbohydrate-containing macromolecules. The traditional methods used for qualitative analysis: paper chromatography and thin-layer chromatography, (TLC), have largely been replaced by rapid quantitative techniques, but, for completeness, are described briefly. Methods for the separation and quantitation of mixtures of oligosaccharides and for the identification and quantitation of individual purified oligosaccharides are described. Rather than providing a listing of all methods which have been devised, only those which have become the more commonly used techniques are described.

# 2. Colorimetric methods

The colorimetric methods which have been devised are mainly used for the gross determination of total carbohydrate content or total reducing sugar content, although specific assay methods have been developed for the quantitation of an individual oligosaccharide from a mixture of compounds via the use of specific enzymes. Both total carbohydrate and reducing sugar contents are important features of oligosaccharide analysis since it is possible to obtain a value for the size of an oligosaccharide or, for a mixture of related oligosaccharides, a mean value for the ratio of total residues to terminal reducing residues. With the development of sugar syrups (1) methods for the characterization of these products have been developed with the various syrups being categorized by their dextrose (i.e. D-glucose) equivalent or DE value. Despite the shortcomings of defining mixtures of oligosaccharides in terms of this one parameter (2) it is still used as the principle method for characterizing oligosaccharide mixtures.

## 2.1 Total sugar assays

A number of assays have been developed which rely on the action of concentrated (or near concentrated) sulfuric acid causing hydrolysis of all glycosidic linkages and the subsequent dehydration of the monosaccharides released to give derivatives of furfural (e.g. hexoses produce 5-hydroxy-methyl furfuraldehyde). The dehydration products react with a number of compounds such as L-cysteine (3), phenol (4), orcinol (5), and anthrone (6) to give coloured products. Whilst total sugar concentrations are readily obtained for homoglycans, care must be taken in interpreting the results obtained with heteroglycans due to the different colour intensities produced by different monosaccharides. Practical details for the performance of the phenol assay can be found in Chapter 1, *Protocol 2*.

---

**Protocol 1.** Orcinol–sulfuric acid assay

- Sensitivity: 0–20 µg carbohydrate in 200 µl  • Final volume: 1.0 ml

*Reagent*

Ice-cold orcinol (recrystallized from benzene) dissolved in concentrated acid (2 g/litre). This reagent should be prepared fresh each day but can be stored at 4 °C for up to one week.

*Method*

1. To precooled (to 4 °C) samples, standards, and controls (200 µl) add 800 µl of reagent with care. Mix well.

---

**2.** Heat the solutions at 80 °C for 15 min and cool rapidly to room temperature.

**3.** Determine the absorbance at 420 nm.

*Note*: The original method (4) suggested that absorbances should be determined at 510 nm but the use of 420 nm results in an increased sensitivity (by a factor of two) and a reduced interference from uronic acids and deoxy sugars.

## 2.2 Reducing sugar assays

Traditionally the Lane and Eynon (7) assay was used to determine the content of reducing sugars in a sample and still is the method of choice in some industrial applications. The method involves the reaction of reducing sugars with alkaline cupric salts to give cuprous oxide which can be monitored titrimetrically to give the concentration of various reducing sugars by reference to standard tables. Further modifications to this assay have eliminated the use of tables (8).

A far more convenient assay is that which involves the reaction with alkaline 3,5-dinitrosalicylic acid originally devised by Bernfeld (9) and for which practical details are given in Chapter 1, *Protocol 3*. The major disadvantages of this assay are its insensitivity at low carbohydrate concentrations (10) and some workers report different responses obtained for equimolar amounts of oligosaccharides (11). For this reason the less convenient Nelson–Somogyi (12) assay is preferred in some instances (13). This assay is based on the reduction of cupric salts to cuprous salts which further reduce the arsenomolybdate complex to molybdenum blue, the intensity of which is determined spectrophotometrically. It is however less sensitive than the 3,5-dinitrosalicylic acid assay. The assay of reducing groups in oligosaccharide homologues with 2,2′-bicinchonite has been recently proposed (14).

A number of other assays have been reported such as those using alkaline ferricyanide (15) or alkaline picric acid (16). The reaction based on the production of coloured soluble formazan salt from the dye tetrazolium blue (17), as a result of an essential solvent extraction stage, can be applied to samples containing particulate matter.

---

**Protocol 2.** Tetrazolium blue assay

- Sensitivity: 0–10 µg glucose equivalent in 100 µl
- Final volume: 1.0 ml

*Reagent*

Add 3 vol. of 0.3 M NaOH to 1 vol. of an aqueous suspension of tetrazolium blue (1% w/v) and stir until completely dissolved. Dilute with 5 vol. distilled water and store at 5 °C in the dark.

---

**Protocol 2.** *Continued*

*Method*

1. To samples, standards, and controls (100 μl) add 900 μl of reagent and mix well.

2. Heat the solution at 100 °C for 30 sec and cool rapidly to room temperature.

3. Add 1 ml of toluene and shake until no more colour is extracted fro the aqueous layer.

4. Determine the absorbance at 570 nm.

*Note*: The sensitivity of the assay can be adjusted to accommodate small or large colour changes by halving or doubling the volume of toluene used for the extraction stage. Extended heating of reaction mixture results in alkaline hydrolysis of oligosaccharides and increased absorbance values.

## 2.3 Automated assay system

One of the major uses of colorimetric assays is the monitoring of chromatographic columns. Traditionally this was done by collecting fractions of effluent stream and performing a series of manual assays. To a large extent this has now been replaced by either non-specific detectors (see Section 5.6) or automated assay systems of various degrees of specificity. The pioneering work on assay automation was performed using air segmented systems based on glass tubing of approximately 1.6–2.4 mm i.d. and Technicon AA1 type equipment. This allows mixing of reagents via peristaltic pumps equipped with tubing of different sizes and construction to facilitate accurate mixing of the majority of reagents including certain organic solvents and concentrated minerals acids. Increasingly the assay systems now being devised use non-segmented flow-through narrow bore (0.5 mm i.d. and smaller) Teflon tubing but such systems require pumps capable of pumping solvents and corrosive liquids at the higher pressures resulting from the use of narrow bore tubing.

Several systems have been devised for the detection of reducing sugars (see Chapter 1) but these are only applicable to the detection of the smaller oligosaccharides due to the response being related to the molar concentration of the oligosaccharide and not the concentration by weight. Since the response of maltohexaose is only 18% of that obtained for the same weight of D-glucose, reducing sugar assays are never used for the detection of oligsaccharides above hexasaccharides, and ideally are best suited to the detection of disaccharides and trisaccharides. The automated 3,5-dinitro-salicylic acid assay is shown in *Figure 1* by way of an example.

The detection of total sugar content relies on the use of reagents such as concentrated acid to hydrolyse the oligosaccharides to monosaccharides and to produce a suitable chromogen (e.g. 5-hydroxymethyl furfuraldehyde) and consequently the development in suitable automated assay systems has not received the same degree of attention. However systems have been devised for the estimation of total sugar content based on the L-cysteine (18) (see *Figure 2*) or orcinol assays (19) for neutral oligosaccharides and the carbazole (18) assay for uronic acid-containing oligosaccharides using segmented (18) or non-segmented (19) flow systems. The combination of such systems and chromatographic columns has led to the development of fully automated oligosaccharide analysers based on ion-exchange (19) or gel permeation (20) chromatography. Oligosaccharides were labelled with a solution containing 5% 4-aminobenzoic acid hydrazide in 0.5 M HCl with a 2.4 M NaOH solution in a ratio of 1 : 2 (v/v) (21). This method was used as post-column reaction pumping the reagent in a ratio of 1 : 2 (v/v) to a HPLC column eluent and detecting the complex at 400 nm. The reaction conditions (reaction time 200 sec and reaction temperature 105 °C) were optimized, giving a detection limit of 20–50 pmol for mono- and oligosaccharides (22).

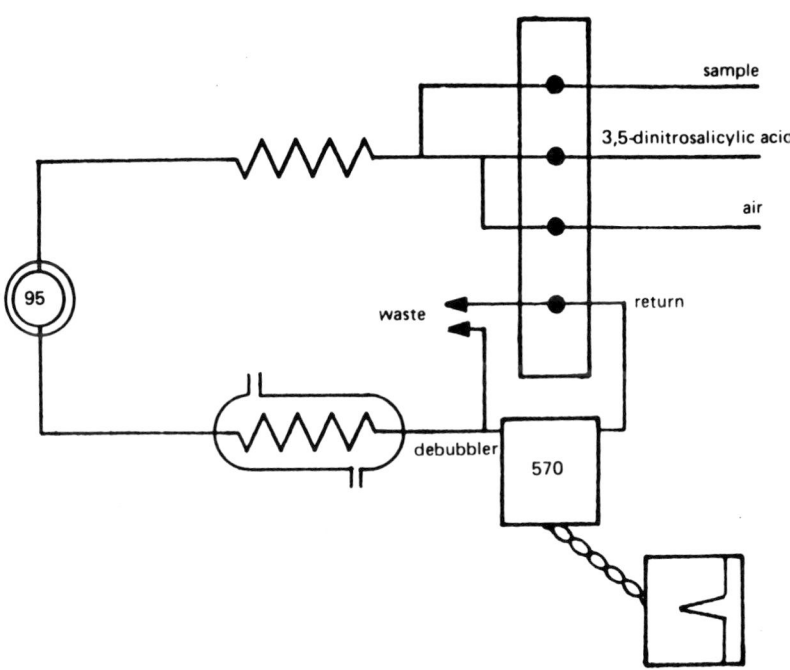

**Figure 1.** The automated 3,5-dinitrosalicylic acid assay. (DNS reagent consisting of 1.0 g of 3,5-dinitrosalicylic acid and 300 g of potassium sodium tartrate, per litre of 0.4 M NaOH 0.73 ml/min; sample, 0–3 mg D-glucose equivalent/ml water or buffer, 0.20 ml/min; air, 0.73 ml/min; waste, 0.73 ml/min.)

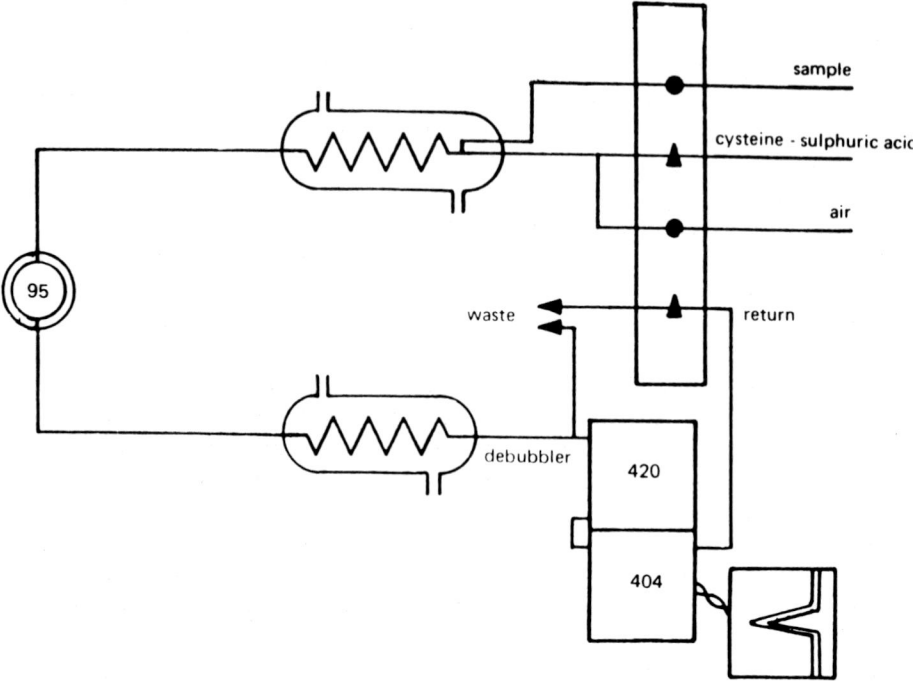

**Figure 2**. The automated L-cysteine–sulfuric acid assay. (L-Cysteine hydrochloride reagent, 700 mg/ml 86% sulfuric acid, 0.53 ml/min; sample, 0–50 µg/ml water or buffer, 0.1 ml/min; air, 0.23 ml/min; waste 0.32 ml/min.)

## 3. Thin-layer chromatography

TLC is not widely used for oligosaccharide analysis, but can be useful as a rapid technique for monitoring, for example, the hydrolysis of starch to give a particular oligosaccharide spectrum. It is also useful when it is not possible to perform more complex studies, as for example during field studies. The technique is essentially qualitative in its execution, although semi-quantitative data can be obtained using scanning densitometric analysis of the visualized chromatogram. Such results must however be interpreted with care since the degree of colour produced by the various visualization techniques is affected by parameters such as temperature and duration of heating, coverage of spray reagent, traces of salt or eluants, etc. all of which can vary on a single chromatogram as well as between different chromatograms.

Two systems which can be used to provide initial data on the relative abundance of the lower members of homologous series of oligosaccharides are described briefly. Both use the multiple-ascents technique, whereby the chromatogram is eluted in the normal manner, dried, and re-eluted with the same eluant a number of times, in order to improve the resolution between components. Full practical details for analysis can be found in Chapter 1.

Kieselgel G has been used to separate the individual members of a series of non-reducing oligofructosides, up to a hexasaccharide, derived from plant extracts by applying the mixture in 70% ethanol and eluting the chromatogram with chloroform/acetic acid/water (3 : 3.5 : 0.5) at ambient temperature with three ascents being used for optimum resolution (23). Individual members of the series of oligosaccharides derived from the enzymatic hydrolysis of starch, up to maltodecaose, can be separated using silica gel G and an eluent of butanol/ethanol/water (5 : 3 : 2) with three ascents for optimum resolution (24) (see *Figure 3*).

Detection of components on a thin-layer chromatogram can readily be achieved using a non-selective charring technique by heating the chromato-gram after spraying it with a reagent prepared from concentrated sulfuric acid in ethannol (1 : 9). This method can be improved such that increased contrast and sensitivity is obtained if ceric sulfate (3% w/v) is added to the sulfuric acid/ethanol spray (24). Selective detection can be obtained using spray reagents containing diphenylamine aridine phosphate (25) or *o*-anisaldehyde (26) which give different colours with different carbohydrates and thereby assist in identifying overlapping or unresolved peaks. The use of reagents such as tetrazolium blue and its derivatives which only visualize reducing

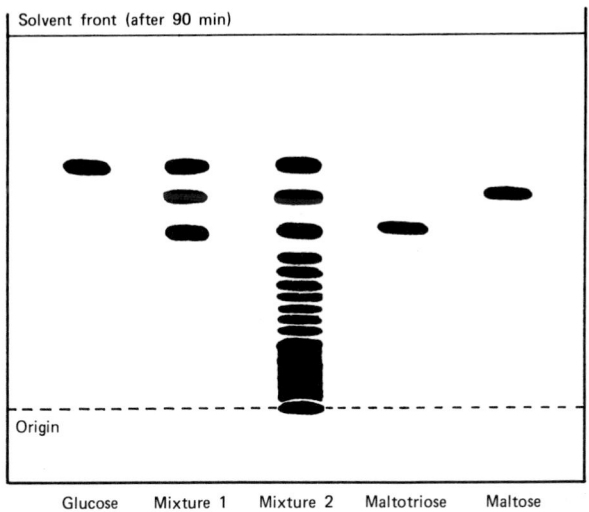

**Figure 3.** TLC separation of standard malto-oligosaccharides on silica gel G using butanol/ethanol/water (5:3:2) eluent. (Individual components, 4 mg/ml, 5 µl loaded; mixture 1 = D-glucose, maltose, and maltotriose, 2 mg each/ml, 10 µl loaded; mixture 2 = starch hydrolysate 50 mg/ml, 5 µl loaded.)

sugars, are of litte value for oligosaccharide analysis due to the diminishing response with increasing molecular weight of oligosaccharides.

Amino-bonded high performance TLC silica plates, and pyridine/ethyl acetate/acetic acid/water (6 : 2 : 1 : 3) as eluent, has been used for directly immunostaining of free oligosaccharides (27).

# 4. Low pressure column chromatography

The distinction between low pressure (traditional) column chromatography and high performance liquid chromatography (HPLC) is not as clear cut as many workers would have you believe, with definitions being based on particle size (25 μm is a typical division) or the capital cost of equipment. Consequently some of the discussion in this section can equally apply to Section 5 and vice versa. Low pressure column chromatography is typified by the use of compressible column packings which require the use of low pumping pressures (hydrostatic pressure of peristaltic pumps frequently being used) and extended analysis times in the region of 2–18 hours. Detection is frequently performed using automated assay systems (Section 2.3) or by manual assay following collection of eluate fractions. Many of the historical developments and applications with respect to oligosaccharide analysis have been reviewed (28).

## 4.1 Ion-exchange chromatography

### 4.1.1 Types of systems available

Ion-exchange chromatography has almost universally replaced the traditional adsorption chromatography based on charcoal-celite mixtures developed in the early 1950s which was capable of separating series of oligosaccharides up to a degree of polymerization (DP) of 8–10 using ethanol or butanol gradients in water as eluent (29). The use of ion-exchange resins is preferred due to their increased selectivity, lower operational back pressures, increased reproducibility, and the ability to use eluents which do not decrease the solubility of the higher oligosaccharides as is the case with mixtures of alcohols in water.

Anion-exchange resins, most commonly in the carbonate or bicarbonate form, although chloride and hydroxyl forms can be used, will fractionate oligosaccharides in order of decreasing molecular size. The use of water as eluent allows fractionation of neutral sugars up to DP about 6 but the possibiltiy of interconversion of terminal residues under the effects of the basic ion-exchange resins has prevented a full exploitation of the method. The use of acetic acid, sodium acetate, formic acid, or lithium chloride eluents allows the fractionation of acid-containing oligosaccharides such as uronic acid- or aldonic acid-containing oligosaccharides with, for example, up to DP 14 being fractionated for the series of xylonic acids prepared from a birch xylan hydrolysate using Dowex 1-X2.

Cation-exchange resins have also been used for separation of oligo-saccharides in an attempt to overcome the effects of sugar interconversions by anion-exchange resins. Cation-exchange resins in the lithium, barium, or potassium form have been used to give results similar to those obtained with anion-exchange resins but with the aid of less complex eluents. The effect of the counterion is considerable and the correct choice is essential for optimum resolution of a given oligosaccharide mixture (30).

Use of water/alcohol mixtures to elute ion-exchange columns results in a partition mechanism being responsible for the separation, with equilibria being established between the phase within the resin and the bulk solution. With anion-exchange resins there is a higher proportion of water within the resin than in the bulk solution which favours retention of the higher oligosaccharides such as a reversal of elution order can be achieved compared to anion-exchange chromatography using water as sole eluent, with mono-saccharides eluting before disaccharides and the higher oligosaccharides following. This results in incomplete recovery of some mixtures of higher oligosaccharides and restriction of the method to mixtures containing DP less than 10. The most common resin used is the sulfate form of styrene-divinylbenzene resins (31) although quaternary ammonium salts derived from polymeric carbohydrate resins (i.e. cross-linked dextran) can be used (32). The use of cationic resins in this partition-type mode is of little use in oligosaccharide analysis due to the high alcohol concentrations required for elution. This restricts the analysis to disaccharides only.

If borate buffers are used to form complexes with carbohydrates the different affinities of the negatively charged complexes for the ion-exchange resins can be exploited as a means of separating oligosaccharides as an extension to the normal ion-exchange chromatographic separation of mono-saccharides (19) (see *Figure 4*). The extended analysis times required and the

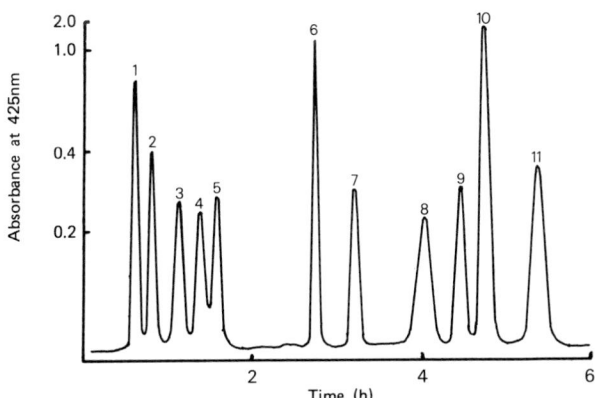

**Figure 4.** Ion-exchange chromatography of monosaccharides and disaccharides using Jeol LC-R-3 ion-exchange resin (1, sucrose; 2, cellobiose; 3, maltose; 4, lactose; 5, rhamnose; 6, ribose; 7, mannose; 8, fucose; 9, galactose; 10, xylose; 11, glucose).

alkaline conditions can cause interconversion of the terminal residues of reducing sugars. By alteration of the composition of the eluting buffers the system is ideally suited for separation of families of structurally related disaccharides and trisaccharides such that structural features of, for example, reversion products of enzymically degraded starch can be identified (33) (see *Figure 5*).

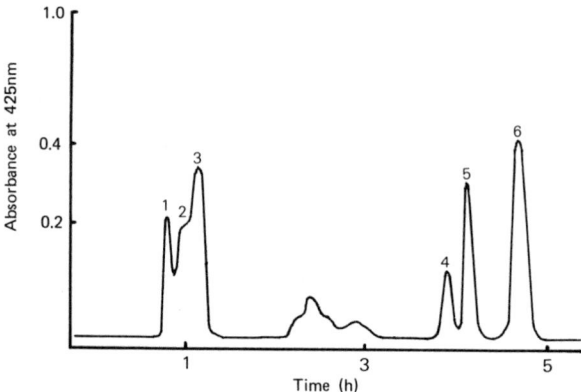

**Figure 5.** Ion-exchange chromatography of starch-derived oligosaccharides showing structural features enhanced by borate ion complexation using Jeol LC-R-3 resin (1, panose; 2, maltotriose; 3, maltose; 4, isopanose; 5, isomaltose; 6, D-glucose).

---

**Protocol 3.** Ion-exchange column chromotagraphy

1. Convert the ion-exchange resin to the desired form and solvate following the manufacturers' recommendations.

2. Fill a glass column (diameter in the range of 0.6–1.0 cm, and length 10–30 cm) drawn to a neck at the lower end and containing a small plug of glass wool with buffer. Pour the resin in and allow it to settle.

*Note*: The use of proprietary columns and fittings such as the Omnifit® range (Omnifit Ltd.) greatly assists in the production of a chromatographic column which can be pumped under increased pressure or which flows under hydrostatic pressure.

3. Pump the resin with buffer or eluent until a stable packed bed is obtained.

4. Introduce the sample onto the column by use of proprietary system injector. Alternatively it can be introduced by removal of the eluent head from above the resin, allowing the sample to enter the resin (with the aid of air pressure from a syringe attached via a rubber bung to the

top of the column if necessary), washing the sample into the resin with a small amount of eluent, and replacing the eluent head.

5. Pump the column with eluent and monitor either by automated assay (Section 2.3) or by collection of fractions and manual assay (Section 2.1).

## 4.2 Gel permeation chromatography

Gel permeation chromatography (GPC) is defined as the separation of compounds according to their molecular weight, or more correctly their hydrodynamic volume. It has been developed into a very useful technique for oligosaccharide analysis. Tridimensional cross-linked gels, based on dextran (e.g Sephadex G-10 or G-25 from Pharmacia) or polyacrylamide (e.g. Bio-Gel P-2, P-4, P-6 from Bio-Rad Laboratories) have the degree of cross-linking controlled such that the resultant pores are too small to allow high molecular weight species to penetrate. They are excluded from the gel and therefore elute in a volume equal to the void (dead) volume ($V_O$) of the column. Very small molecules penetrate the gel freely and are eluted in a volume ($V_T$) equal to the sum of the void volume and the volume of the solvent within the gel matrix available to small molecules (i.e. the pore volume, $V_p$). Thus

$$V_T = V_O + V_P$$

Molecules of intermediate size can penetrate some of the pores of the gel matrix and the degree of penetration, which is related to the distribution of pore sizes and the hydrodynamic volume of these intermediate size molecules is reflected in the elution volume ($V_e$) which lies within the range $V_O$ and $V_T$ (see *Figure 6*). The distribution of pore sizes within the gel matrix is such that over a considerable range of elution volumes the relationship:

$$V_e \propto \log (\text{molecular weight})$$

can be applied. At the extremities of the range of elution volume this relationship does not apply, but it is still possible to obtain some data provided a calibration curve (see *Figure 7*) is constructed to cover the range of elution volume involved, using materials having similar known hydrodynamic volumes. Examples have been published (34, 35) which show that homo-logous series of oligosaccharides have different gradients for their calibration curves due to the different hydrodynamic volumes for oligosaccharides having the same molecular weight. Therefore it is important to exercise care in interpreting the molecular weight of an unknown oligosaccharide when using a gel permeation column which has been calibrated with, for example, the malto-oligosaccharide series (derived from the controlled hydrolysis of amylose) or the cello-oligosaccharide series (as shown in *Figure 7*).

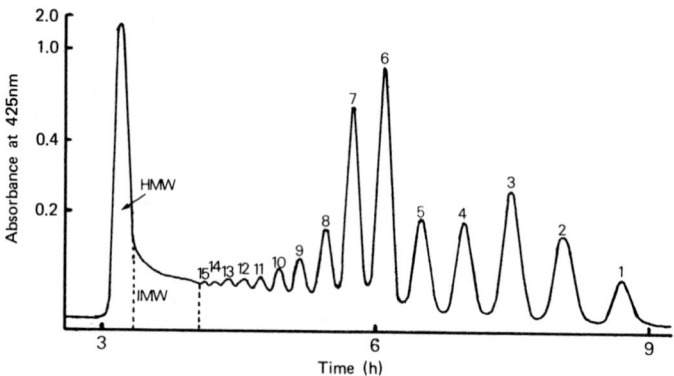

**Figure 6.** Gel permeation chromatographic fractionation of starch-derived oligo-saccharides on Bio-Gel P-2. (1,2,3 etc., D-glucose, maltose + isomaltose, D-glucotri-saccharides, etc., IMW, intermediate molecular weight material DP 15 to approximately 20; HMW, high molecular weight above DP approximately 20.)

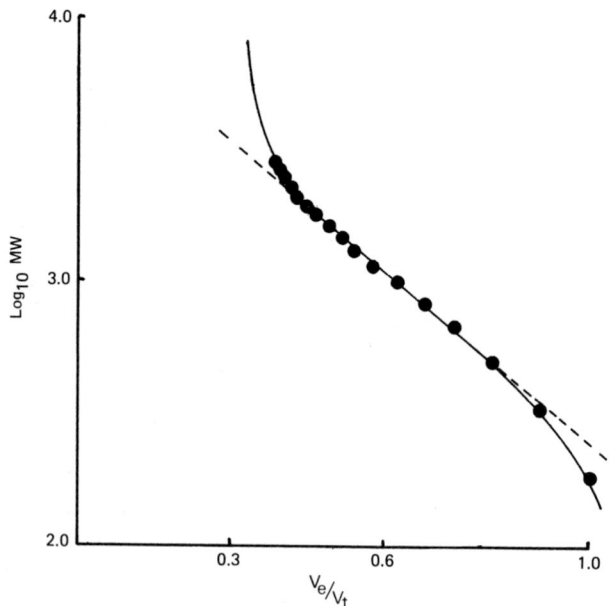

**Figure 7.** Calibration curve for Bio-Gel P-2 derived from the cello-oligosaccharide series obtained by the controlled hydrolysis of cellulose.

In an ideal system the only effects which contribute to the separation of oligosaccharides are steric effects but it has now become well-recognized that elution volumes greater than unity can be obtained for both carbohydrate-(36) and polyacrylamide-(18) based permeation supports. It is recommended that eluents containing, for example, 0.1 M NaCl should be used to prevent adsorption phenomena affecting the separation. The use of polyacrylamide-based gels is also recommended for optimum resolution at high temperatures (e.g. 65 °C) (20) or when samples of biological origin are used (18) in order to prevent elution of extraneous carbohydrate material due to disruption of the gel matrix. Such a system can be operated using automated assay detection systems for up to one year with no loss of resolution (20).

The use of oligosaccharide fractionation by gel permeation is now recommended as a superior method for the characterization of starch hydrolysates (2) particularly in the determination of dextrose equivalent (37) based on the actual composition of oligosaccharides present rather than on a single gross determination (see Section 2).

Gel permeation gives very good results for both analytical and preparative separations (38) and has been frequently combined with HPLC (39). The effective size of oligosaccharides from glycoprotein, expressed in glucose units using Bio-Gel P-4 (< 400 mesh), has been summarized (40).

## 4.3 Affinity chromotagraphy

Affinity chromatography is a unique separation method which allows the isolation of particular molecules according to the biospecific interaction with an immobilized ligand. This technique, using immobilized carbohydrate-binding proteins (i.e. lectins (41) and antibodies (42)) permits separation of carbohydrate complex mixtures, based on the stereochemistry (43). The elution is performed with buffers in controlled temperature (water-jacked columns are often used).

Fractionation and structural analysis of oligosaccharides by means of immobilized lectin columns has been recently reviewed (44).

# 5. High performance liquid chromatography

HPLC, characterized by small particle sizes (< 25 μm), narrow bore columns, high inlet pressures, and short analysis times, uses many of the same separation principles described for low pressure column chromatography. HPLC can be considered a valid alternative to gas chromatography for quantitative and qualitative analysis. It is possible to inject the sample without a prior derivatization, obtaining in short time analysis a high resolved chromatogram. Since its inception in the mid 1970s the number of methods and types of packing material have expanded rapidly. More recently the bonded-phase columns represent the most used separation systems. Much of the earlier work on oligosaccharide analysis has been reviewed (45) and the theory, separation modes, and instrumentation described (46), and the developments are such that a number of the low pressure chromatographic systems are now being replaced, particularly in the food industry for the analysis of the simpler mixtures. Details of the experimental approach can be found in Chapter 1, Section 6.

## 5.1 Adsorption chromatography

Adsorption, or normal phase, chromatography relies on the surface hydroxyl groups of silica (and to a lesser extent alumina) which can interact with solutes and affect a separation on account of the different strengths of interaction. Alumina causes a strong adsorbment and its basic character can induce, in some cases, epimerization. The separation of neutral oligosaccharides cannot be carried out conveniently by this method although limited separations can be achieved in water (47) or using solvent mixtures (48). The method is, however, well suited to the analysis of derivatives of oligosaccharides of low DP using non-aqueous eluents (see *Table 1*) particularly when ultrasensitive (10–100 ng level) detection is required (see Section 5.6).

**Table 1.** Derivatives used for the adsorption chromatography of oligosaccharides

| Derivative | Eluent | Reference |
|---|---|---|
| Benzoate | Hexane/ethyl acetate (85:15) | 49 |
| 4-Nitrobenzoate | Hexane/acetonitrile/chloroform (65:20:15) | 50 |
| Benzyloxime-perbenzoate | Hexane/dioxane (80:20) | 51 |
| Phenyldimethylsilyl ethers | Hexane/ethyl acetate (197:3) | 52 |

## 5.2 Reversed-phase chromatography

Reversed-phase chromatography packings are characterized by hydrocarbon chains bonded to the surface of the silica matrix. Whilst chain lengths can range from 1 to 22 carbon atoms the most popular are the 18 carbon atom chain (octadecylsilyl silica), the eight carbon chain (octylsilyl silica), and the six carbon chain (hexylsilyl silica) (53). The essential criterion responsible for separation is the interaction of the packing with polar materials. Using aqueous solutions or solvents of medium polarity the more polar species elute first and as the polarity decreases the more tightly bound less polar species are eluted. Using this kind of packing it is possible to separate both derivatized and underivatized sugars. Separations of underivatized oligosaccharides were obtained at room temperature, using pure water as eluent (54, 55). Using low temperatures it is possible to achieve an improvement in the resolution. Broad peaks poorly resolved are obtained if anomers are present in the mixture. To overcome this shortcoming a derivatization is necessary by reducing the terminal aldehyde group to the alcohol or by reductive amination.

The use of high concentrations of organic solvents in the mixtures of water and solvent used as eluent gives rise to problems of solubility of oligosaccharides. Nevertheless oligosaccharides up to DP 30 contained in wood extracts have been fractionated in under 30 minutes using a gradient of 70–62.5% acetonitrate in water (56) although the separation between the first five members of the series (DP 1–5) was very poor. Use of fully acetylated oligosaccharides overcomes problems of solubility and has resulted in the fractionation of malto-oligosaccharides up to DP 30 in about 150 minutes (57) using an exponential gradient of acetonitrile (10–70%) in water. Another advantage is the possibility after preparative HPLC to recover the original oligosaccharides by treatment with mild base allowing further analysis.

Perbenzoylated derivatives of high mannose-type oligosaccharides were used to improve the chromatographic properties on reverse-phase columns, using a 15 minutes linear gradient from acetonitrile/water (4:1 v/v) to pure acetonitrile (58).

## 5.3 Bonded-phase chromatography

By far the most frequently used systems for separation of oligosaccharides are those using chemically bonded phases which fractionate materials on the basis of their relative affinities for the mobile phase and the bonded phase. Oligosaccharides are eluted in order of increasing molecular weight. Supports containing a variety of cyano- or amino-bonded phases have been used in the past but the two most important types of column are those containing the aminopropyl-bonded phase (59) and hybrid phase of cyano- and amino-derivatives, i.e. the Partisil PAC column (60) with the aminopropyl-bonded column being regarded by some workers as the ultimate in column technology for oligosaccharide analysis (61).

As eluent, nonpolar organic solvents or an aqueous/organic mixture (such as aqueous methanol or aqueous acetonitrile) are usually used. Separation of series of oligosaccharides from, for example, hydrolysed starch can readily be achieved with up to DP 10 being separated in 15–20 minutes using acetonitrile/water eluents containing 35–40% water (see *Figure 8*). Increasing the water content to 45% can increase the number of detectable oligosaccharides up to about DP 15. However very high molecular weight materials cannot be analysed due to prolonged retention times and solubility problems in the acetonitrile/water eluents and for a full analysis separation by gel permeation of ion exchange is also required.

In some cases (45) interactions are possible between the reducing sugar and the amino group of the stationary phase. Prolonged use of bonded-amine columns is accompanied by a loss of performance through loss of the bonded phase, and to overcome this shortcoming and to introduce different degrees of selectivity into the system the use of *in situ* modification of the silica has been developed (62). Instead of using an amine phase bonded to the silica support a diamine or polyamine is added to the eluent which results in a dynamic equilibrium between an amine phase coating the silica and that in the

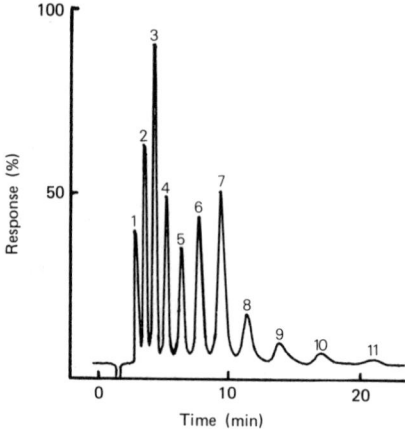

**Figure 8.** Separation of starch-derived oligosaccharides on a Spherisorb S5 NH₂ column (1,2,3, etc., refer to the DP of the oligosaccharide).

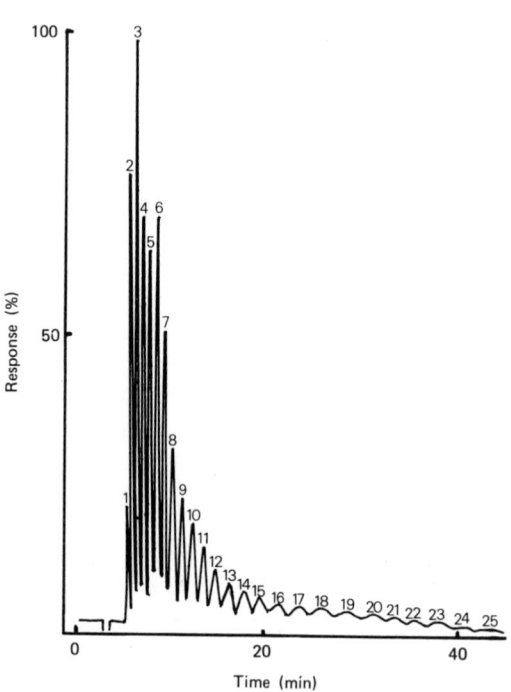

**Figure 9.** Separation of starch-derived oligosaccharides on an *in situ* modified silica column using 1,4-diaminobutane (0.01% v/v) as modifier. (1,2,3, etc. refer to the DP of the oligosaccharide.)

eluent. Separations are similar to those obtained using bonded-amine columns with separations of up to 20–25 being achieved in 45 minutes (see *Figure 9*) using eluents containing 50% water in acetonitrile to which the modifier has been added at the 0.01% (v/v) level. Even at this low modifier concentration it is imperative to use a presaturation column in the system prior to the injector to protect the analytical column packing from dissolution. Whilst diamine modifiers give the optimum resolution between oligosaccharides of different DP polyamine modifiers give improved selectivity and separation of oligosaccharides of the same DP such that the presence of, for example, isomaltose in starch derived oligosaccharides can be determined (61). Recently stationary phases similar to the aminopropyl silica have been developed that give a higher packing stability (63, 64).

## 5.4 Ion-exchange chromatography

High performance ion-exchange chromatography utilizes the same mechanisms for separation described in Section 4.1.

Anion-exchange HPLC has been recently developed performing an exceptional resolving power for complex oligosaccharides. These analyses are carried out at high pH coupled with pulsed amperometric detection (PAD) (see Section 5.8), allowing separation of oligo- and polysaccharides up to DP ≥ 50 (see *Figure 10*) (65). The separation depends on the molecular size, sugar composition, and kind of linkages between the monosaccharide units.

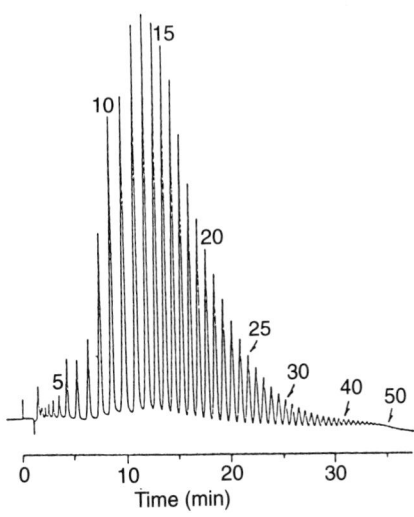

**Figure 10.** Chromatogram of (1 → 4)-α-D-glucans (short chain amylose EX-1 (DP ≈ 17)). The number on each peak indicates its DP. Chromatographic conditions: column, HPIC-AS6 (250 mm × 4 mm i.d.): eluent and gradient program as shown in *Table 1*; flow rate, 1 ml/min; detector, PAD II; meter scale, 10 K nA; temperature, ambient.

Analysis of linear and branched glucose oligosaccharides was investigated (66). Anion-exchange chromatography has also been used to separate a variety of positional isomers of neutral, sialylated, and phosphorylated oligosaccharides in a wide range of pH values (pH ~ 4–13) (67, 68). A very interesting application is given by the possibility of injecting crude enzyme digests of complex oligosaccharides (69). This method does not damage the phase and permits a quick and easy monitoring of the enzymatic digestion course and to isolate the reaction intermediates. Anion-exchange chromatography has been used recently to analyse oligosaccharides contained in complex biological samples (22, 70).

The separation of oligosaccharides by cation-exchange chromatography is based on the size exclusion and ligand-exchange mechanism. Most cation-exchange resins are prepared from cross-linked polystyrene and silica-based ion exchangers. The use of 4% and 8% cross-linked cation-exchange resins in the calcium (71, 72) or silver (73) form have been used to provide a rapid separation of oligosaccharides of up to DP 6–8 for calcium counterions (72) or DP 8–12 for silver counterions (73) depending on the size of the chromatographic column and the time of analysis. The major drawbacks to the use of such systems include the compressibility of the gel matrix (depending on the amount of cross-linking), extended analysis times, efficiency losses of the order to 50% for a doubling of the flow rate, the need for high temperature (85 °C) operation, and the need for specialized regeneration and re-packing of contaminated columns. However these drawbacks are offset by the ability to obtain a total analysis of material applied to the column (see *Figure 11*) and the use of water as the only eluent (74). Another way to overcome the losses in efficiency is the use of 2% cross-linked stationary phase (H⁺ form) with low flow rates and very low back pressure (75). In this case the size exclusion mechanism is predominant. The use of 0.01 M sulfuric acid enables the

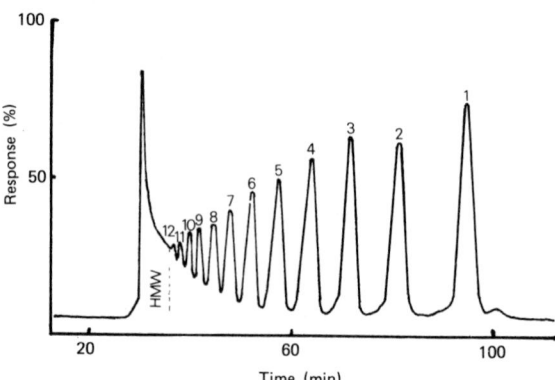

**Figure 11.** Separation of starch-derived oligosaccharides by high performance ion-exchange chromatography using 4% cross-linked cation-exchange resin with silver counterions (1,2,3, etc. refer to the DP of the oligosaccharide; HMW, high molecular weight material above DP 12).

constant regeneration of the $H^+$ form stationary phase. The effects of the temperature and the flow rate on the resolution has been investigated (76). The size fractionation of oligosaccharides up to DP 20 has been performed using a cation-exchange resin column ($Na^+$ form) using water/acetonitrile as mobile phase, in the presence of variable concentrations of sodium acetate or triethylammonium buffer (77). The separation efficiency has been tested at different temperatures and mobile phase pH.

## 5.5 Gel permeation chromatography

The development of non-compressible matrices for GPC analysis of water soluble materials has not reached the degree of sophistication available for the organic eluent compatible matrices. Consequently there has been no direct replacement for the cross-linked polysaccharide or polyacrylamide materials used for traditional gel permeation analysis of oligosaccharides. Some advances have, however, been made with the development of silica matrices deactivated by the chemical bonding of an organic ether stationary phase to provide a hydrophilic surface. The currently available materials have fractionation ranges which extend down to molecular weights of about 2000 (i.e. DP 10–12) whilst unmodified silica with 6 nm pore size can extend the fractionation range down to molecular weights of about 1000 (i.e. DP 5–6). Even with the modified materials adsorption effects are present and elution with ionic buffers (0.1 M) is recommended within the pH range 2–7. Products of starch hydrolysis has been separated using 0.15 M NaCl as the mobile phase (78). Oligosaccharides up to DP 6 from partially degraded hyaluronic acid have been resolved but a complicated three-column arrangement was required (79).

Water-compatible hydroxylated polyether-based matrices have been developed (for example the TSK PW and TSK HW series) which overcome some of the disadvantages of silica-based materials and have fractionation ranges which are comparable to the cross-linked polysaccharide and poly-acrylamide gels (80). Such materials are less rigid than the silica-based materials and therefore require lower operating pressures. Their lower selectivity compared to silica-based matrices is offset by their increased stability towards alkaline pH (up to pH 12). Whilst analysis times are of the order of one third to one tenth that of traditional low pressure gel permeation analysis the separations are inferior to those currently obtainable by ion-exchange chromatography (Section 5.4) and consequently little emphasis is placed on high performance gel permeation chromatography for oligo-saccharide fractionation.

## 5.6 Affinity HPLC

The affinity principle (see Section 4.3) has been used in HPLC to separate oligosaccharide molecules (81, 82). The ligand is covalently bonded with a

10 μm macroporous silica matrix. This improvement provides better resolution and shorter analysis times. Varying the temperature it is possible to obtain excellent chromatographic separation.

## 5.7 Preparative HPLC

The HPLC wide bore columns (2.0–2.5 cm i.d.) have been used to isolate milligram to gram quantities of malto-oligosaccharides from starch hydro-lysates (83). Very good results have been achieved using reverse-phase, bonded, and cation-exchange ($H^+$ and $Ag^+$ form) packings. In order to obtain retention times for each oligosaccharide, comparable to those obtained with the analytical columns, the flow rates for the preparative column ($F_P$) have been calculated using the following equation:

$$Fp = Fa[i.d.p/i.d.a]^2 . Lp/La$$

where $Fa$ is the analytical flow rate, i.d.p and i.d.a are the inner diameters of the preparative and analytical columns, and $Lp$ and $La$ are the column lengths.

## 5.8 Detection

Detection of oligosaccharides eluting from HPLC columns is the biggest challenge and the weakest link in the analysis of oligosaccharides. For underivatized oligosaccharides the most common detectors are: refractive index, UV, and electrochemical detectors. Non-specific detectors such as the refractive index detector are used routinely (64, 74, 76, 83). Ultra-sensitive refractive index detectors have detection limits of less than 200 nanograms of monosaccharide (83). A severe restriction is the high sensitivity to the temperature changes and to the mobile phase composition. It is necessary to use isocratic elution (i.e. single non-gradient eluents). Low wavelength UV detection (below 210 nm) (54) has been shown to have comparable sensitivity whilst allowing the use of limited gradient elution (84). Practically insensitive to the temperature and mobile phase changes (if the solvents used do not adsorb in the UV).

Electrochemical detection by means of a pulsed amperometric detector (PAD) has been considerably developed in the last few years, coupled in particular with anion-exchange (69, 70) and affinity (81, 82) HPLC. This technique is very sensitive (detection of 10–100 pmol) (67). A complete study of linear and branched glucose oligosaccharides using anion-exchange HPLC and PAD has been recently carried out looking at the variation of detector response at different degrees of polymerization (85). An electrochemical quantification of underivatized oligosaccharides has also been performed (68). The main limit of PAD is the necessity to operate at relatively high pH, and that can influence the choice of the separation conditions.

A mass detector has been developed which allows gradient elution to be performed and has been used for monosaccharide analysis (86) but the use of high water contents in the eluents needed for oligosaccharide analysis results in instability of the evaporation system which is an integral part of the detection system.

The problem of removing all traces of eluent after chromatographic separation are being resolved (e.g. termospray interface) and are allowing the mass spectrometer to be developed as an ultra-sensitive HPLC detector (87) which could have many applications in oligosaccharide analysis (53).

To obtain an improved sensitivity pre- or post-column derivatizations have been developed. Aromatic or heterocyclic substituents are added to the oligosaccharide structure to allow the UV or fluorescence detection. Oligosaccharides can be labelled with 2-aminopyridine (UV and fluorescent absorber) by reductive amination (77, 88), or with ethyl 4-aminobenzoate (UV absorber) (63, 89). This last derivative allows the quantification of picomolar amounts of oligosaccharides (58). Another alternative method in which the reducing carbohydrates are labelled for UV detection with 1-(4-methoxy)phenyl-3-methyl-5-pyrazolone allows sensitivity close to the sub-nanomole level (90).

Oligosaccharides have been $^3$H-labelled at the reducing terminal by reduction with NaBH$_4$ for radioactivity monitoring (91). The reduction conditions has been studied (92) to obtain the best detector limit (0.3 pmol).

Post-column derivatization labels the saccharides with a chromophore (see Section 2) for variable wavelength or fluorescence detectors. Whilst the use of post-column reaction systems have been developed for monosaccharide analysis (see Chapter 1) such systems have a limited applicability to oligosaccharide analysis due to the reliance on reducing sugar assays. Attempts to use strong cation-exchange resins (in the protonated form) to hydrolyse the glycosidic bonds in oligosaccharides to give complete conversion to monosaccharides after chromatographic separation have been reported (93) with the resulting monosaccharides being detected as reducing compounds. The method of post-column reaction for reducing carbohydrates, based on 4-aminobenzoylhydrazide followed by variable wavelength detection at 400 nm (see Section 2.3), has been compared with the PAD (22). Post-column reactions has been developed in order to detect reducing and non-reducing sugars at trace levels (94).

A wide number of detection systems have been recently reviewed (45).

# 6. Gas chromatography

Despite the expansion of HPLC techniques, gas chromatography (HGLC) still continues to have a place in oligosaccharide analysis for both structural studies, to determine the component monosaccharide residues (95) and position of inter-residue glycosidic bonds (see Chapter 3), and the routine analysis and quantitation of oligosaccharides. Whilst the determination of oligosaccharides by GLC is normally restricted to the analysis of disaccharides as an extension of monosaccharide analysis (see Chapter 1) methods have been reported for the fractionation of oligosaccharides up to DP 6–7 (96–97).

The efficacy of packed column GLC analysis has been improved using the high resolution as chromatography (HRGC), with higher resolution and reproducibility. The analysis times are shorter and the chromatograms show sharp peaks for the larger oligosaccharides too. The use of high-temperature columns and more volatile derivatives has extended separation of permethylated isomalto-oligosaccharides up to DP 10 with a sensitivity of one nanogram (< 1 pmol) (98, 99). The effect of GLC analysis parameters (i.e. carrier flow rate, split ratio, and nature of the derivatizing agent) on the separation of legume oligosaccharides on HRGC was recently studied (100). Using long elution times and temperatures higher than 400 °C the thermal degradation of saccharides occurs (98, 99).

The choice between HPLC and GLC is not clear cut with GLC being at least ten times more sensitive and producing shorter separation times than HPLC techniques. To offset these advantages the additional time required to prepare the sample and produce the required derivative (see Chapter 1) with no partially derivatized contaminants makes single analyses very unattractive. The final decision as to whether to use HPLC or GLC frequently depends not on the relative merits of the method but on the nature of a particular sample (101). Typical samples which are better suited to GLC analysis are those containing trace amounts of oligosaccharides as in plant tissues or where the low DP oligosaccharides are present in small amounts relative to high contents of disaccharides and monosaccharides such as found in high DE glucose syrups (96) or adulterated honey (102).

# 7. Supercritical fluid chromatography

Supercritical fluid chromatography (SFC) is a relatively new technique which combines the advantages of GLC and HPLC. In fact, using $CO_2$ as the mobile phase above its critical temperature, it is possible to analyse high molecular weight or thermo-labile compounds. Another great advantage of this technique is the compatibility with universal detectors for organic compounds (e.g. flame ionization detector). The possibiltiy of SFC-MS (mass spectrometry) coupling has been discussed (103). Small internal diameter capillary columns are used (50–100 μm) with a stationary film (similar to the bonded-phase GLC columns) of 0.05–0.02 μm. The low solubility of carbohydrates in the $CO_2$ fluid is the biggest shortcoming. For this reason derivatization is necessary (trimethylsilyl or permethyl derivatization) (104–106). Silylated malto-oligosaccharides can be separated by SFC from DP 1 to beyond 20 (104) in pressure programmed conditions. In this case double peaks, due to the two anomers at each DP value have been observed. Resolution decreases in the course of the program due to increased mobile phase pressure.

# 8. Capillary electrophoresis

Capillary electrophoresis (CE) is a powerful new technique that provides short time and high efficiency separations of complex mixtures of ions. This is due to the application of electrophoretic separation to a fused silica capillary. The two ends of the capillary are immersed in two separated electrolyte reservoirs containing a high voltage electrode. Variable parameters to improve the separation are the buffer composition and the pH.

Detection is usually performed with on-line systems (UV or fluorometric detectors) and derivatization of the oligosaccharides is necessary. It is very important that the sample components possess charged moieties in their structures. For this reason the neutral oligosaccharides are converted to primary amines by reductive amination. This derivatization followed by the reaction with a fluorogenic reagent (3-(4-carboxybenzoyl)-2-quinolinecarbox-aldehyde or 3-benzoyl-2-naphthaldehyde) has extended the detection limit down to the attomole level (107). The capillary zone electrophoresis (CZE) is the simplest form of CE. The capillary is only filled with the buffer solution (phosphate or borate buffer). The separation is due to variations in the molecular size/electric charge ratios of the mixture components. Direct CZE separates the oligosaccharide derivatives according to the increasing degree of polymerization up to 20 (see *Figure 12*) but cannot distinguish oligo-saccharides having the same DP CZE as borate complexes (108) provides a

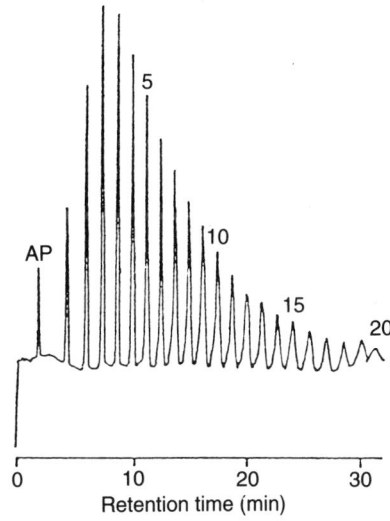

**Figure 12.** Separation of reductively pyridylaminated isomalto-oligosaccharides by direct CZE. Capillary, fused silica coated with polyacrylamide (Bio-Rad, 25 μm i.d., 20 cm); carrier, 100 mM phosphate buffer (pH 2.5); applied voltage, 8 kV; detection, UV absorption at 240 nm. AP, 2-aminopyridine (excess reagent).

separation according to the structural differences of the outer mono-saccharide residues. The two systems used to analyse the same sample are demonstrated to be complementary to each other (109).

## 9. Mass spectrometry

While mass spectrometry has traditionally been of major use when coupled to GLC columns (more recent is the possibility to interface HPLC and SFC to mass spectrometers) to provide structural information on oligosaccharides and polysaccharides (see Chapter 3) the development of several new ionization techniques now allows the direct analysis of underivatized oligosaccharides (110). Field desorbtion, in which ions are produced by thin wire emitters on which the sample is deposited, is a very soft ionization technique capable of producing molecular ions or simplified spectra (111). Other soft ionization techniques, that has been recently developed, are fast atom bombardment (FAB) and liquid secondary ionization (LSI). Fast atom bombardment (FAB), produces ions via bombardment of the sample with high energy rare gas atoms. It direct coupling with micro HPLC for oligosaccharide analysis has been studied (112). Structural information can be obtained in addition to molecular weight for underivatized oligosaccharides of DP up to 20 and beyond. Analysis by FAB of neutral underivatized oligosaccharides showed a low sensitivity (1–50 nmol of sample required) (113).

Also liquid secondary ionization mass spectrometry has been used to analyse native neutral homoglycan (114). This technique has shown many advantages like, short measuring time and direct access to molecular masses and high sensitivity. These characteristics are maintained when derivatized oligosaccharides are submitted to LSI mass spectrometry with a detection limit of 100 fmol for maltopentaose derivatized by reductive amination using phosphatidylethanolamine dipalmitoate) (115).

## 10. Nuclear magnetic resonance spectroscopy

Nuclear magnetic resonance (NMR) spectroscopy has been developed into a very useful non-destruction technique for the determination of oligosaccharide structures.

$^{13}$C-NMR not only gives information on the anomeric configuration on the carbohydrate residues but also provides information on the composition of component monosaccharides, their sequence, and the overall conformation of the molecule (see Chapter 5). Methylated and charged agarose oligo-saccharides have been analysed by $^{13}$C-NMR (116). The primary analysis of the structures of manno-oligosaccharides has been carried out (117).

$^{1}$H-NMR whilst being a more sensitive technique, does not provide as much structural information due to incomplete separation of the proton resonance signals obtained for carbohydrate molecules. It can, however, be used to quantitate individual oligosaccharide contents in situations where the unique selectivity of the method overcomes interference problems encountered when using other spectroscopic methods of great sensitivity but inferior selectivity. Determination of substituent groups on carbohydrate residues can also be made without the problems encountered with the traditional degradation techniques (118). Recently $^1$H-NMR has been used to characterize enzymically derived oligosaccharides (119, 120).

The degree of polymerization of reducing oligosaccharides can be determined by using $^2$H-NMR (121).

Another recent development which allows the complete structural determination of oligosaccharides is two-dimensional nuclear magnetic resonance (2D NMR) (122). Some peculiar problems (hidden resonace) can be in this way resolved. 2D NMR can be homonuclear (123, 124) (same nulcear types $^1$H-$^1$H) or heteronuclear (125) (different nuclear types $^1$H-$^{13}$C). Combining the data of the 2D signals, unambiguous information can be obtained about the type of residue and its position in the carbohydrate chain, also without derivatization or reducing of the reducing end. This method allows the study of oligsaccharide mixtures (126).

It is outside the scope of this chapter to describe the various methods available for recording spectra and the methods by which spectra can be interpreted. These subjects have been reviewed in depth with particular emphasis being given to oligosaccharide analysis (127, 128).

With the advent of Fourier transform techniques development is underway to prepare on-line HPLC detectors (129) which will provide valuable structural information on components in mixtures rather than the more normal determination of net properties of a component. This non-destructive detector will allow its use in conjunction with other detectors or fraction collectors if preparative scale chromatography is used.

# 11. Infra-red spectroscopy

Infra-red spectra of carbohydrates are complex and the method is usually restricted to the identification of specific structural features, particularly when monitoring of chemical reactions is required with interpretation of spectra normally comparison with published spectra or spectra obtained for a standard reference compound. The advent of infra-red spectroscopy has led to the development of analysers capable of detecting total carbohydrate contents in aqueous solutions and complex mixtures and formulations (130, 131). By choice of selected resonance frequencies it is possible to introduce a degree of selectivity into the determination such that the concentration of D-glucose in starch suspensions or specific oligosaccharides such as lactose can be measured.

# References

1. Dziedzic, S. Z. and Kearsley, M. W. (ed.) (1984). *Glucose syrups: science technology*. Elsevier Applied Science, London.
2. Kennedy, J. F., Noy, R. J., Stead, J. A., and White, C. A. (1985). *Starch/ Stärke*, **37**, 343.
3. Dische, Z., Shettles, L. B., and Osnos, M. (1949), *Arch. Biochem.*, **22**, 169.
4. Dubois, M., Gilles, K. A., Hamilton, J. K., Rebers, P. A., and Smith, F. (1956). *Anal. Chem.*, **28**, 350.
5. Svennerholm, L. (1956). *J. Neurochem.*, **1**, 42.
6. Roe, J. H. (1955). *J. Biol. Chem.*, **212**, 335.
7. Lane, J. H. and Eynon, L. (1923). *J. Chem. Soc. Ind.*, **42**, 32T.
8. Egan, H., Kirk, R. S., and Sawyer, R. (1981). In *Pearson's chemical analysis of foods* (8th edn), p. 150. Churchill Livingstone, London.
9. Bernfeld, P. (1955). In *Methods in enzymology* (ed.) S. P.Colowick and N. D. Kaplan, Vol. 1, p. 149. Academic Press, London and New York.
10. Lindner, W. A., Dennison, C., and Quicke, G. V. (1983). *Biotechnol. Bioeng.*, **25**, 377.
11. Robyt, J. F and Whelan, W. J. (1972). *Anal. Biochem.*, **45**, 510.
12. Somogyi, M. (1952) *J. Biol. Chem.*, **195**, 19.
13. Breuil, C. and Saddler, J. N. (1985). *Enzyme Microbol. Technol.*, **7**, 327.
14. Doner, L. W. and Irwin, P. L. (1992). *Anal. Biochem.*, **202**, 50.
15. Kidby, D. K. and Davidson, D. J. (1973). *Anal. Chem.*, **55**, 321.
16. Osborne, D. R. and Voogt, P. (1978). *The analysis of nutrients in food*. Academic Press, London.
17. Mullings, R. and Parish, J. H. (1984). *Enzyme Microb. Technol.*, **6**, 491.
18. White, C. A. and Kennedy, J. F. (1979). *Clin. Chim. Acta*, 95, 369.
19. Kennedy, J. F. and Fox, J. E. (1980). *Methods Carbohydr. Chem.*, **8**, 3.
20. Kennedy, J. F. and Fox, J. E. (1980). *Methods Carbohydr. Chem.*, **8**, 13.
21. Yránty, P., Brinkman, U. A. T., and Frei, R. W. (1985). *Anal. Biochem.*, **57**, 224.
22. Peelen, G. O. H., de Jong, J. G. N., and Wevers, R. A. (1991). *Anal. Biochem.*, **198**, 334.
23. Buffa, M., Congiu, G., Lombard, A., and Tourn, M. L. (1980). *J. Chromatogr. Anal. Biochem.*, **200**, 309.
24. Kennedy, J. F., White, C. A., and Riddiford, C. L. (1979). *Starch/Stärke*, **31**, 235.
25. Damonte, A., Lombard, A., and Tourn, M. L. (1980). *J. Chromatogr.*, **60**, 203.
26. Stahl, E. and Kaltenbach, U. (1961). *J. Chromatogr.*, **5**, 351.
27. Magnani, J. L. (1987). In *Methods in enzymology*. Vol. 138, p.208.
28. Heyraud, A. and Rinaudo, M. (1981) *J. Liquid Chromatogr.*, **4**, 175.
29. French, D., Robyt, J. F., Weintraub, M., and Knock, P. (1966). *J. Chromatogr.*, **24**, 68.
30. Barker, S. A., Hatt, B. W., Kennedy, J. F., and Somers, P. J. (1969). *Carbohydr. Res.*, **9**, 327.
31. Haylicek, J. and Samuelson, Ö. (1972). *Carbohydr. Res.*, **22**, 307.

32. Jonsson, P. and Samuelson, Ö. (1967). *J. Chromatogr.*, **26**, 194.
33. Kennedy, J. F., Cabral, J. M. S., and Kalogerakis, B. (1985). *Enzyme Microb. Technol.*, **7**, 22.
34. White, C. A., Kennedy, J. F., Lombard, A., and Rossetti, V. (1985). *Br. Polymer J.*, **17**, 327.
35. Kennedy, J. F., Stevenson, D. L., and White, C. A. (1985). *Cellulose Chem. Technol.*, **19**, 505.
36. Kennedy, J. F. (1972). *J. Chromatogr.*, **69**, 325.
37. Kiser, D. L. and Hagy, R. L. (1979). In *Liquid chromatographic analysis of food and beverages* (ed. G. Charalambous), Vol. 2, p. 263. Academic Press, New York.
38. Djordjevic, S. P., Batley, M., and Remond, J. W. (1986). *J. Chromatogr.*, **354**, 507.
39. Goso-Kato, K., Iwase, H., Ishihara, K., and Hotta, K. (1986). *J. Chromatogr. Biomed. Appl.*, **380**, 374.
40. Kobata, A., Yamashita, K., and Takasaki, S. (1989). In *Methods in enzymology*, Vol. 138, p. 84.
41. Kobata, A. and Yamashita, K. (1987). In *Methods in enzymology*. Vol. 179, p. 46.
42. Dakour, J., Lundblad, A., and Zopf, D. (1988). *Arch. Biochem. Biophys.*, **264**, 203.
43. Smith, D. F. and Torres, B. V. (1989). In *Methods in enzymology*. Vol. 179, p.30.
44. Kobata, A. and Endo, T. (1992). *J. Chromatogr.*, **597**, 111.
45. Ben-Bassat, A. A. and Grushka, E. (1991). *J. Liquid Chromoatogr.*, **14**, 1051.
46. Macrae, R. (ed.) (1982). *HPLC in food anlaysis*. Academic Press, London.
47. Rocca, J. F. and Rouchousse, J. (1976). *J. Chromatogr.*, **117**, 216.
48. Iwata, S., Narui, T., Takahashi, K., and Shibata, S. (1984). *Carbohydr. Res.*, **133**, 157.
49. White, C. A., Kennedy, J. F., and Golding, B. T. (1979). *Carbohydr. Res.*, **76**, 1.
50. Nachtmann, F. and Budna, K. W. (1977). *J. Chromatogr.*, **136**, 279.
51. Thompson, R. M. (1978). *J. Chromatogr.*, **166**, 201.
52. White, C. A., Vass, S. W., Kennedy, J. F., and Large, D. G. (1983). *J. Chromatogr.*, **264**, 99.
53. Silvestro, L., Viano, G., Naggi, A., Torri, G., Da Col, R., and Baiocchi, C. (1992). *J. Chromatogr.*, **591**, 225.
54. Chaturvedi, P. and Sharma, C. B. (1988). *Biochim. Biophys. Acta*, **967**, 115.
55. Bock, K., Duus, J. Ø., Norman, B., and Pedersen, S. (1991). *Carbohydr. Res.*, **221**, 219.
56. Noel, D., Hanai, T., and D'Amboise, M. (1979). *J. Liquid Chromatogr.*, **2**, 1325.
57. Wells, G. B. and Lester, R. L. (1979). *Anal. Biochem.*, **97**, 184.
58. Daniel, P. F. (1987). In *Methods in enzymology*. Vol. 138, p. 94.
59. Schwarzenbach, R. (1976). *J. Chromatogr.*, **117**, 206.
60. Rabel, F. M., Caputo, A. G., and Butts, E. T. (1976). *J. Chromatogr.*, **126**, 731.
61. Folkes, D. J. and Taylor, P. W. (1982). In *HPLC in food analysis* (ed. R. Macrae), p. 149. Academic Press, London.

62. White, C. A., Corran, P. H., and Kennedy, J. F. (1980). *Carbohydr. Res.*, **87**, 165.
63. Akiyama, T. (1991). *J. Chromatogr.*, **588**, 53.
64. Koizumi, K., Utamura, T., Kubota, Y., and Hizukuri, S. (1987). *J. Chromatogr.*, **409**, 396.
65. Koizumi, K., Kubota, Y., Tanimoto, T., and Okada, Y. (1989). *J. Chromatogr.*, **464**, 365.
66. Ammersaal, R. N., Delgado, G. A., Tenbarge, F. L., and Friedman, R. B. (1991). *Carbohydr. Res.*, **215**, 179.
67. Townsend, R. R., Hardy, M. R., Hindsgaul, O., and Lee, Y. C. (1988). *Anal. Biochem.*, **174**, 459.
68. Townsend, R. R., Hardy, M. R., Hindsgaul, O., and Lee, Y. C. (1989). In *Methods in enzymology*. Vol. 179, p. 65.
69. Hernandez, L. M., Ballou, L., and Ballou, C. L. (1990). *Carbohydr. Res.*, **203**, 1.
70. Reddy, G. P. and Bush, A. C. (1991). *Anal. Biochem.*, **198**, 278.
71. Ladisch, M. R. and Tsao, G. T. (1978). *J. Chromatogr.*, **16**, 85.
72. Fitt, L. E., Hassler, S., and Just, D. E. (1980). *J. Chromatogr.*, **187**, 381.
73. Scobell, H. D. and Brobst, K. M. (1981). *J. Chromatogr.*, **212**, 51.
74. Scott, F. .W. and Hatina, G. (1988). *J. Food Sci.*, **53**, 264.
75. Hicks, K. B. and Hotchkiss, A. T., Jr. (1988). *J. Chromatogr.*, **441**, 382.
76. Derler, H., Hörmeyer, H. F., and Bonn, G. (1988). *J. Chromatogr. Anal. Biochem.*, **440**, 281.
77. Seto, Y. and Shinohara, T. (1989). *J. Chromatogr.*, **464**, 323.
78. Brooks, J. R. and Griffin, V. K. (1987). *Cereal Chem.*, **64**, 253.
79. Kundsen, P. J., Erikson, P. B., Fenger, M., and Florentz, K. (1980). *J. Chromatogr.*, **187**, 373.
80. Thurl, S., Offermanns, J., Müller-Werner, B., and Sawatzki, G. (1991). *J. Chromatogr. Biomed. Appl.*, **568**, 291.
81. Zopf, D., Ohlson, S., Wang, W., and Lundblad, A. (1989). In *Methods in enzymology*. Vol. 179, pp. 55–
82. Wang, W., Lundgren, T., Lindh, F., Nilsson, B., Grönberg, G., Brown, J. P., Mentzer-Dibert, H., and Zopf, D. (1992). *Arch. Biochem. Biophys.*, **292**, 433.
83. Hicks, K. B. and Sondey, S. M. (1987). *J. Chromatogr.*, **389**, 183.
84. Binder, H. (1980). *J. Chromatogr.*, **189**, 414.
85. Ammeraal, R. N., Delgado, G. A., Tenbarge, F. L., and Friedman, R. B. (1991). *Carbohydr. Res.*, **215**, 179.
86. Macrae, R. and Dick, J. (1981). *J. Chromatogr.*, **210**, 138.
87. Chapman, J. R. (1985). *Inst. Food Sci. Technol. Proc.*, **18**, 59.
88. El Rassi, Z., Tedford, D., An, J., and Mort, A. (1991). *Carbohydr. Res.*, **215**, 25.
89. Webb, J. W., Jiang, K., Gillence-Castro, B. L., Tarentino, A. L., Plummer, T. H., Byrd, J. C., Fisher, S. J., and Burlingame, A. L. (1988). *Anal. Biochem.*, **169**, 337.
90. Kakehi, K.; Suzuki, S., Honda, S., and Lee, Y. C. (1991). *Anal. Biochem.*, **199**, 256.
91. Hall, N. A. and Patrick, A. D. (1989). *Anal. Biochem.*, **178**, 378.
92. Guo, Y. and Conrad, H. E. (1988). *Anal. Biochem.*, **168**, 54.

93. Vrátny, P. and Ouhrabková, J. (1980). *J. Chromatogr.*, **191**, 313.
94. Kraemer, M. and Engelhardt, H. (1992). *J. High Resol. Chromatogr.*, **15**, 24.
95. Helleur, R. J., Budgell, D. R., and Hates, E. R. (1987). *Anal. Chim. Acta*, **192**, 367.
96. Folkes, D. J. and Brookes, A. (1984). In *Glucose syrups: science and technology* (ed. S. Z. Dziedzic and M. W. Kearsey), p. 197. Elsevier Applied Science, London.
97. Traitler, H., Del Vedovo, S., and Schewizer, T. F. (1984). *J. High Resol. Chromatogr. Chromatogr. Comm.*, **7**, 558.
98. Karlsson, H. and Hannson, G. C. (1988). *J. High Resol. Chromatogr. Chromatogr. Comm.*, **11**, 821.
99. Gabius, S., Hellmann, K.-P., Hellmann, T., Brinck, U., and Gabius, H.-J. (1989). *Anal. Biochem.*, **182**, 447.
100. Degen, P. H., Risser, F., and Lauber, J. W. (1992). *J. Chromatogr.*, **623**, 191.
101. Brobst, K. M. and Scobell, H. D. (1982). *Starch/Stärke*, **34**, 117.
102. Doner, L. W., White, J. W., and Phillips, J. G. (1979). *J. Assoc. Off. Anal. Chem.*, **62**, 186.
103. Smith, R. D., Kalinoski, H. T., and Udseth, H. R. (1987). *Mass Spectrom. Rev.*, **6**, 445.
104. Chester, T. L., Pinkston, J. D., and Owens, G. D. (1989). *Carbohydr. Res.*, **194**, 273.
105. Leroy, Y., Lemoine, J., Ricart, G., Michalski, J.-C., Monteuil, J., and Fournet, B. (1990). *Anal. Biochem.*, **184**, 235.
106. Chester, T. L. and Innins, D. P. (1986). *J. High Resol. Chromatogr. Chomatogr. Comm.*, **9**, 209.
107. Liu, J., Shirota, O., and Novotny, M. (1991). *J. Chromatogr.*, **559**, 223.
108. Honda, S., Suzuki, S., Nose, A., Yamamoto, K., and Kakehi, K. (1991). *Carbohydr. Res.*, **215**, 193.
109. Honda, S., Makino, A., Suzuki, S., and Kakehi, K. (1990). *Anal. Biochem.*, **191**, 228.
110. Niessen, W. M. A., Van der Hoeven, R. A. M. and Van der Greef, J. (1992). *Org. Mass Spectrom.*, **27**, 341.
111. Schutten, H.-R. and Lehmann, W. D. (1980). *Trends Biochem. Sci.*, **5**, 142.
112. Ito, Y., Takeuchi, T., Ishii, D., Goto, M., and Mizuno, T. (1987). *J. Chromatogr.*, **391**, 296.
113. Poulter, L. and Burlingame, A. L. (1990). In *Methods in enzymology*, Vol. 193, p. 661.
114. Stahl, B., Steup, M., Karas, M,. and Hillenkamp, F. (1991). *Anal. Chem.*, **63**, 1463.
115. Lawson, A. M., Chai, W., Cashmore, G. C., Stoll, M. S., Hounsell, E. F., and Feizi, T. (1990). *Carbohydr. Res.*, **200**, 47.
116. Lahaye, M., Yaphe, W., Phan Viet, M. T., and Rochas, C. (1989). *Carbohydr. Res.*, **190**, 249.
117. Cumming, D. A., Hellerqvist, C., and Touster, O. (1988). *Carbohydr. Res.*, **179**, 369.
118. Kennedy, J. F., Stevenson, D. L., White, C. A., Tolly, M. S., and Bradshaw, I. J. (1984). *Br. Polymer J.*, **16**, 5.

119. Hoffmann, R. A., Leeflang, B. R., de Barse, M. M. J., Kamerling, J. P., and Vliegenthart, J. F. G. (1991). *Carbohydr. Res.*, **221**, 63.
120. Gruppen, H., Hoffmann, R. A., Kormelink, F. J. M., Voragen, A. G. J., Kamerling, J. P., and Vliegenthart, J. F. G. (1992). *Carbohydr. Res.*, **233**, 45.
121. Goux, W. J. (1988). *Carbohydr. Res.*, **173**, 292.
122. Koerner, T. A. W., Prestegard, J. H., and Yu, R. K. (1987). In *Methods in enzymology* Vol. 138, pp. 38–
123. Dabrowski, J., Ejchart, A., Kordowicz, M., and Handfland, P. (1987). *Magn. Res. Chem.*, **25**, 338.
124. Platzer, N., Davoust, D., Lhermitte, M., Bauvy, C., Meter, D. M., and Derappe, C. (1989). *Carbohydr. Res.*, **191**, 191.
125. Breg, G., Romijn, D., Vliegenthart, J. F. G., Strecker, G., and Montreuil, J. (1988). *Carbohydr. Res.*, **183**, 19.
126. Colquhoun, I. J., de Ruiter, G. A., Schols, H. A., and Voragen, A. G. J. (1990). *Carbohydr. Res.*, **221**, 63.
127. Coxon, B. (1980). In *Developments in food carbohydrates—2*, (ed. C. K. Lee) p. 351. Applied Science, London.
128. Rathbone, E. B. (1985). In *Analysis of food carbohydrates* (ed. G. C. Birch), p. 149. Elsevier Applied Science, London.
129. Dorn, H. C. (1984). *Anal. Chem.*, **56**, 747A.
130. Kennedy, J. F., White, C. A., and Browne, A. J. (1985). *Food Chem.*, **16**, 115.
131. Kennedy, J. F., White, C. A., and Browne, A. J. (1985). *Food Chem.*, **18**, 95.

# 3

# Neutral polysaccharides

JOHN H. PAZUR

## 1. Introduction

Polysaccharides are polymers of monosaccharide units joined together by glycosidic bonds which are formed by the elimination of the elements of water between the hemiacetal hydroxyl group of one residue and a primary or secondary hydroxyl group of an adjacent residue (1). Polysaccharides may consist of a small or a large number of residues and in structure may be linear, branched, or occasionally cyclic polymers. These polymers may be homopolysaccharides composed of a single type of residue or may be heteropolysaccharides composed of two or more different types of residues. Originally polysaccharides were named to reflect biological sources or properties of the polymer and many such names have remained in common usage. At the present time, it is the recommendation of the Joint Commission on Biochemical Nomenclature that polysaccharide nomenclature follow the general principles established for organic nomenclature (2). Glycan is the name recommended for a polysaccharide by the Commission. A glycan or polysaccharide consisting of a single type of monosaccharide is named by replacing the 'ose' ending of the constituent monosaccharide unit with the suffix 'an' as for example, glucan for a polysaccharide of glucose. In the case of a glycan with more than one type of residue the name is determined from the nature of the structural unit of the main chain and the prefix names of the other units as for example, fuco-manno-galactan. In the case of a glycan with no obvious main chain, the compound is named by listing the prefix names of all the constituent units alphabetically followed by the general term glycan, as for example galacto-gluco-manno-glycan. However both the common name and the systematic name are acceptable.

In living cells, polysaccharides may be combined covalently with members of other classes of compounds notably proteins and lipids. Many of these combinations as well as free polysaccharides have important biological functions and the group has become known as complex carbohydrates (3). Naturally occurring polysaccharides vary greatly in molecular size ranging from several thousands to millions of daltons. The molecular structure and architecture is diverse but can be divided into three major types, a linear structure in which the units are joined by the same type of glycosidic linkage to form long linear chains, a substitued linear structure in which mono-saccharides or short side chains of oligosaccharide units are attached to the main chain, and a branched structure in which long side chains are attached randomly to the main chain and to side chains. Polysaccharides with structures intermediate to the above basic types are known and are important constituents of biological materials.

For the elucidation of the complete molecular structure of a polysaccharide the structural features that need to be determined are: the types of monosaccharide residues, the D or L configuration of the residues, the number of residues per molecule, the positions of glycosidic linkages between residues, the ring structure of the residues, the sequence of monosaccharide residues in the chains, and the anomeric configuration of the glycosidic bonds.

The analytical techniques used in the structural analysis of polysaccharides are many and varied in types. The common methods used singly or in combination are methylation analysis, periodate oxidation, enzymic assays, immunological reactivities, paper chromatography, gas-liquid chromato-graphy, high performance liquid chromatography, mass spectrometry, polarimetry, and nuclear magnetic resonance spectrometry. Brief statements on the basic principles of the methods, detailed descriptions of experimental protocols, and descriptions of recent researches on complex polysaccharides are presented in this chapter.

## 2. Extraction of polysaccharides

Polysaccharides are the most abundant compounds of biological materials and in structure may be simple, or complex, and in function highly specific or very diverse. Some of these compounds are the major constituents of foods and serve as energy sources for sustaining life, some serve as structural elements providing shape and size to the living entity, some function as barriers to passage of metabolites and foreign substances into living cells, some impart immunological properties, and some regulate reactions of metabolic cycles of the living cells.

In *Table 1* are listed a few of the important polysaccharides of micro-organisms, marine plants, terrestrial plants, and animals. The polysaccharides occur in nature as members of heterogeneous mixtures of cellular constituents. Accordingly to investigate properties and reactions of polysaccharides, it is first necessary to isolate the desired polysaccharide and to purify the compound to homogeneity. With regard to criteria of purity, there is no single measure of purity that can be applied to all polysaccharides. It is generally agreed that a polysaccharide is pure if it may be isolated by two different procedures and the resulting preparations possess the same chemical, physical, and biological properties.

**Table 1.** Some naturally occurring polysaccharides

| Micro-organisms | Marine plants | Terrestial plants | Animals |
|---|---|---|---|
| Dextran | Agar | Starch | Glycogen |
| Levan | Alginic acid | Cellulose | Chondroitin sulfate |
| Xanthan | Carrageenin | Xylan | Heparin |
| Mannan | Furcellaran | Pectin | Hyaluronic acid |
| Nigeran | Laminaran | Gum guar | Chitin |
| Cell wall polysaccharides | | Gum arabic | |

**Table 2.** Common monosaccharide constituents of polysaccharides

| Type | Compound |
|---|---|
| Pentoses | D-Xylose, L-arabinose |
| Hexoses | D-Glucose, D-mannose, D-galactose, L-galactose, D-fructose |
| Hexosamines | N-Acetylglucosamine, N-acetylgalactosamine |
| Uronic acids | D-Glucuronic acid, D-galacturonic acid, D-mannuronic acid |
| Deoxyhexoses | L-Rhamnose, L-fucose, 6-deoxy-L-talose |

There are many possible stereoisomers of pentoses and hexoses that can occur but only a few of the isomers are constituents of polysaccharides. These monosaccharides are listed in *Table 2*. Because of the variations in the types of residues and linkages between residues there is an enormous number of polysaccharides in nature. Due to this diverse structure of polysaccharides, the isolation procedures vary greatly. A compilation of procedures for the isolation of polysaccharides has been published recently (4). As already mentioned several criteria of purity should be used to check the purity of polysaccharide preparations. These are:

- constancy in monosaccharide composition
- constancy in quantitative values of unique structural constituents
- constancy in the ratio of monosaccharide constituents
- uniform sedimentation rate on ultracentrifugation
- uniform behaviour on gel filtration or ion-exchange chromatography

Suitable colorimetric analytical procedures must be devised to measure the property under consideration. For sedimentation methods an ultracentrifuge will be required. Polysaccharides which sediment as a single symmetrical peak on ultracentrifugation are most likely homogeneous in molecular size but those which yield multi-peaks are heterogeneous. The behaviour of the polysaccharides on filtration on gels or on ion-exchange resins yields information about the purity of a polysaccharide preparation. In these procedures the preparation is passed through a column of the gel or resin and the eluates from the column are analysed for carbohydrate. A symmetrical elution peak is indicative of homogeneity. The technique of gel filtration has become a popular method for determining molecular size as well as for assessing homogeneity. By using polysaccharides of known molecular size as markers, molecular weights of new polysaccharides can be determined. Ultrafiltration through calibrated membranes is a relatively new method that can be used to assess homogeneity in polysaccharide preparations.

# 3. Identification of monosaccharides

In the structural analysis of a polysaccharide it is first necessary to determine the types of the monosaccharide residues which constitute the compounds. With the advent of chromatographic methods, the determination of the monosaccharide composition of a polysaccharide has been considerably simplified. For chromatographic analysis of the monosaccharides, the polysaccharide must first be hydrolysed to its constituent units. A satisfactory method for the hydrolysis of common polysaccharides is as follows.

---

**Protocol 1.** Hydrolysis of a polysaccharide

1. Dissolve a sample of 2–5 mg of the polysaccharide in 0.1–0.25 ml of 2 M HCl. Seal tube and heat at 100 °C for periods of 2–5 h.

2. Analyse the hydrolysate for monosaccharide components by a chromatographic method using paper, gel, ion exchanger, GLC, or HPLC.

3. Analyse suitable standards of monosaccharides or derivatives by the same chromatographic method.

---

The selection of a chromatographic method to be used will depend on the availability of equipment, standards, and polysaccharide supply. Very useful procedures are paper chromatography (5), gas-liquid chromatography (6), high performance liquid chromatography (7), automated ion-exchange chromatography (8), and recently developed HPAE-PAD chromatography (9). The unequivocal identification of the monosaccharides will involve the recovery of the monosaccharides from the hydrolysate in sufficient quantities for measurement of physical constants and the preparation of suitable derivatives.

# 4. Determination of D and L configuration

The D or L configuration of a monosaccharide can be determined by measurements of specific properties of the monosaccharides isolated from acid hydrolysates of a polysaccharide. It is essential that the monosaccharide be isolated in pure form. The specific rotation, circular dichroism absorption, enzymic susceptibility, or chemical reactivity with optically active reagents can then be used. By comparison of the results with the isolated compounds and that of known monosaccharides the D or L configuration of the unknown monosaccharide can be established. Some of the methods for determining the D or L configuration of monosacchride units of a polysaccharide are illustrated for a polymer of 6-deoxytalose, rhamnose, galactose, and glucuronic acid from *Streptococcus bovis* (10). The molecular structure of this polysaccharide has been determined by use of a combination of analytical methods. The structure of the repeating unit is shown in Formula 1.

$$\rightarrow 3)\text{-6-deoxy-L-Tal}p\text{-}(1\rightarrow 3)\text{-D-Gal}p\text{-}(1\rightarrow 3)\text{-L-Rha}p\text{-}(1\rightarrow 2)\text{-L-Rha}p\text{-}(1\rightarrow$$

$$4$$

$$\uparrow$$

$$1$$

$$\text{D-GlcA}p$$

[Formula 1]

---

**Protocol 2.** D and L configuration

A. *Optical rotation*

1. Hydrolyse 100 mg of polysaccharide in 2 ml of 0.2 M HCl in a boiling water-bath for 3 h.

2. Analyse the hydrolysate by qualitative paper chromatography and identify the monosaccharides.

3. Isolate the monosaccharides by preparative paper chromatography (see Chapter 1). From 100 mg of the polysaccharide 12 mg of 6-deoxytalose, 10 mg of rhamnose, 15 mg of galactose, and 8 mg of glucuronic acid may be obtained.

4. Measure the optical rotation of each monosaccharide in a 2 dcm polarimeter tube with 1.4 ml capacity in a Rudolph polarimeter.

5. Calculate the specific rotations from these data: 6-deoxytalose $\alpha_d = -17°$ (c.0.01, water); rhamnose $\alpha_d = +8°$ (c.0.009, water); galactose $\alpha_d = +77°$ (c.0.005, water); and glucuronic acid $\alpha_d = +35.3°$ (c.0.01, water).

B. *Enzymic method with glucose and galactose oxidase*

1. Reduce the carboxyl group of the glucuronic acid of the polysaccharide to a primary alcohol group by reacting 20 mg of polysaccharide with 20 mg carbodiimide in 2 ml of water for 6 h, and then reacting the

**Protocol 2.** *Continued*

product with 10 mg of sodium borohydride for 18 h (11). Acidify with acetic acid and remove the borates by evaporation from methanol. Dialyse for 24 h and lyophilize to dryness.

2. Hydrolyse 5 mg of the reduced polysaccharide in 0.25 ml of 0.1 M HCl for 2 h in a boiling water-bath.

3. Separate the monosaccharides in the hydrolysate by paper chromatography (see Chapter 1, Section 4).

4. Spray the dried chromatogram lightly with a glucose oxidase/peroxidase solution (0.1% glucose oxidase and 0.01% peroxidase in 0.1 M acetate buffer of pH 5.2) and then with 0.2% alcoholic *O*-tolidine solution (12).

5. Spray a duplicate chromatogram with a solution of galactose oxidase and peroxidase in buffer of pH 7, and then with *O*-tolidine, and observe for blue colour appearance on the chromatogram (13).

C. *Configuration from chiral glycosides and GLC*

1. React 0.5–1 mg of monosaccharide with 0.5 ml of 2-octanol or 2-butanol and add a drop of trifluroacetic acid in a small ampoule.

2. Seal ampoule and heat overnight in an oil-bath at 130 °C.

3. Evaporate to dryness in vacuum at 55 °C.

4. Prepare the acetate or the silyl derivative by dissolving the residue in acetic anhydride and heating at 100 °C for 20 min, or treating the residue with hexamethyl disilazine and chlorotrimethyl silane/pyridine for 30 min at room temperature.

5. Remove excess reagent, dissolve in chloroform, and perform GLC analysis.

Comparison of the specific rotation values of the isolated compounds with the values recorded in the literature shows that the deoxy sugars are of the L configuration and the galactose and glucuronic acid are of the D configuration. In the enzymic method reference D-glucose and the product of reduced glucuronic acid migrated with identical $R_f$ values and both yielded intense blue on paper chromatograms (12). Reference L-glucose did not yield a positive test with the glucose oxidase. Thus, the glucose obtained by reduction of the glucuronic acid from the polysaccharide is shown to be of the D configuration and therefore the glucuronic acid is also of the D configuration. D-Galactose and the product with the same $R_f$ value from the hydrolysate of the polysaccharide give a blue colour on the chromatogram sprayed with galactose oxidase but L-galactose did not (13). The galactose from the polysaccharide is also of D configuration.

A method for the determination of D and L configuration for a number of monosaccharides (glucose, galactose, xylose, fucose) using the chiral property has been developed (14) and described recently (15). It involves reaction of the monosaccharides with optically active alcohols to yield the corresponding gycosides, followed by acetylation or silylation, and analysis of products by GLC. This method has not been used with the *S. bovis* polysaccharide.

# 5. Degree of polymerization

After the identification of the constituent residues in a polysaccharide it is necessary to determine the number of residues per molecule. Since the biosynthesis of a polysaccharide is under secondary genetic control, polymers with different numbers of residues will be synthesized in living cells resulting in heterogeneity in molecular size. Other factors such as self-association of the chains will contribute to the polydisperse nature of carbohydrate polymers. The molecular weight values determined for polysaccharides are therefore average molecular weights. Several types of methods have been employed for determining molecular weights. The older methods based on osmotic pressure, viscosity, light scattering, and streaming biofringance measurements are not used very often. Newer methods based on gel filtration, ultracentrifugation, and ultrafiltration are now in use (16, 17).

Ultracentrifugation can be used if the molecular weight of the polysaccharide is reasonably large. Procedures based on the use of an analytical ultracentrifuge or preparative ultracentrifuge have been devised. In the procedure utilizing the analytical ultracentrifuge an elaborate optical arrangement is needed, and other properties of the polysaccharide, specifically diffusion coefficient, partial specific volume, and the density of the solvent must also be measured. Such methods have been employed with the glucose polymer of the starch and glycogen types. Density gradient centrifugation is advantageous to use with some polymers. Gradients are prepared from solutions of increasing concentrations of inert compounds such as sucrose, glycerol, or caesium chloride (18, 19).

---

**Protocol 3.** Density gradient centrifugation

1. Place a sample of about 2 mg of polysaccharide on top of the gradient column (5–40% glycerol).

2. Insert the tubes in a swinging-bucket rotor and centrifuge at 42 000 r.p.m. for 10–16 h.

3. Fractionate the column in 0.2 ml fractions by means of a mechanical fractionator.

4. Analyse the eluates for the polysaccharide component by a suitable colorimetric method.

---

**Protocol 3.** *Continued*

5. Subject the polysaccharides of known molecular weights to identical centrifugation conditions.

6. Measure the distance that the polysaccharides sediment from the top meniscus to the middle of the polysaccharide peak.

7. From the empirical relationship $(D_1/D_2) = (M_1/M_2)^{2/3}$ calculate the molecular weight of the polysaccharide (18). $D_1$ and $M_1$ are distance and molecular weight for the unknown and $D_2$ and $M_2$ for the standard.

In the density gradient centrifugation procedure, the polysaccharide can be easily recovered following centrifugation by collecting the fractions from the gradient tube, analysing for carbohydrate, and combining the carbohydrate containing fractions. The solution is then dialysed to remove gradient material and taken to dryness by lyophilization. The method has been applied for the isolation and characterization of unique polysaccharides in the cell wall of streptococcal organisms.

The molecular weights of polysaccharides of low degree of polymerization can be determined by methods based on the quantitative determination of a functional group of the polysaccharide and appropriate standard curves. The reducing group of polysaccharides can be measured by colorimetric procedures (20) or can be reacted with radioactive reagents (21). The colour intensity and radioactivity values of the reaction products can be correlated with molecular weights.

The carbohydrate residues which constitute the non-reducing end of a polysaccharide can be quantitatively determined and the values used to calculate molecular weights. In this method the polysaccharide is methylated completely by the procedure outlined (see *Protocol 4*) and hydrolysed to the constituent units which are converted to partially methylated alditol acetates. The non-reducing end residue is identified from the nature of the methylated products. For linear polysaccharides the degree of polymerization is calculated from the yield of the methylated derivative from the terminal residue. For branched molecules the average chain length of terminal chains can be determined. An alternate method is to use periodate oxidation which yields data from which similar types of calculations can be made. In the oxidation reaction, formic acid is produced and measured quantitatively by a titration method with standard base.

Recently, gel filtration methods based on the use of gels of various pore sizes have become available and are being used for the determination of molecular weight of polysaccharides and degradation products (22, 23). In the filtration methods, standard polysaccharides of known molecular weight are needed for calibration curves. Also it is necessary to establish that polysaccharides of unknown molecular weight behave in an identical manner as the standards in the gel permeation. Ultrafiltration methods through membranes which retain molecules of specified sizes have also become available. Such methods may be advantageous to use for the determination of molecular size of some polysaccharides.

# 6. Position of glycosidic linkages

## 6.1 Methylation

Methylation analysis has been an important method in structural analysis of polysaccharides for many years (24). With the introduction of microtechniques for methylation (25) and subsequent analysis of hydrolytic products by gas-liquid chromatography and mass spectrometry (26) the utility of the method has been greatly enhanced. Briefly, the method involves complete methylation of a polysaccharide, hydrolysis to a mixture of partially methylated monosaccharides, reduction of the methylated monosaccharides to alditols, acetylation of the alditols, and identification of the partially methylated alditol acetates by gas-liquid chromatography and mass spectrometry. The types of the glycosidic linkages are deduced from the nature of the partially methylated alditol acetates. Methylation analysis does not give information on the anomeric configuration of the glycosidic linkages nor on the sequence of the monosaccharide residues in the polysaccharide. The latter determinations must be done by other methods.

In methylation analysis it is essential that a complete methylation of all of the hydroxyl groups of a polysaccharide be achieved. In early studies of polysaccharide structure the methylations were performed with dimethyl sulfate in sodium hydroxide or with methyl iodide and a silver oxide catalyst and such reagents quite often give incomplete methylation. Prolonged reaction times and corresponding delays in completion of the analysis are encountered. Further, large samples of polysaccharides were needed for these methylations in order to obtain sufficient derivatives for analysis by fractional distillation methods.

A method superior to the older methods for methylating polysaccharides has been evolved using a strong base, methylsulfinyl methyl sodium, to ionize free hydroxyl groups of the polymer and using methyl iodide for methylating these groups (25). In this reaction sequence alkoxide ions are generated readily from free hydroxyl groups by the strong base and methylation of these ions occurs rapidly. Dimethyl sulfoxide is used as the solvent for the reagents and the reactants. A reaction sequence for the methylation of a 4-substituted monosaccharide residue of a polysaccharide is shown in Equation 1.

[Equation 1]

A satisfactory method for assessing the completeness of methylation of all free hydroxyl groups is not available. Generally neutral polysaccharides are completely methylated by the micro method described in *Protocol 4*B and special precautions are not needed. However, polysaccharides containing uronic acid or hexosamine residues are more difficult to methylate and may yield secondary products. For example, a uronic acid residue can yield a ketal derivative by reaction of the carboxyl group with the methylsulfinyl methyl sodium and the ketals may undergo elimination reactions yielding undesirable side products. Hexosamine residues with *N*-acetyl groups yield the *N*-methyl *N*-acetamido, *O*-methyl derivative of the hexosamine on methylation. Such derivatives may require special analytical techniques for identification. The preparation of reagents and procedural steps in the methylation analysis are recorded in *Protocol 4*.

---

**Protocol 4**. Methylation analysis

A. *Preparation of methylsulfinyl methyl sodium from dimethyl sulfoxide (DMSO) and sodium hydride*

1. Thoroughly dry DMSO by stirring 500 ml of reagent grade DMSO with excess powdered calcium hydride for several hours at 65 °C. Distil DMSO under nitrogen and at reduced pressure.

2. Transfer the DMSO into brown bottles stoppered tightly with serum caps. Store over molecular sieves in the cold room.

3. For use, thaw the frozen solvent and remove a sample with a dry glass syringe. Transfer into the desired vessel.

4. Prepare the methylsulfinyl methyl sodium by adding 30 ml of dried DMSO into a three-necked 250 ml flask containing 1.5 g of 57% suspension of sodium hydride in mineral oil mixed with 10 ml of anhydrous ether.

5. Evacuate the flask and flush with nitrogen three times under vacuum. Add the DMSO from a dry syringe through the serum cap.

6. Attach a glass syringe to the flask through the rubber cap to facilitate the escape of hydrogen.

7. Sonicate the reaction mixture in the flask in a water-bath of 50 °C for 2–3 h. The synthesis of methylsulfinyl methyl sodium is complete when a transparent green colour appears in the solution in the flask.

8. Transfer the dimethylsulfinyl methyl sodium to several 5 ml serum bottles which are capped with serum caps and flushed with nitrogen. Store in the frozen state at 4 °C in the dark.

9. For methylations, thaw the methylsulfinyl methyl sodium and remove the sample through the rubber cap by means of a dry glass

syringe. Add the solution directly from the syringe into the flask containing the polysaccharide.

B. *Methylation procedure*

1. Place a sample of 2–5 mg of the thoroughly dry polysaccharide into a dry 25 ml round-bottom flask.

2. Fit the flask with a rubber serum cap and evacuate for 1 h to further dry the polysaccharide.

3. Continue the evacuation and flush the flask with nitrogen three times.

4. Add 1 ml of dry DMSO from a dry glass syringe and sonicate the reaction mixture for 20 min at room temperature.

5. Add 0.4 ml of methylsulfinyl methyl sodium solution using a dry glass syringe.

6. Liquify the resulting gel by sonication for 20 min at 25 °C.

7. Add 0.3 ml of dry methyl iodide and sonicate the mixture for 15 min at 25 °C.

8. Maintain the reactants at room temperature for an additional 1–6 h.

9. Remove the serum cap from the flask and add a small amount of water to neutralize excess methylsulfinyl methyl sodium.

10. Transfer the resulting suspension into 12 ml clinical centrifuge tubes and extract the methylated polysaccharide in the suspension with 4 ml of chloroform.

11. Wash the chloroform layer with 3 ml of water and centrifuge at high speed in a clinical centrifuge. Remove and discard the top water layer and wash the chloroform layer three more times with 3 ml aliquots of water, discarding the water layer after each washing.

12. After the last washing dry inside of tubes with Kimwipe and add anhydrous magensium sulfate to remove last traces of water.

13. Remove the magnesium sulfate by filtration through glass filter in a Pasteur pipette.

14. Collect the chloroform filtrate with the methylated polysaccharides in a 10 ml ampoule.

15. Evaporate the chloroform under a stream of nitrogen at 40 °C until approximately 1 ml of solution of the methylated polysaccharide remains.

16. Purify the derivative by chromatography on a column of 2 g of Sephadex LH-20 which has been washed with a solvent of chloroform/acetone mixture (2:1 v/v).

17. Place the sample of methylated polsaccharide on the column and elute the derivative with the chloforom/acetone mixture. The

**Protocol 4.** *Continued*

    methylated polysaccharide can be detected in the Sephadex as a light pink band.

**18.** Collect this band and transfer to a 10 ml ampoule.

**19.** Remove the solvent by evaporation under a stream of nitrogen at 40 °C.

C. *The hydrolysis of methylated polysaccharides*

**1.** Suspend the methylated polysaccharide in the 10 ml ampoule in 1 ml of 90% formic acid and sonicate for 15 min at room temperature.

**2.** Flush the ampoule with nitrogen, then seal by heat. Heat in an oven at 105 °C for 1.5 h.

**3.** Open the ampoule and evaporate the formic acid by heating to 40 °C.

**4.** Add 2 ml of 0.15 M $H_2SO_4$ to the ampoule containing the dry methylated polysaccharide.

**5.** Flush the ampoule with nitrogen, re-seal, and re-heat for 12–18 h at 105 °C in an oven.

**6.** Open the ampoule and add barium carbonate until the acid is neutralized.

**7.** Add 0.1 M NaOH to aid in flocculation of the precipitate.

**8.** Transfer the mixture into 12 ml clinical centrifuge tubes and centrifuge.

**9.** Recover the supernatant by filtration into 25 ml Erlenmeyer flasks.

D. *Reduction of partially methylated monosaccharides*

**1.** Reduce partially methylated monosaccharides with 1 mg of sodium borohydride for 12 h at room temperature.

**2.** Neutralize the mixture by acidification with Dowex 50 ($H^+$) ion-exchange resin and evaporate the solvent in a rotary evaporator at a temperature below 40 °C and at reduced pressure.

**3.** Dissolve the residue in redistilled methanol and transfer into a small round-bottom flask.

**4.** Remove the methanol under suction by heating at 40 °C in a water-bath.

**5.** Add methanol and evaporate three additional times in order to insure that all the borate is removed as volatile trimethyl borate.

**6.** Transfer the residue to 10 ml ampoule with methanol. Evaporate the methanol.

---

**E.** *Acetylation with acetic anhydride in pyridine*

**1.** Add 0.5 ml of equal parts of acetic anhydride and pyridine, flush the ampoule with nitrogen, seal, and heat at 105 °C for 2 h.

**2.** Open the ampoule and evaporate the acetic anhydride and pyridine in a stream of nitrogen at 40 °C.

**3.** Add ether to the reaction mixture and evaporate the solvent.

**4.** Repeat this treatment several times to remove all traces of acetic anhydride and pyridine.

**5.** Finally dissolve the sample in a small amount of chloroform and use aliquots of the chloroform solution for analysis by GLC and MS.

---

The analysis of the methylated alditol acetates in a hydrolysate of methylated polysaccharide is achieved with a coupled system of gas–liquid chromatography and mass spectrometry. Many types of commercial units are available for this purpose. A Varian 1400 Aerograph coupled to a DuPont 21–490 Mass Spectrometer and a Carlo Erba GC coupled to Kratos MS 25 Spectrometer are in use in this laboratory. The Varian GC unit is equipped with a six foot stainless steel column packed with an adsorbent of OV-138, OV-225, or SP2230. It gives satisfactory separations of most partially methylated acetylated monosaccharide derivatives. The Carlo Erba GC is equipped with a 60 metre J&W Scientific DB 5 capillary column. This unit has much higher resolving power and more sensitivity than the Varian–DuPont unit. The desired type of separation will also require the selection of proper temperatures, proper flow rate for the carrier gas, and optimum sample size. Both MS assemblies operate by the electron impact pathway.

The partially methylated alditol acetates when introduced into the mass spectrometer collide with the high energy electrons and undergo characteristic fragmentation. The ion fragments from different derivatives vary in masses and accordingly are accelerated in a magnetic field of the spectrometer at different rates. The ions impinge on the face of a target disc and lose their charge initiating a flow of current which is recorded by the instrument. Extensive studies have been conducted on the identification of the major fragments from partially methylated alditol acetates (27). It has been found that the fragmentation of the derivatives occurs in accord with a number of empirical rules deduced from experimental observations (see also Chapter 1, Section 8).

(a) Fragments which arise from methylated alditol acetates of stereoisomers with the same substitution patterns are identical in m/z values and in relative abundance.

(b) Ion fragments produced from the methylated alditol acetates on impact with electrons are of two types, primary fragments which arise from an initial fission of a carbon–carbon bond and secondary fragments which arise by loss of some functional groups of atoms from the primary fragments.

(c) Primary fragments are produced by fission of methylated alditol acetates between contiguous carbon atoms which carry methoxyl groups.

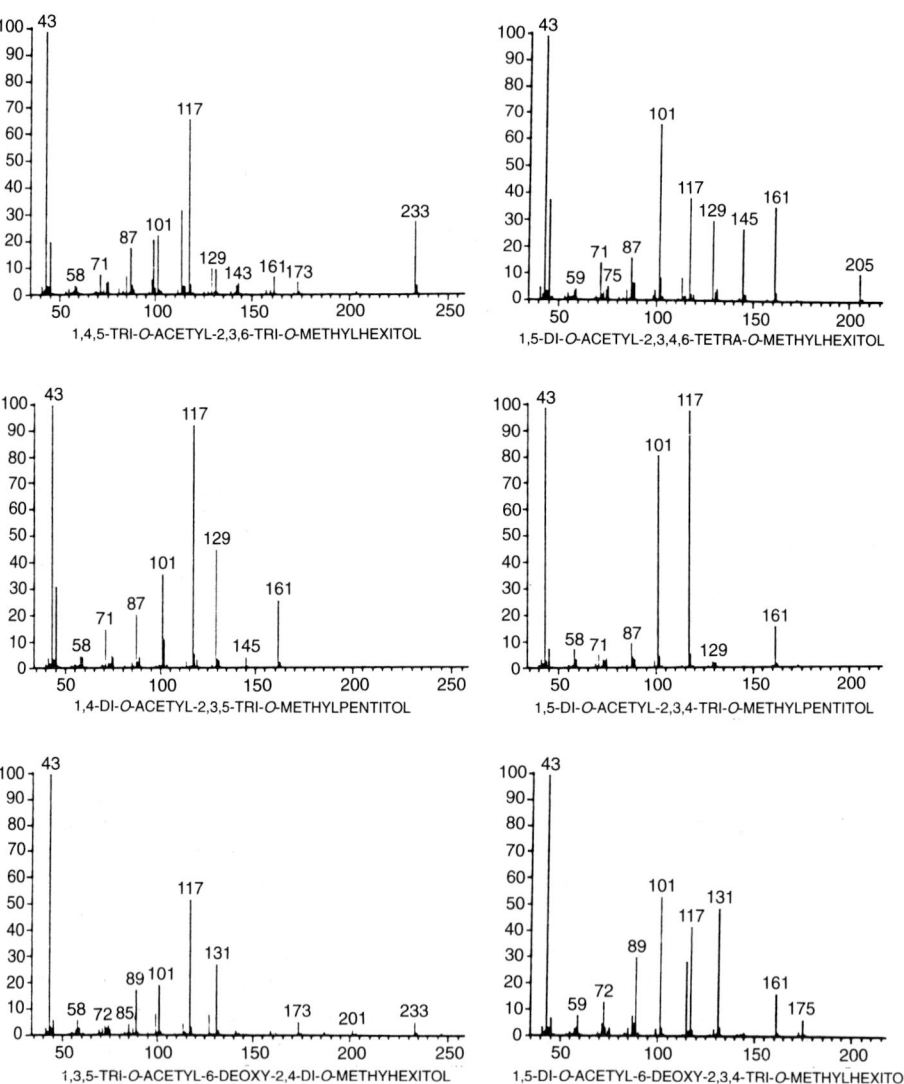

**Figure 1.** Electron impact mass spectrum of some partially methylated alditol acetates.

(d) Fission can also occur between contiguous carbons with acetyl and methoxyl groups.

(e) Secondary fragments are produced by loss of structural groups from primary fragments.

Comparison of spectra of unknowns with standards is necessary for definitive identification. The spectra for a few derivatives are recorded in *Figure 1*. A more complete list is given in ref. 27. *Figure 2* shows the type of separation by GLC of the methylated products on methylation analysis of a polysaccharide of glucose and galactose from the cell wall of *Streptococcus faecalis* (28).

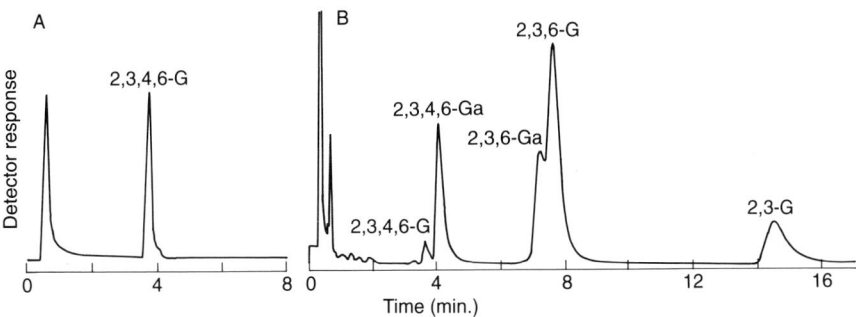

**Figure 2.** A photograph of the GLC patterns for the methylated alditol acetates from glucose (A) and from a streptococcal diheteroglycan of glucose and galactose (B): 2,3,4,6-G = 1,5-di-*O*-acetyl-2,3,4,6-tetra-*O*-methyl glucitol, 2,3,4,6-Ga = 1,5-di-*O*-acetyl-2,3,4,6-tetra-*O*-methyl galactitol, 2,3,6-Ga = 1,4,5-tri-*O*-acetyl-2,3,6-tri-*O*-methyl galactitol, 2,3,6-G = 1,4,5-tri-*O*-acetyl-2,3,6-tri-*O*-methyl glucitol, and 2,3-G = 1,4,5,6-tetra-*O*-acetyl-2,3-di-*O*-methyl methyl glucitol.

## 6.2 Periodate oxidation

Periodate oxidation of polysaccharides is a valuable analytical technique that has been used in the structural characterization and sequence determination of polysaccharides (29). The reactions which occur on periodate oxidation of monosaccharides and oligosaccharides have been discussed in Chapters 1 and 2 and these types of reactions occur during the oxidation of polysaccharides, some of which are shown in Equation 2. Briefly, carbohydrate residues containing glycol groups on adjacent carbon atoms are oxidized to dialdehydes. If residues contained hydroxyl groups on three adjacent carbons formic acid will also be produced. The newly formed aldehyde groups may form hemiacetal bonds with non-oxidized hydroxyl groups, thereby complicating interpretations of results (30). The formation of one such product is

$$R_2O \xrightarrow{\text{CH}_2\text{OH}} \text{OR}_1 \xrightarrow{\text{NaIO}_4} R_2O \xrightarrow{\text{CH}_2\text{OH}} \text{OR}_1 \xrightarrow{\text{NaBH}_4} R_2O \xrightarrow{\text{CH}_2\text{OH}} \text{OR}_1$$

(structures: starting sugar with CH₂OH, R₂O, OR₁, OH, OH → NaIO₄ → dialdehyde with OHC OHC → NaBH₄ → reduced with HOH₂C, CH₂OH)

$$\xrightarrow[\text{hydrolysis}]{\text{mild acid}} \begin{array}{c} \text{CH}_2\text{OH} \\ | \\ \text{HCOH} \\ | \\ \text{HCOR}_2 \\ | \\ \text{CH}_2\text{OH} \end{array} + \begin{array}{c} \text{CHO} \\ | \\ \text{CH}_2\text{OH} \end{array} + \text{R}_1\text{OH}$$

$$\xrightarrow[\text{hydrolysis}]{\text{acid}} \begin{array}{c} \text{CH}_2\text{OH} \\ | \\ \text{HCOH} \\ | \\ \text{HCOH} \\ | \\ \text{CH}_2\text{OH} \end{array} + \begin{array}{c} \text{CHO} \\ | \\ \text{CH}_2\text{OH} \end{array} + \text{R}_2\text{OH}$$

[Equation 2]

[Equation 3]

diagrammed in Equation 3. In practice, the oxidized residues are reduced with sodium borohydride yielding acetals which are easily removed by acid hydrolysis. Additional information on structure can be obtained by a second oxidation of these products. This procedure has become known as the Smith degradation.

Another analytical scheme for using periodate oxidation for polysaccharide structural studies is known as the Barry degradation. This method has been described very well in a methods article (31). In the Barry degradation the periodate oxidized polysaccharide is reacted with phenylhydrazine in dilute acetic acid. Phenylhydrazones are formed with the aldehyde groups generated by the oxidation. The phenylhydrazone containing units are released by hydrolysis. Structural data from Barry degradations have been obtained from gum arabic, galactan, arabinogalactan, and nigeran. In some cases polysaccharides can be degraded by the Barry method with a sequential removal of monosaccharide residues and the sequence of carbohydrate residues in the polysaccharide becomes known.

An example of the application of the periodate method has been its use in the determination of the structure of a polysaccharide of glucose and galactose from the cell wall of *S. faecalis*. Formula 2 shows the structure of the repeating unit deduced for the polysaccharide from methylation analysis, periodate oxidation, and acetolysis. This is an unusual polysaccharide with many lactose side chains (28). There are only a few compounds in nature with lactose units in the structure suggesting a specialized function for a lactose unit. Some of the methylation data obtained with the native polysaccharide have already been shown (see *Figure 1*). The products of periodate oxidation followed by reduction and hydrolysis can be readily identified by paper chromatography as shown in *Figure 3*. These products are glycol, with $R_f$ value greater than glycerol but not labelled in the figure, glycerol ($R_f$ 0.76),

$$\rightarrow 4)\text{-}\beta\text{-}D\text{-}Glc}p\text{-}(1 \rightarrow 4)\text{-}\beta\text{-}D\text{-}Glc}p\text{-}(1 \rightarrow 4)\text{-}\beta\text{-}D\text{-}Gal}p\text{-}(1 \rightarrow$$

$$6$$
$$\uparrow$$
$$1$$
$$\beta\text{-}D\text{-}Glc}p$$
$$4$$
$$\uparrow$$
$$1$$
$$\beta\text{-}D\text{-}Gal}p$$

[Formula 2]

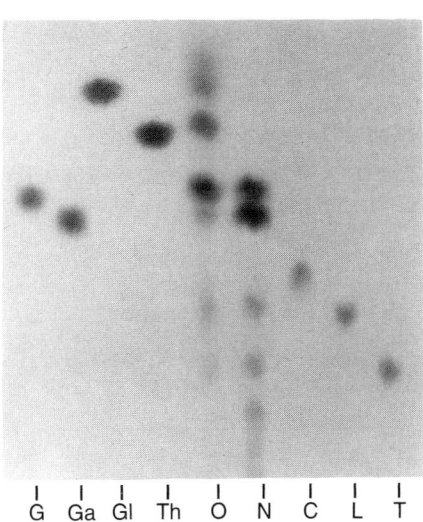

$$\begin{array}{ccccccccc} | & | & | & | & | & | & | & | & | \\ G & Ga & Gl & Th & O & N & C & L & T \end{array}$$

**Figure 3.** A paper chromatogram of the products in acid hydrolysates of native (N) and periodate oxidized and borohydride reduced (O) polysaccharide of glucose and galactose from *S. faecalis*: G = glucose, Ga = galactose, Gl = glycerol, Th = threitol, C = cellobiose, L = lactose, T = trisaccharide of glucose with $\alpha(1 \rightarrow 4)$ and $\alpha(1 \rightarrow 6)$ linkages (29).

threitol ($R_f$ 0.67), glucose ($R_f$ 0.55), and a trisaccharide ($R_f$ 0.25). Trace amounts of galactose ($R_f$ 0.50) and some oligosaccharides were also present. The native polysaccharide yielded only glucose, galactose, and an oligosaccharide series. The presence of lactose side chains as shown in Formula 2 was confirmed by results of acetylosis experiments (28).

---

**Protocol 5.** Periodate oxidation of a polysaccharide

1. Dissolve 50 mg of the polysaccharide in 50 ml of 0.02 M sodium periodate, adjusted to pH 4.5.

2. Cover the reaction flask with aluminium foil and place the flask and reaction mixture in a cold room at 4 °C for 18 h. At the end of this time destroy the excess periodate by addition of a few millilitres of ethylene glycol.

3. Remove the low molecular weight substances by dialysis against distilled water for 48 h.

4. Lyophilize the oxidized polysaccharide to dryness.

5. Dissolve 20 mg of the periodate oxidized polysaccharide in 10 ml of water.

6. Add 5 mg of sodium borohydride and maintain the reaction mixture for 24 h at room temperature.

7. Dialyse the reaction mixture against distilled water for 24 h.

8. Lyophilize the sample to dryness.

9. Dissolve 10 mg of the periodate oxidized and borohydride reduced polysaccharide in 0.5 ml of 0.02 M HCl in a small tube, stopper the tube, and heat the reaction mixture in a boiling water-bath for 20 min.

10. Hydrolyse 2 mg of the native polysaccharide in 0.1 ml of 0.1 M HCl for 3 h in a boiling water-bath.

11. Analyse both hydrolysates for reducing sugars by paper chromatography using two ascents of the solvent system of n-butyl alcohol/pyridine/water (6:4:3 by vol.) as described in Chapter 1, Section 4.

12. Stain the finished chromatogram with silver nitrate and sodium hydroxide reagents and photograph the chromatogram.

---

## 6.3 Acetolysis

Acetolysis of polysaccharides results in the complete acetylation of the free hydroxyl groups of the polysaccharide and the selective cleavage of glycosidic bonds (32). There are wide differences in the rate constants for the cleavage of various glycosidic bonds. By proper selection of the acetolysis conditions it is possible to cleave polysaccharides into specific homogeneous fragments. The $(1 \rightarrow 6)$ glycosidic linkage is highly susceptible to acetolysis whereas the $(1 \rightarrow 2)$ and the $(1 \rightarrow 3)$ linkages are comparatively resistant.

Acetolysis reactions are performed as described in *Protocol 6*. On the basis of detailed studies (32), a mechanism proposed for acetolysis reactions in Equation 4 involves the formation of an acyclic intermediate followed by cleavage of glycosidic bonds and release of the acetolysis fragments.

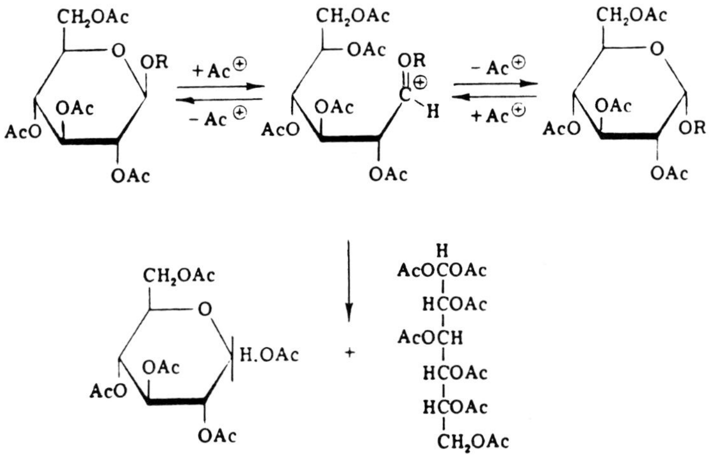

[Equation 4]

The acetolysis method has been especially useful for investigating the structure of polysaccharides from micro-organisms. The method has been used in an extensive series of studies designed to elucidate the structure of yeast mannans (33). The sequence of linkages in a yeast mannan was deduced from the nature of acetolysis products identified by gel chromatography methods.

---

**Protocol 6.** Acetolysis method

1. Dissolve 10 mg of the polysaccharide in 1 ml of the acetolysis solution of acetic anhydride/glacial acetic acid/concentrated $H_2SO$ (10:10:1 by vol.).

2. Heat the solution in a stoppered reaction vessel for 3 h at 40 °C.

3. Add pyridine to the mixture to stop the reaction and evaporate the solvents in a stream of nitrogen at 40 °C.

4. Deacetylate the products by dissolving the sample in methanol containing a catalytic amount of barium methoxide. Maintain the reaction mixture at room temperature for 15 min.

5. Neutralize the base by adding dry ice and then evaporate the solvent.

---

---

**Protocol 6.** *Continued*

6. Suspend the residues in a small amount of water and remove the barium carbonate by centrifugation.

7. Concentrate the supernatant to a small volume (0.5–1 ml) and analyse for carbohydrate types by paper chromatography in a solvent of *n*-butyl alcohol/pyridine/water (6:4:3 by vol.).

8. Isolate the fragments by preparative paper chromatography and subject the compounds to methylation analysis.

---

## 6.4 Methanolysis

Simultaneous hydrolysis and methanolysis of polysaccharides to yield methyl glycosides can be effected at elevated temperatures in anhydrous methanolic hydrogen chloride (34). On prolonged heating the polysaccharides are converted to the methylglycosides of the constituent monosaccharides but on short heating methylglycosides of oligosaccharides may be produced. The reaction is carried out at temperatures in the range of 80–100 °C and for periods of 1–5 hours.

---

**Protocol 7.** Methanolysis

1. Dissolve 0.5 mg of the polysaccharide in 1 ml of anhydrous methanolic hydrogen chloride and heat at 85 °C for 5 h.

2. Neutralize the reaction mixture with powdered silver carbonate and centrifuge the suspension to remove the precipitated silver salt.

3. Transfer the supernatant to a vial and remove the solvent with a stream of nitrogen.

4. The residue consists of a mixture of carbohydrate derivatives from the polysaccharide and is used in subsequent analyses.

5. If necessary reacetylation of the amino groups of the polysaccharide is effected in acetic anhydride and pyridine at room temperature for 6 h.

---

The methylglycosides are converted to the trimethylsilyl derivatives or the acetyl derivatives for GLC analysis. The trimethylsilyl and the acetyl derivative can also be prepared from the reduced initial products. Such products are obtained by first removing the methyl group from the glycoside by acid hydrolysis and reducing the aldehyde group with sodium borohydride. The derivatives are prepared on the reduced products. Derivatives of the reduced monosaccharides are advantageous to use since α- and β-anomers cannot be formed because of the destruction of the asymmetry at carbon-1 by the reductions. As a result one peak is obtained for individual compounds on gas-liquid chromatography analysis.

## 6.5 Hydrolysis by acids

The rate constants for the hydrolysis of the glycosidic bonds in polysaccharides vary greatly and are affected by factors such as the type of glycosidic linkages, the ring structure of the residues, and the anomeric configuration of the linkages. *Table 3* contains data for a few types of disaccharides containing different types of glycosidic linkages taken in part from ref. 35. The types of the monosaccharide residues constituting the carbohydrate also affect the rate constants for the hydrolysis of glycosidic bonds. Thus linkages between uronic acid residues or amino sugar residues to neutral monosaccharides are more resistant to hydrolysis than similar linkages between other residues. In some cases it may be necessary to modify uronic acid or amino sugar residues by chemical reactions before effecting the hydrolysis of the linkage.

Partial acid hydrolysis by acids has been an extremely useful technique in structural analysis of polysaccharides. By partial acid hydrolysis, poly-saccharides can be cleaved randomly into a mixture of oligosaccharides or may be cleaved preferentially at a specific type of bond to give a high yield of product with uniform structure. The oligosaccharides are isolated and then subjected to appropriate analysis. If a sufficient number of oligosaccharides is prepared which contain overlapping segments of the polysaccharide, the sequence of residues and glycosidic linkages in the polysaccharide can be deduced.

Monosaccharides with furanose rings are very acid labile and this type of linkage can be readily cleaved by mild acid hydrolysis (see *Table 3*). Analysis

**Table 3.** Rates of hydrolysis of disaccharides in 0.1 M HCl at 99 °C

| Disaccharide | Structure | Relative rate |
|---|---|---|
| Maltose | $\alpha$-D-Glc (1 → 4)-D-Glc | 100 |
| Cellobiose | $\beta$-D-Glc (1 → 4)-D-Glc | 40 |
| Isomaltose | $\alpha$-D-Glc (1 → 6)-D-Glc | 31 |
| Gentiobiose | $\beta$-D-Glc (1 → 6)-D-Glc | 44 |
| Kojibiose | $\alpha$-D-Glc (1 → 2)-D-Glc | 106 |
| Sophorose | $\beta$-D-Glc (1 → 2)-D-Glc | 62 |
| Nigerose | $\alpha$-D-Glc (1 → 3)-D-Glc | 87 |
| Laminarabiose | $\beta$-D-Glc (1 → 3)-D-Glc | 57 |
| Trehalose | $\alpha$-D-Glc (1 → 1)-$\alpha$-D-Glc | 5 |
| Turanose | $\alpha$-D-Glc (1 → 3)-D-Fru | 73 |
| Lactose | $\beta$-D-Gal (1 → 4)-D-Glc | 102 |
| Melibiose | $\alpha$-D-Gal (1 → 6)-D-Glc | 95 |
| GlcA-Gal | $\beta$-D-GlcA (1 → 6)-D-Gal | 1 |
| Ara-GlcA | $\beta$-L-Ara$_f$ (1 → 4)-D-GlcA | 44 500 |
| Sucrose | $\beta$-D-Fru$_f$ (2 → 1)-$\alpha$-D-Glc | 89 000 |

of the residual fragment can yield valuable structural data. In some cases if the polysaccharide is resistant to hydrolysis the methylated polysaccharide is hydrolysable in trifluoroacetic acid.

## 6.6 Hydrolysis with enzymes

The hydrolysis of polysaccharides can be achieved with appropriate hydrolytic enzymes. Such hydrolysates will probably contain a series of oligosaccharides and these can be isolated and analysed. Enzymes which catalyse the hydrolysis of polysaccharides are of two types, the exohydrolases which hydrolyse terminal glycosidic linkages, and the endohydrolases which hydrolyse internal glycosidic bonds located at specific positions. The endohydrolases generally effect a progressive shortening of the complete polysaccharide chain with long segments of residues being produced in the initial stages of hydrolysis and short oligosaccharide segments in later stages. Some endohydrolases possess a specificity for a particular type of linkage and as a result relatively uniform fragments are produced.

For the hydrolysis of heteropolysaccharides the ideal situation is to have a series of exoglycosidases available, each with specificity for a single type of terminal saccharide residue and each capable of removing those residues quantitiatively. The series of enzymes could be employed sequentially for liberating terminal residues that would be identified at each step. It has not yet been possible to obtain exoglycosidases needed for the complete hydrolysis of a heteropolysaccharide. However it has been possible to use glycosidases for sequencing the carbohydrate moiety of the glycoprotein, fetuin (36). The monosaccharide sequence of the oligosaccharide is shown by Formula 3. The enzymes employed for the sequential hydrolysis were *β*-galactosidase, *N*-acetyl-*β*-glucosaminidase, *β*-mannosidase, and α-mannosidase. The identification of the oligosaccharide fragments was by chromatography on a Bio-Gel P-4 column. The *β*-galactosidase has also been used in structural studies on *S. faecalis* cell wall polysaccharide (28).

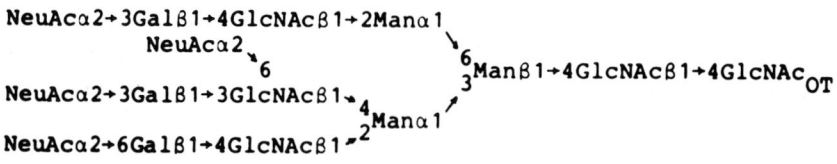

[Formula 3]

---

**Protocol 8.** An antigenic polysaccharide of glucose and galactose

1. Dissolve 60 mg of the polysaccharide in 1.5 ml of water and mix with 0.75 ml of 4% almond *β*-glycosidase solution in 0.1 M phosphate buffer (pH 6.8).

2. Maintain the enzymic digest at room temperature for a period of 44 h.

3. Examine the digest periodically for the liberation of reducing sugars by paper chromatography.

4. At the end of 44 h inactivate the enzyme solution with heat and precipitate the enzymes with an equal volume of 10% trichloroacetic acid.

5. Remove the precipitate by centrifugation, collect the supernatant, and dialyse for 48 h against distilled water.

6. Dry the enzyme treated polysaccharide by lyophilization.

7. Subject 2 mg of the enzyme treated polysaccharide to methylation analysis.

8. Identify and quantitate the products by GLC and MS.

**Table 4.** Moles of methylated monosaccharides from the native and enzymically modified polysaccharide from cell wall of *S. faecalis*

| Compound | Native | Modified | Difference |
|---|---|---|---|
| 2,3,4,6-Tetramethyl glucose | 4.0 | 15.9 | +11.9 |
| 2,3,4,6-Tetramethyl galactose | 17.5 | 5.0 | −12.5 |
| 2,3,6-Trimethyl galactose | 17.6 | 17.8 | +0.2 |
| 2,3,6-Trimethyl glucose | 35.0 | 21.9 | −13.1 |
| 2,3-Dimethyl glucose | 17.1 | 16.8 | −0.3 |
| Galactose (release) | | | (+12.5) |

A comparison of the quantitative data for the native and enzymically modified polysaccharide is recorded in *Table 4*.

## 6.7 Degradation reactions

### 6.7.1 Degradation with alkali

The degradative action of alkalis on polysaccharides has been investigated for many years but the exact nature of the reactions has not yet been determined (37). Variations in the types of degradation products obtained from polysaccharides with different glycosidic linkages have been noted. It is reasonably well established that alkaline degradation begins at the residue with the reducing group and this residue is ultimately eliminated. In concentrated alkali, residues with free hydroxyl groups are also degraded by an oxidative route. The presence of substituents on hydroxyl groups of such residues markedly affects the degradative reactions and in some cases arrests the degradative process. This observation has been used to devise a method for determining the position of branch linkages of branched polysaccharides.

It has been found that treatment of polysaccharides with alkali brings about many changes including isomerization, oxidation, and molecular rearrangements and the fragmentation of the polysaccharide chains. The low molecular weight products are numerous and varied in types and may consist of formic acid, acetic acid, glycolic acid, lactic acid, saccharinic acids, and carbon dioxide. Only a limited amount of structural information can be obtained by this method because of the complex nature of the degradation.

### 6.7.2 β-Elimination reactions

Carbohydrate residues with substituent groups located in the β-position to electron withdrawing groups undergo β-elimination reactions on treatment with a base (32) and information on glycosidic linkages can be obtained. The substitutents which can be eliminated include alkoxyl and glycosyl groups, and the electron withdrawing groups may be aldehydic, carbonyl, carboxylic esters, amides, or sulfones. In order for the elimination reaction to occur a hydrogen must be present in the α-position to the electron withdrawing group. The β-elimination reactions have been especially useful in linkage studies on polysaccharides containing uronic acid residues. Reactions of this type are also utilized in structural studies on the carbohydrate moiety of glycoproteins with O-glycosidic linkages.

In order for β-elimination reactions to occur with acidic polysaccharides, the carboxyl group must be esterified to form an electron withdrawing group and the carbohydrate residue must possess a substituent at the β-position to the ester grouping. The reaction sequence for the elimination is shown in Equation 5. The initial reaction of the sequence is the formation of a double bond between carbons-4 and 5 of the uronic acid followed by the elimination of the substituent at carbon-4. Subsequently mild acid hydrolysis is used to release the substituent at carbon-1 of the uronic acid residue. Carbohydrate moieties which are attached to carbon-1 or 4 will be eliminated. Both types of moieties can be isolated and used for further structural studies. The position of attachment of the original uronic acid units can be determined by

[Equation 5]

methylation of the carbohydrate fragments and identifying the new methyl derivative which is produced. GLC results of uronic acid elimination reactions and subsequent analysis for a bacterial polysaccharide with the structure shown in Formula 1 are presented in *Figure 4*. The results show that the uronic acid units are attached to carbon-4 of the rhamnose residues of the main chain and after elimination of the glucuronic acid this rhamnose moiety appears methylated at positions 2 and 4 (38). Such a derivative was not present in the methylation mixture of the native polysaccharide.

The β-elimination reaction can be coupled with other degradative reactions and used for obtaining additional information on the structure of a polysaccharide. The oxidation of the β-eliminated product can be effected at the free hydroxyl group converting the hydroxyl group to a carbonyl group by oxidation with ruthenium tetraoxide. The product of methylation, β-elimination, oxidation, and reduction was subjected to remethylation, hydrolysis, reduction, and acetylation (39). An analysis by gas–liquid chromatography–mass spectrometry showed that 3-*O*-acetyl-1,2,4,5,6-penta-*O*-methyl galactitol, 1,5-di-*O*-acetyl-2,3,4-tri-*O*-methyl rhamnitol, and 1,3,5-tri-*O*-acetyl-2,4-di-*O*-methyl-6-deoxyltalitol were present. The mono-saccharide sequence of this polysaccharide (see Formula 1) is thereby substantiated.

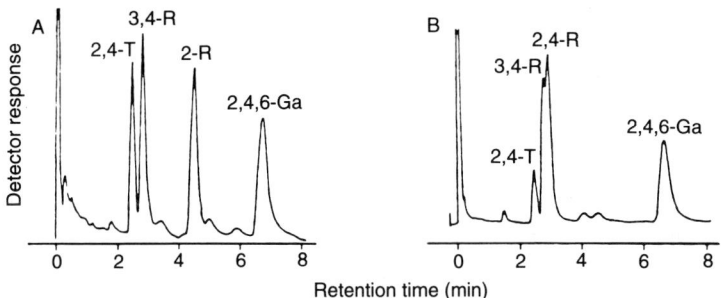

**Figure 4.** A photograph of GLC patterns for the methylated alditol acetates from a native tetraheteropolysaccharide (A) and the fragment after removal of the D-glucuronic acid residues (B): 2,4-T = 1,3,5-tri-*O*-acetyl-2,4-di-*O*-methyl-6-deoxytalitol; 3,4-R = 1,2,5-tri-*O*-acetyl-3,4-di-*O*-methyl rhamnitol; 2-R = 1,3,4,5-tetra-*O*-acetyl-2-*O*-methyl rhamnitol; 2,4,6-Ga = 1,3,5-tri-*O*-acetyl-2,4,6-tri-*O*-methyl galactitol; 2,4-R = 1,3,5-tri-*O*-acetyl-2,4-di-*O*-methyl rhamnitol (38).

**Protocol 9.** Elimination of glucuronic acid and oxidation with ruthenium tetraoxide

A. *Esterification and methylation*

1. Methylate the hydroxyl groups and esterify the carboxyl groups of 12 mg of streptococcal polysaccharide by the procedure in *Protocol 4*.

**Protocol 9.** *Continued*

2. Extract the methylated and modified polysaccharide with 10 ml portions of chloroform and wash the combined chloroform extracts three times with water.

3. Divide the extract into two equal portions. Use one portion to locate the position of $\beta$-elimination (38) and the other portion for oxidation with ruthenium tetraoxide (39).

B. *Hydrolysis and identfication of products*

1. Evaporate one portion to dryness.

2. Subject the residue obtained above to mild hydrolysis in 2 ml of 10% acetic acid in a sealed ampoule under nitrogen for 1 h at 100 °C.

3. Dry the reaction mixture by lyophilization and purify the polymeric residue by gel filtration on Sephadex LH-20.

4. Evaporate the solvent and remethylate by the standard procedure.

5. Hydrolyse and convert the products to the alditol acetates as described in *Protocol 4C*.

   The products identified by GLC in the mixture were 1,3,5-tri-*O*-acetyl-2,4-di-*O*-methyl-6-deoxytalitol, 1,2,5-tri-*O*-acetyl-3,4-di-*O*-methyl rhamnitol, 1,3,5-tri-*O*-acetyl-2,4-di-*O*-methyl rhamnitol, and 1,3,5-tri-*O*-acetyl-2,4,6-tri-*O*-methyl galactitol shown in *Figure 4*.

C. *Oxidation with ruthenium tetraoxide*

1. Dissolve the other portion of the methylated and modified polysaccharide in 3 ml of methylene dichloride in a 50 ml round-bottom flask and add 13 ml of a 0.1 M solution of ruthenium tetraoxide in methylene chloride from a dry glass syringe.

2. Add 16 ml of saturated aqueous solution of sodium *meta*-periodate.

3. Stopper the flask and shake vigorously until both solvent phases remain yellow in colour indicating that the reaction is complete.

4. Remove the aqueous layer containing the periodate.

5. Shake the organic phase with 0.2 ml of isopropyl alcohol to convert the ruthenium tetraoxide to ruthenium dioxide.

6. Remove the precipitate of ruthenium dioxide and other impurities by washing the organic phase several times with 8 ml portions of water.

7. Extract the combined aqueous washings with 8 ml of methylene chloride and combine the methylene chloride phase and the original organic phase.

8. Filter through glass wool and evaporate the solvent under nitrogen at 40 °C.

9. Dissolve the material in 3 ml of methylene chloride and add 2 ml of freshly prepared 1.0 M sodium methoxide solution. Maintain the reaction mixture at room temperature for 4 h to liberate the oligosaccharide fragments from the polysaccharide.

10. Acidify to pH 4 by adding 90% acetic acid. Remove the solvents by evaporation under nitrogen at 45 °C.

11. Hydrolyse the modified polysaccharide in 8 ml of 50% acetic acid for 1 h at 105 °C and lyophilize to dryness.

12. Partition the residue between methylene chloride and water phases.

13. Wash the organic phase three times with 3 ml of water and dry over $MgSO_4$. Remove the $MgSO_4$ by centrifugation.

14. Filter the methylene chloride layer through glass fibre paper and evaporate the solvent under nitrogen.

15. Reduce the methylated oligosaccharide by dissolving the sample in 5 ml of a mixture of dioxane and ethanol (8:3 v/v) and adding 50 mg of sodium borohydride. Maintain the reaction mixture for 24 h at room temperature in which time the pH increases from 8.0 to 10.5.

16. Acidify by adding Dowex 50 (hydrogen ion form) until the pH drops to 4.0.

17. Remove the resin by filtration through glass fibre paper and wash the resin several times with methanol. Combine the washings and the filtrate from the glass fibre paper and remove the solvent under reduced pressure at 40 °C.

18. Remove the borate by repeated evaporation of methanol from the residue.

19. Dry the methylated and reduced oligosaccharide in a vacuum desiccator overnight and remethylate by the procedure described in *Protocol 4*B.

Analysis by GLC and MS establishes the presence of 3-*O*-acetyl-1,2,4,5, 6-penta-*O*-methyl galactitol, 1,5-di-*O*-acetyl-2,3,4-tri-*O*-methyl rhamnitol, and 1,3,5-tri-*O*-acetyl-2,4-di-*O*-methyl-6-deoxytalitol, and verifies the sequence of residues and types of glycosidic linkages of the polysaccharide.

### 6.7.3 Degradation by deamination

The most common hexosamines found in biological materials are 2-acetamido-2-deoxy-D-glucose (*N*-acetylglucosamine) and 2-acetamido-2-deoxy-D-galactose (*N*-acetylgalactosamine). Less common is 2-acetamido-2-deoxy-D-mannose (*N*-acetylmannosamine) which occurs as a constituent unit of some polysaccharides and as a moiety of sialic acids in glycoproteins.

Recently several reports have appeared describing amino sugar residues of unusual structure in microbial cell walls. Such polymers often function as immunological substances. Generally the *N*-acetyl amino sugars occur as constituents of heteropolysaccharides but chitin, a polymer of only *N*-acetylglucosamine, is a notable exception. For procedures used in structural investigations of amino sugar-containing polysaccharides (32) the amino sugar is deacetylated in the initial step of the analysis. The deacetylation is performed under mild basic conditions to minimize removal of other labile groups and prevent extensive degradation reactions (40). Barium hydroxide in water, sodium ethoxide in ethanol, or dry hydrazine have been used in the past, and the use of trifluoroacetic acid and trifluoroacetic anhydride has been introduced recently. The deamination reaction involves treatment of amino sugar with nitrous acid. In this reaction, a 2,5 anhydro ring and a free aldehyde group at carbon-1 is formed. The reaction sequence is illustrated in Equation 6.

[Equation 6]

In the reaction the hexosamine molecule undergoes a rearrangement because of the attachment of the intermediate diazonium ion on the pyranose ring oxygen. Very often an inversion of configuration occurs at carbon-2. If the amino sugars are constituents of polysaccharides, the glycosidic bonds are also cleaved simultaneously thereby yielding polysaccharide fragments useful in further structural studies.

# 7. Ring structure of monosaccharides

In solution monosaccharides occur in the open chain, the pyranose ring, and the furanose ring structures. These structural forms are in equilibrium and the concentration of an individual form is characteristic of the monosaccharide. When monosaccharides are joined together by glycosidic linkages to form polysaccharides, the ring structures become fixed for all residues except the residue at the reducing end. In nature polysaccharides may contain a single type of ring structure or a mixture of ring types. With the latter polysaccharides the type of ring structure is more difficult to deduce.

In the elucidation of the complete structure of a polysaccharide it is necessary to establish the type of ring structure of each monosaccharide unit. Residues which possess a furanose ring and which are terminal units are removed easily by mild acid hydrolysis. Following hydrolysis the monosaccharide which is released is isolated and characterized by the usual methods. If residues with furanose rings are internal residues then a fragmentation of the polysaccharides will occur at the locations of such residues. In this process the furanose residues will become reducing units of the fragments. Such polysaccharide fragments are very useful for structural studies.

Monosaccharides with different ring structures will yield characteristic products on methylation analysis. Thus on methylation analysis the derivatives from hexose residues with furanose rings will contain methoxyl groups at carbon-5 and not at 4 while residues with a pyranose ring structure will contain methoxyl groups at carbon-4 and not at 5. The methylation technique and the identification of the methyl derivatives can be performed as described earlier. In such an analysis the monosaccharide units with pyranose rings which are substituted at carbon-4 would yield the same derivative as residues with a furanose ring that are substituted at carbon-5. To differentiate such structures additional analysis with different alkylation reagents can be performed utilizing methylating and ethylating reagents.

# 8. α and β configuration of glycosidic linkages

There are several methods for determining α and β configuration of the glycosidic linkages and these are enzymic assays, NMR spectroscopy, and selective oxidations.

## 8.1 Enzymic assays

The susceptibility of polysaccharides to hydrolysis by enzymes of known specificiity is the method of choice for determining the configuration of the glycosidic linkages. However enzymes of known specificity and sufficient purity should be used but are not always available. Some polysaccharides contain glycosidic linkages with only one type of anomeric configuration but others contain both types of anomeric linkages. Consequently more than one type of enzyme will need to be employed in determination of the configuration of the linkages in the latter polysaccharides. The ideal situation is to have a spectrum of enzymes with a single type of specificity; one to remove residues from the terminal end of a polysaccharide, another to remove the second layer of residues, and so on. From such results the configuration of all the linkages can be established.

## 8.2 Nuclear magnetic resonance spectroscopy

The $^1$H-NMR and $^{13}$C-NMR spectra of polysaccharides have been recorded and the data used to identify anomeric configuration of the glycosidic linkages of polysaccharides (41, 42). For such measurements reference derivatives of known structure are needed for comparisons. Chemical shifts in the spectra are attributable to the orientation of groupings in the molecule. Different shifts in the signals may be obtained for α- and β-anomeric linkages and assignment of configuration of linkages in an unknown polysaccharide can be made.

## 8.3 Oxidation with chromium trioxide

The configuration of the glycosidic linkages of some polysacchrides may be determined by chromium trioxide oxidation of the acetylated polysaccharide followed by deacetylation and methylation analysis. It has been shown with model compounds that data from the oxidation with chromium trioxide will differentiate the α and β configuration. Thus glycosides in which the aglycone moieties occupy the equatorial position are oxidized much faster than glycosides in which the aglycone occupies the axial position. The difference has been attributed to the ease of formation of the ketoester by cleavage at the bridge oxygen of β-anomeric compounds. A polysaccharide with α-glycosidic linkages is only slowly oxidized and hence on methylation analysis yields essentially the same types of derivatives as those from the native polysaccharide. A polysaccharide with β-glycosidic linkages is oxidized rapidly and residues with this linkage will be degraded. As a result gas–liquid chromatography and mass spectrum patterns for the methylated fragments are markedly different.

The chromium trioxide oxidation method has been used for determining the configuration of the glycosidic linkages in a polysaccharide in rhamnose and glucose from *S. bovis* (43), the structure of the repeating unit is shown in Formula 4. The oxidation results establish the glycosidic linkages to be α since the oxidized and native glycan yielded the same derivatives on methylation analysis.

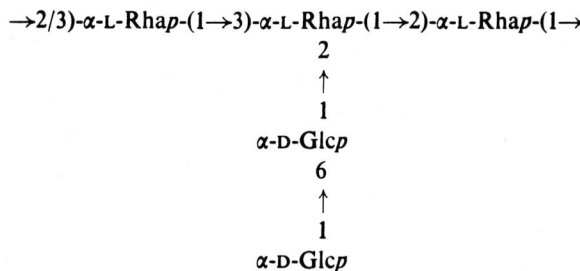

→2/3)-α-L-Rha*p*-(1→3)-α-L-Rha*p*-(1→2)-α-L-Rha*p*-(1→
                                                            2
                                                            ↑
                                                            1
                                                    α-D-Glc*p*
                                                            6
                                                            ↑
                                                            1
                                                    α-D-Glc*p*

[Formula 4]

---

**Protocol 10.** Oxidation with chromium trioxide

**1.** Dissolve 5 mg of the polysaccharide in 1 ml of *N,N'*-diomethyl-formamide, and mix with 1 ml of an equal part mixture of pyridine and acetic anhydride.

3: Neutral polysaccharides

2. Acetylate the polysaccharide at room temperature overnight.

3. Recover the products by chromatography on a column of Sephadex LH-20, with acetone as the eluting solvent.

4. Identify the fractions containing the polysaccharide acetate by a suitable colorimetric test.

5. Combine and evaporate these fractions under a stream of nitrogen.

6. Subject the derivative to oxidation by adding 25 mg of finely powdered chromium trioxide in 0.2 ml of glacial acetic acid. Maintain the reaction mixture at 50 °C for 1 h.

7. Reisolate the polysaccharide by chromatography on Sephadex LH-20 as described above.

8. Methylate the chromium trioxide treated polysaccharide as described in *Protocol 4*B.

9. Hydrolyse, reduce, and acetylate the product and analyse by GLC.

# 9. Immunological methods

That polysaccharides can stimulate the immune system to produce antibodies was discovered many years ago. The early work was concerned with the antigenicity of polysaccharides from the cell wall of micro-organisms (44) and the development of serological methods for identifying pathogenic groups of bacteria (45). It is now established that polysaccharides (46, 47), glycoproteins (48), and glycoconjugates (49) initiate an immune response in which antibodies specific for certain carbohydrate residues called haptens are synthesized. Antibodies are proteins of unique structure with special arrangements of the polypeptide chains which combine with structural groups of the immunizing substance and neutralize infectious organisms, viral particles, and toxic compounds. An immunizing substance is called an antigen which forms a precipitin complex with the antibody and the complex is eliminated. Recent work has shown that the antibody molecule consists of two light and two heavy polypeptide chains arranged in a structure in which segments of a light and heavy chain form a combining site at which a structural group of the antigen is bound (50, 51).

The precipitin complex formed by antibody and antigen can be observed by capillary precipitin tests, agar diffusion, or production of a coloured product in enzyme linked immunoassays. A photograph of an agar diffusion plate is reproduced in *Figure 5* showing the reaction of an antigenic polysaccharide (see Formula 2) and its homologous antiserum (well 1 and well 2). Also diffusion patterns with antigen dissolved in 0.1 M HCl prior to heating (well 1 and 3), heated at 100 °C for three hours (well 1 and well 4) are shown. Well 5 contains a water blank. Quite clearly the antigenicity of the polysaccharide is

**Figure 5**. Agar diffusion patterns of anti-lac antibodies against native and acid hydrolysed polysaccharide from *S. faecalis*: 1 = anti-lac antibodies, 2 = native polysaccharide, 3 = polysaccharide in 1 M HCl prior to heating, 4 = polysaccharide hydrolysed in 1 M HCl for 3 hours at 100 °C, and 5 = blank of $H_2O$.

destroyed by hydrolysis. The laboratory techniques for the agar diffusion test, preparation of vaccines, and the immunization procedure for antibody production are recorded in *Protocol 11*.

---

**Protocol 11.** The immunological technique

A. *Agar diffusion*

1. Layer 3 ml of 1% agarose solution in a microscope slide. Allow the agarose to gel.

2. Cut holes in the gel 10 μl in size with a standard punch.

3. Introduce 10 μl of the antigen or antibodies into individual wells.

4. Place the slide in a Petri dish in a moist atmosphere.

5. Observe the gel for formation of precipitin bands for the next 36 h. Photograph the slide showing bands for a permanent record.

B. *Vaccine*

1. Grow *S. faecalis* strain N in 10 ml of a 3% solution of Todd–Hewitt broth at 37 °C for 24 h.

2. On successive days inoculate 75 ml of Todd–Hewitt broth with the 10 ml culture, and then 500 ml of broth with the 75 ml culture. Grow all cultures at 37°C for 24 h.

3. Collect the cells from the 500 ml culture by centrifugation at 4°C for 15 min.

---

4. Wash the pellet three times with 100 ml of 0.2% solution of formaldehyde in saline (0.9% NaCl). Recover the cells after each washing by centrifugation.

5. Stir the cells in 100 ml of 0.2% formaldehyde in saline solution at 4 °C for 24 h.

6. Recover the cells by centrifugation and suspend the cells in 80 ml of sterile saline solution.

7. Perform viability tests with the formaldehyde treated cells to check that the cells are non-viable.

8. Measure the absorption at 600 nm. Absorption values from 1.0 to 1.5 should be obtained.

9. Store the vaccine at 4 °C.

C. *Immunization*

1. Rabbits are immunized by intravenous injection of 0.3 ml of the suspension of the formalinized cells daily for four days, followed by a rest period of three days.

2. This schedule is repeated for four additional weeks.

3. After a rest period of three weeks, a second cycle of immunization is performed following the above schedule.

4. Blood samples are obtained from the rabbits weekly from the ear vein and the serum is tested for potency utilizing the polysaccharide as the antigen.

5. After the ninth week of vaccination, the blood samples of the rabbits exhibit a high titre against the polysaccharide.

6. Immunization is continued for two additional cycles. Blood samples are taken weekly.

The anti-lactose antibodies produced in the above experiment were purified by affinity chromatography on an immunoadsorbent. *Protocol 12* contains descriptions for the preparation of the adsorbent and the affinity chromatography method.

**Protocol 12.** Immunoadsorbent and affinity chromatography

A. *Preparation of Lactosyl–Sepharose 4B*

1. Swell 12 g of cyanogen bromide activated Sepharose 4B in dilute acid by washing the Sepharose on a glass sintered filter with 700 ml of 0.001 M HCl.

**Protocol 12.** *Continued*

2. Wash the activated Sepharose with 200 ml of 0.1 M NaHCO$_3$ adjusted to pH 9.0 with NaOH.

3. Dissolve 170 μmol of *p*-aminophenyl *β*-lactoside in 20 ml of 0.1 M NaHCO$_3$ (pH 9.0) and mix each solution with 12 g of the swollen and moist Sepharose.

4. Shake the mixture for 24 h on a wrist action shaker at 4 °C.

5. Recover the reaction product (Lactosyl–Sepharose) on a glass sintered filter and wash with 150 ml of 0.1 M NaHCO$_3$ (pH 9.0).

6. Shake the reaction product with 100 ml of 1 M ethanolamine (pH 9.0) for 2 h at room temperature.

7. Wash with 0.1 M phosphate buffer (pH 7.2) in saline. The adsorbent will contain *p*-aminophenyl-lactosyl units linked via the amino groups to the imidocarbonate groups of the activated Sepharose.

B. *The technique of affinity chromatography of antibodies*

1. Pour the Lactosyl–Sepharose in a 30 × 2.5 cm column and equilibrate with 0.1 M, phosphate buffer (pH 7.2) in saline at room temperature.

2. Introduce a sample of 2 ml of antiserum containing anti-lactose antibodies on a column of Lactosyl–Sepharose washed well with 0.1 M phosphate buffer, pH 7.2.

3. Wash the column and sample with 0.1 M phosphate buffer (pH 7.2) in saline.

4. Monitor the eluate with an UV analyser and flow cell.

5. When the serum protein has passed through the column, elute the adsorbed antibodies with 10 ml of 0.5 M lactose solution.

6. Collect these fractions and precipitate the antibodies with an equal volume of saturated ammonium sulfate. Recover the antibodies by centrifugation.

7. Dissolve the antibodies in 0.2 ml of 0.1 M phosphate buffer (pH 7.2) in saline.

8. Combine such anti-lactose antibody preparations from several affinity runs.

9. Dialyse against 0.1 M phosphate buffer (pH 7.2) in saline for 72 h at 4 °C.

10. Concentrate by lyophilization and store the antibodies at −20 °C until used for other experiments.

Many types of anti-carbohydrate antibodies have been isolated in pure form by affinity chromatography methods (48). The specificity of these antibodies for carbohydrate units of the antigens has been established by the elution behaviour of the antibodies from carbohydrate Sepharose adsorbents, periodate oxidation, hapten inhibition, and reactivity of the antibodies with carbohydrate units attached to other carrier proteins. The elution pattern for the anti-lactose antibodies is reproduced in *Figure 6*. Oxidation of the antigen with periodate destroys the carbohydrate units and also destroys the antigenicity. Hapten inhibition techniques are used for identifying the immunodeterminant group of a polysaccharide which combines with antibodies. Some results of hapten inhibition experiments with antibodies specific for lactose are shown in *Figure 7*. It is noted in the figure that lactose and aminophenyl lactoside are potent inhibitors of the precipitin reaction at the concentrations tested inhibiting about 90% of the preciptin reaction. On the

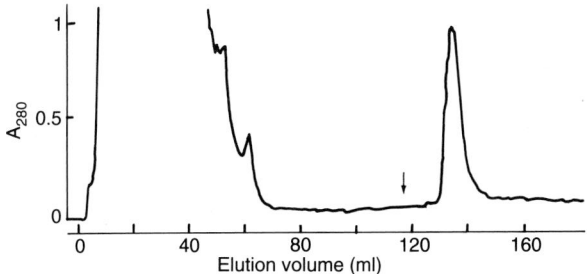

**Figure 6**. Affinity chromatography of anti-lac antibodies on Lactosyl–Sepharose and 0.5 M lactose introduced at arrow (51).

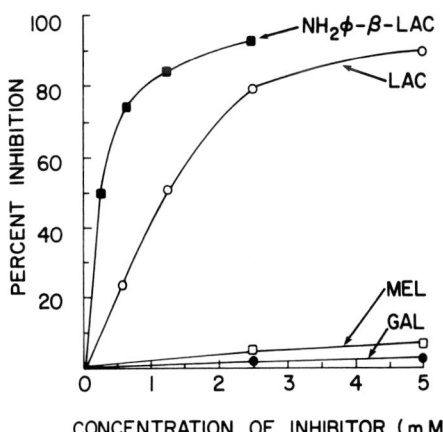

**Figure 7**. Percent inhibition of anti-lac antibodies (47), *p*-aminophenyl-*β*-lactoside (NH₂φ-*β*-LAC), lactose (LAC), melibiose (MEL), and galactose (GAL).

**107**

other hand, galactose and melibiose (a galactose–glucose disaccharide) are not inhibitors. Therefore the determinant group of this antigen is a lactose unit. This type of inhibition experiment is applicable to all other antigen–antibody systems.

# 10. Recent researches on complex polysaccharides

## 10.1 Anti-gum arabic antibodies

Bacterial cell wall polysaccharides have been known to be antigenic for a long time but only recently it has been shown conclusively that plant polysaccharides can also induce antibody synthesis (52). Gum arabic, an exudate of subtropical shrubs, is a complex polysaccharide of galactose, rhamnose, arabinose, and glucuronic acid with an intricate arrangement of these residues joined together by a variety of glycosidic bonds (53). Immunization of rabbits with a vaccine of gum arabic and Freunds adjuvant stimulates the immune system to produce antibodies specific for certain structural groups of the polysaccharide. Two sets of antibodies with specificity for different structural groups of gum arabic have been detected in anti-gum arabic serum by agar diffusion (52). *Figure 8* shows agar diffusion patterns (well 1 and well 2) for the two types of antibodies (set I and set II). Gum arabic can be hydrolysed in dilute acid with the release of arabinose which is in the furanose ring. The slightly hydrolysed gum arabic reacts only with set II antibodies (well 1 and well 3). On prolonged acid hydrolysis, the polysaccharide is no longer reactive with the antibodies (well 1 and well 4). The two sets of antibodies can be separated and isolated by affinity chromatography by a two column method on different adsorbents (54). Electrofocusing and agar diffusion

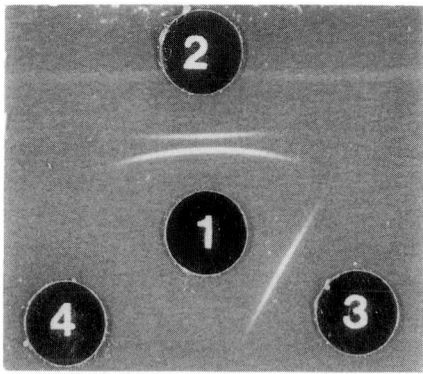

**Figure 8.** Agar diffusion patterns of native and acid modified gun arabic and antibodies: 1 = antibodies, 2 = native gum arabic, 3 = gum arabic hydrolysed in 0.01 M HCl, 4 = gum arabic hydrolysed in 1 M HCl (53).

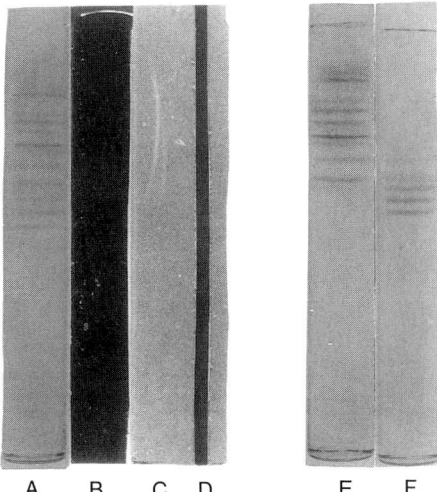

A    B    C    D         E    F

**Figure 9.** Gel isoelectric focusing and agar diffusion of anti-gum arabic antibodies: gel A = gel stained for protein, B = gel unstained and embedded in agar, C = precipitin complex arcs, D = trough of 2% solution of gum arabic, E = set I antibodies, F = set II antibodies.

results with the complete serum and the native gum arabic are shown in *Figure 9*. The isofocusing gel (a) which was stained for proteins shows the presence of 12 major protein bands. The unstained gel (B) embedded in agar gives two precipitin arcs (one for set I antibodies and the other for set II antibodies) with the antigen. All the protein isomers in the serum yield a precipitin complex (C) with the native gum arabic (D). The antibodies of set I and set II purified by affinity chromatography also reacted with the gum arabic but yielded single precipitin arcs (54). The purified antibody gels stained for protein show that set I antibodies consist of eight isoantibodies (E) and set II antibodies consist of four isoantibodies (F).

Two oligosaccharides were isolated from acid and enzymic hydrolysis of gum arabic and their structures were established by methylation analysis. The structures of the side chains of gum arabic which contain these oligosaccharides are shown in Formula 5. Hapten inhibition experiments by a micro method (55) showed that the disaccharide units from the gum arabic with the structures (Ara(1 → 4)GlcA and GlcA(1 → 6)Gal) inhibit the precipitin reaction with the complete anti-gum arabic serum. Purified set I antibodies were inhibited by the former and set II antibodies were inhibited in the latter. An agar diffusion plate showing the inhibition of set II antibodies by the disaccharide is pictured in *Figure 10*. It can be seen in the figure that the antibodies with the disaccharide yield much reduced precipitin bands with the antigen.

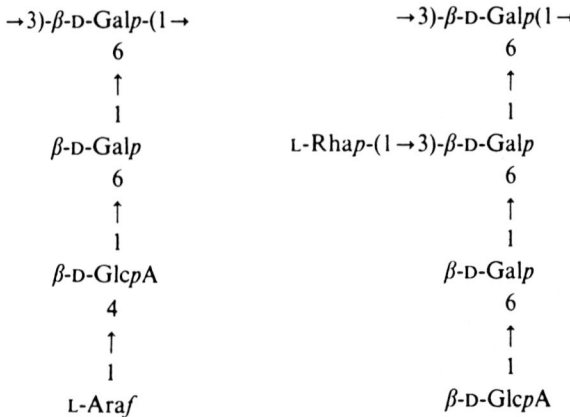

→3)-β-D-Galp-(1→
6
↑
1
β-D-Galp
6
↑
1
β-D-GlcpA
4
↑
1
L-Araf

→3)-β-D-Galp(1→
6
↑
1
L-Rhap-(1→3)-β-D-Galp
6
↑
1
β-D-Galp
6
↑
1
β-D-GlcpA

[Formula 5]

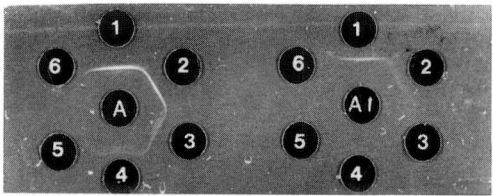

**Figure 10.** Inhibition of mildly hydrolysed gum arabic and antibodies of set II by an oligosaccharide from gum arabic: A = antibodies, Ai = antibodies plus inhibitor, 1 to 6 decreasing concentrations (200 μg to 6.25 μg) of antigen (53).

Immunodeterminant groups of polysaccharides are very often located at the non-reducing ends of the chains. Ara(1 → 4)GlcA is known to be a terminal unit of gum arabic. Evidence that GlcA(1 → 6)Gal is also a terminal unit has now been obtained as follows. The glucuronic acid residues of gum arabic were reduced and the reduced and native samples were subjected to methylation analysis. A portion of the gas chromatogram of the methylated samples is reproduced in *Figure 11*. It will be noted in *Figure 11* that the reduced sample yields derivatives 2,3,4-trimethyl glucose and 2,3,4,6-tetramethyl glucose which were not present in the methylation mixture of the native gum arabic. The glucose methylated derivatives arise from the reduced glucuronic acid. The presence of the tetramethyl derivative shows that glucuronic acid units were terminal residues of some gum arabic chains. The trimethyl derivative arises from glucuronic acid in internal positions.

**110**

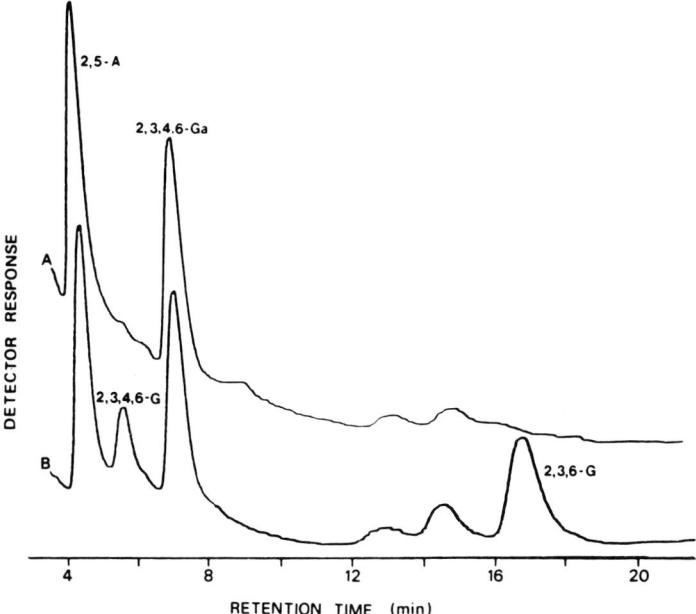

**Figure 11.** Comparable portion of the GLC patterns for partially methylated and acetylated derivatives from native gum arabic (A) and reduced gum arabic (B): 2,5-A = 1,3,4-tri-*O*-acetyl-2,5-di-*O*-methyl arabinitol; 2,3,4,6-G = 1,5-di-*O*-acetyl-2,3,4,6-tetra-*O*-methyl glucitol; 2,3,4,6-Ga = 1,5-di-*O*-acetyl-2,3,4,6-tetra-*O*-methyl galactitol; 2,3,6-G = 1,4,5-tri-*O*-acetyl-2,3,6-tri-*O*-methyl glucitol (53).

## 10.2 Reductive cleavage of a polysaccharide from *Klebsiella*

The elucidation of the complete structure of a polysaccharide requires the determination of many structural features including the identity of mono-saccharide units, sequence of the monosaccharides, ring forms of units, positions of glycosidic linkages, and anomeric configuration of the glycosidic linkages. A new method termed reductive cleavage facilitates the determina-tion of these structural features of polysaccharides (56, 57). The salient feature of the method is that reductive cleavage of a methylated poly-saccharide is carried out in $Et_3SiH$ and $Me_3SiOSO_2CF_3$ and followed by *in situ* acetylation. A cleavage of oxygen bonds of the polysaccharide occurs giving rise to acetyl methylated anhydro alditols. The acetyl methylated anhydride alditols are subsequently analysed by GLC–MS methods. Total or selective cleavage can be effected to yield monosaccharides of oligo-saccharides which can be separated by HPLC methods and characterized by mass spectrometry and $^1$H-NMR spectrometry. The utility of the method was checked with a polysaccharide of *Klebsiella* type 54. The structure deduced for this polysaccharide is shown in Formula 6.

$$→3)-\beta\text{-D-Glc}p\text{-}(1→4)\text{-}\alpha\text{-D-Glc}p\text{A-}(1→3)\text{-}\alpha\text{-L-Fuc}p\text{-}(1→$$

$$4$$
$$↑$$
$$1$$
$$\beta\text{-D-Glc}p$$

[Formula 6]

---

**Protocol 13.** Reductive cleavage of a polysaccharide

1. Methylate a few milligrams of *Klebsiella* type 54 by *Protocol* 4B.

2. Purify by adsorption and elution from C-18 reverse-phase Sep Pak cartridge and chromatography on columns (3 × 39 cm) of Sephadex LH-20 and elution with dichloromethane/methanol (2:1 v/v).

3. Check for complete methylation by $^1$H-NMR spectroscopy.

4. Reduce the ester groups with lithium aluminium hydride.

5. Add the per-*O*-methylated polysaccharide (5 mg) and a small stirring-bar to a Wheaton V-vial equipped with a Teflon lined screw top.

6. Keep the vial and contents under high vacuum for 2 h and add a $CH_2Cl_2$ solution of $Et_3SiH–Me_3SiO_3SCF_3$.

7. Cap and stir for 20 h at room temperature, and add 13 µl of $Ac_2O$ (5 equiv.).

8. Continue stirring for 20 h at room temperature and then quench by the addition of 0.5 ml of saturated aqueous sodium hydrogen-carbonate.

9. Stir the biphasic reaction mixture for 1 h (mild evolution of gas occurs).

10. Remove the aqueous layer, add saturated aqueous $NaHCO_3$ (0.5 ml).

11. Stir for 30 min, remove the aqueous layer, and use the $CH_2Cl_2$ layer for GLC analysis.

---

## 10.3 Pneumococcal capsular $\beta(1 → 2)$ $(1 → 3)$ glucan

The capsular polysaccharide of *Streptococcus pneumonia* type 40 is composed solely of glucose residues and is the type-specific substance of this organism. The polysaccharide is a homopolymer while in other types of this group of organisms heteropolymers are the type-specific substances. The structure of the homopolymer has now been deduced by methylation analysis, periodate oxidation, and NMR spectroscopy (58). The structure is unique and may be described as a comb-like structure. The polysaccharide consists of a main chain of glucose units linked by $\beta(1 \rightarrow 3)$ glycosidic linkages and each glucose of the main chain carries a side chain of single glucose units linked $\beta(1 \rightarrow 2)$ to the main chain. The polymer is appropriately named $\beta(1 \rightarrow 2)$ $(1 \rightarrow 3)$ glucan. The repeating unit for the glucan is shown in Formula 7.

$$\rightarrow 3)\text{-}\beta\text{-D-Glc}p\text{-}(1\rightarrow$$
$$2$$
$$\uparrow$$
$$1$$
$$\beta\text{-D-Glc}p$$

[Formula 7]

---

**Protocol 14.** Pneumococcal capsular glucan

1. Suspend 100 mg of the polysaccharide in 85% formic acid for 20 min at 85 °C. This treatment yields a product which is more soluble.

2. Chromatograph one column (3 × 90 mm) of Bio-Gel P-10 and collect the polysaccharide (82 mg) from the void volume.

3. Perform methylation analysis by standard method.

4. Analyse by GLC on capillary column (SE54) with a program 180°–250° by mass spectrometer (Hewlett Packard 5970). 2,3,4,6-Tetra-*O*-methyl glucose and 4,6-di-*O*-methyl glucose are obtained in comparable amounts but no other products.

5. Run NMR spectra on the purified polysaccharide in deuterium oxide. The signals indicate both substituted glucoses are linked by $\beta$-glycosidic bonds.

6. Dissolve 60 mg of polysaccharide and 200 mg of sodium *meta*-periodate in 15 ml of 0.1 M acetate buffer of pH 6.0 and maintain in the dark for 20 h at 4 °C.

7. Reduce product with 200 mg sodium borohydride overnight.

8. Recover the product.

9. Run NMR spectra.

---

## 10.4 Methylated peracetyl aldonitriles

The structure of the polysaccharide from a cyanobacterium has been investigated by a methylation technique in which the methylated mono-saccharides are converted to methylated peracetyl aldonitriles (PAAN) prior to analysis by GLC–MS (59, 60). The PAAN derivatives are separated by

GLC in capillary columns and subjected to mass spectrometry. The unsymmetrical nature of the aldonitriles aids in the identification of the monosaccharides on the basis of mass spectra. The polysaccharide from the cyanobacterium is a unique polymer composed of six different monosaccharide units namely glucose, fucose, mannose, arabinose, galactose, and galacturonic acid. On the basis of the types of methylated peracetyl aldonitrile derivatives obtained, it is proposed the polysaccharide consists of two types of repeating structural units (60). The structures of the units are shown in Formula 8.

→2 or 3)-GalA-(1→      ⟿3)-Glc-(1→3 or 4)-Fuc-(1→4)-Man-(1⟿$_2$
    ↑                                   ↑
Ac-3GalA                                Ara

[Formula 8]

---

**Protocol 15.** Preparation of methyl peracetyl aldonitrile

1. Methylate the bacterial polysaccharide (see *Protocol 4*) and hydrolyse the product.

2. React the methylated monosaccharides with hydroxylamine hydrochloride in a pyridine solvent at 60 °C for 20 min with vigorous stirring.

3. Add acetic anhydride and heat the reaction mixture for an additional 20 min at 60 °C.

4. Extract the methylated peracetyl aldononitriles with chloroform and use appropriate aliquots for analysis by GLC–MS.

5. The mass spectra data for the methylated peracetylated aldononitrile derivatives of monosaccharides are obtained in a similar manner as with the methylated alditol acetates.

---

## 10.5 Gluco-galacto-glycan containing pyruvic acid

A glycan is extracted from an *Agrobacterium radiobacteri* (ATCC 53271) and is found to be composed of glucose, galactose, and pyruvic acid in the molar ratio of 15:2:2 (61). A detailed study of the structure of the polysaccharide has been made by methylation and ethylation analysis, NMR spectroscopy, periodate oxidation, and deuterium labelling followed by methylation analysis. The results show the polysaccharide to be of complicated structure with a repeating unit of 17 monosaccharide units. The main chain of the polysaccharide is made up of glucose residues linked by (1 → 4) glycosidic bonds interspersed with occasional galactose residues linked to glucose by (1 → 3) bonds. The main chain contains glucose side chains linked to the 6 or 2 position of glucose. Pyruvate units are linked to positions 4 and 6 of the next

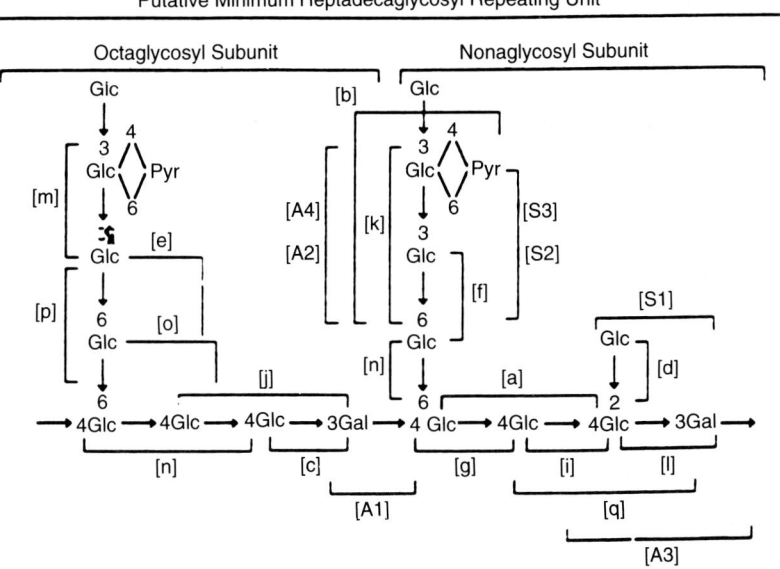

**Figure 12.** Proposed structure of the heptadecasaccharide repeating unit of the polysaccharide from *A. radiobacter* (61).

to the last glucose unit of the side chain as carboxy-ethylidine moieties. The polysaccharide does indeed possess a complex repeating unit diagrammed in *Figure 12*. The biosynthetic pathway and particularly the regulation of the pathway of biosynthesis must be extremely complex.

---

**Protocol 16.** Isolation and characterization of gluco-galacto-glycan

1. Grow *A. radiobacter* (ATCC 53271) in liquid medium (100 ml), pH 7.2 containing D-glucose (2.5%), L-glutamic acid (0.02%), potassium phosphate (0.5%), magnesium sulfate (0.02%), calcium chloride (0.005%), and trace amounts of biotin, pathothenic acid, and manganese.

2. Grow cultures on a rotary shaker (200 r.p.m.) until the D-glucose is consumed (six to eight days).

3. Treat the fermentation broth with 2-propanol (2 vol.) and collect the resulting precipitate which contains the polysaccharide by centrifugation. Reprecipitate a solution of the precipitate in water by the addition of 2-propanol (2 vol.).

**Protocol 16.** *Continued*

4. Centrifuge a water solution of the polysaccharide at 40 000 $g$ for 1 h and precipitate again by the addition of 2-propanol. The polysaccharide preparation contains no protein as determined by the Lowry method.

5. Determine the glycosyl residue composition of the polysaccharide by GLC analysis of the alditol acetate and per-*O*-trimethylsilyl methylglycoside derivatives.

6. Treat the polysaccharide (1 mg for 1 h at 120 °C with 2 M TFA. Remove the TFA under a stream of air and treat the residue for 16 h at 80 °C with S-(+)-2-butanol-HCl. The S-(+)-2-butylglycosides so obtained will be *O*-trimethylsilylated and analysed by GLC.

7. Release glycosyl residues containing *O*-(1-carboxyethylidene) residues from the polysaccharide by treatment with 1 M HCl in dry methanol for 15 h at 80 °C, and examine the *O*-trimethylsilylated products by GLC–MS.

8. Heat 30 mg polysaccharide in 50 mM TFA for 1 h at 80 °C. Dialyse the cooled solution against deionized water and freeze-dry (yield 43 mg), and subject to methylation analysis.

9. Ethylate the partially methylated oligoglycosyl alditols. Separate the partially methylated, partially ethylated oligoglycosyl alditols (3 mg) by HPLC.

10. Dissolve some of the polysaccharide in 50 mM sodium periodate (100 ml) in the dark for 48 h at 20 °C. Destroy the excess periodate by addition of ethylene glycol (3 ml) and dialyse the solution. Add sodium borohydride (400 mg) and keep the solution for 8 h at 20 °C. Destroy the excess borohydride by the addition of glacial acetic acid, dialyse, and freeze-dry the solution. Determine the sequence of residues by methylation analysis.

## 10.6 Cleavage at the glucuronic moiety of a plant arabinoglucuronomannan

Arabinoglucuronomannan is present in tobacco seed cultures and is isolated by standard extraction procedures. The polysaccharide is cleaved at glucuronic acid units and the products used for structural analysis (62). The isolated polymer is permethylated and subjected to the following sequence of reactions: saponification, decarboxylation, acetoxylation, and reductive cleavage with sodium borohydride. A series of 4-*O*-α-D-mannopyranosyl D-xylitol derivatives will be obtained. The oligosaccharide derivatives are separated on silica gel and then by HPLC using a Waters analytical Partisil column and elution with 97:3 benzene/methanol solvent. The derivatives are characterized by [1]H-NMR spectroscopy and GLC–MS. The results are interpreted to show that the main chain of the polysaccharide contain units of 4-linked β-D-glucuronic acid and 2-linked α-D-mannose with side chains of β-L-arabinofuranosyl units attached to position 3 of some of the glucuronic acid, and α-L-arabino-furanosyl units attached to some of the mannose residues. It may be possible to sequence the oligosaccharides by a stepwise degradation method (63).

## 10.7 Mycobacterial cell wall arabinogalactan

The cell wall of *Mycobacterium tuberculosis* contains an unusual poly-saccharide of arabinose and galactose monosaccharide residues occurring in the furanose ring structures (63). The polysaccharide is methylated *in situ* in the cell wall of the organism and the derivatized polysaccharide is isolated by extraction with acetone. The extracted product is made up of about 25% arabinogalactan and 75% peptidoglycan and lipopolysaccharides. The preparation is hydrolysed in 2 M trifluroacetic acid to a series of methylated oligosaccharides. The methylated oligosaccharides are separated by HPLC and subjected to GLC–MS analysis. The structural analysis is done on 43 oligosaccharide fragments obtained by partial acid hydrolysis of the permethylated cell walls. The experimental evidence is interpreted as showing the structure of the polymer to be a main chain of galactofuranose with alternating $(1 \rightarrow 5)$ and $(1 \rightarrow 6)$ glycosidic linkages carrying side chains of arabinfuranose units linked by $(1 \rightarrow 3)$ or $(1 \rightarrow 5)$ linkages forming oligosaccharides of various degrees of polymerization. The arabinose side chains are linked to position 5 of galactofuranose of the main chain. Both monosaccharides are of the L configuration. These studies could lead to a greater understanding of the role of the polysaccharide in the formation of peptidoglycan complex of the cell wall and in development of pathogenesis to the mycobacterium.

---

## Protocol 17. Preparation of arabinogalactan

1. Grow *M. tuberculosis* for two months on Sauton's medium in 2.8 litre Fernbach flasks. Inactivate the bacteria by autoclaving at 80 °C for 2 h and centrifuge to yield 33.5 g of wet cells.

2. Suspend cells in 100 ml Tris buffer and sonicate in a Heat-Systems-Ultrasonicator. Microscopic examination reveals that the majority of bacteria are broken.

3. Recover cell walls by centrifugation at 27 000 *g*, extract overnight with aqueous 2% SDS. Centrifuge at 27 000 *g*. Wash the cell walls several times with water and 80% acetone to remove SDS and lyophilize. These cell walls should contain approximately 25% arabinogalactan.

**Protocol 17.** *Continued*

4. Methylate the cell walls by the Hakomori procedure. Extract the per-*O*-methylated arabinogalactan into acetone.

5. Hydrolyse portions of the acetone soluble per-*O*-methylated arabino-galactan with 2 M TFA at 120 °C for 30 min, reduce, and acetylate.

6. Obtain partially methylated oligoglycosyl alditol fragments with an average size of approximately three units.

7. Dissolve a portion of the mixture of per-*O*-alkylated oligoglycosyl alditols (3–5 mg) in aqueous 30% $CH_3CN$ and separate the derivatives by HPLC phase column.

8. Perform GLC–MS analysis after HPLC purification of the per-*O*-alkylated oligoglycosyl alditol acetates.

## 10.8 Xanthan (manno-glucurono-glucan) from *Xanthomonas campestris*

Xanthan is a microbial polysaccharide composed of glucose, mannose, and glucuronic acid residues with some of the mannose residues carrying pyruvic acid and acetyl groups. The polysaccharide has unique properties and many industrial uses have been developed for the polymer in food, beverage, pharmaceutical, and agro products. It is an exocellular polysaccharide synthesized by *Xanthomonas campestris*. The carbohydrate residues are joined by $\beta(1 \rightarrow 4)$, $\alpha(1 \rightarrow 3)$, or $\beta(1 \rightarrow 2)$ glycosidic linkages and a structural formula for the repeating unit is shown in Formula 9 (65).

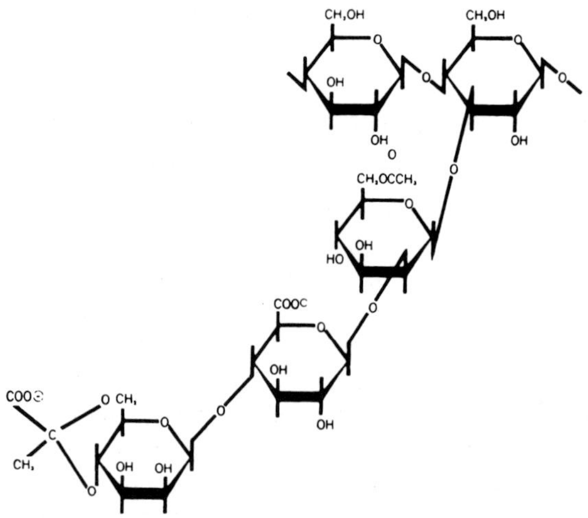

[Formula 9]

Antibodies specific for xanthan have been produced by rabbits immunized with a vaccine of xanthan and Freunds adjuvant (66). The antibodies have been obtained in pure form by affinity chromatography of the serum on a xanthan–Sepharose adsorbent. The antibodies should be useful for devising analytical methods for the quantitative analysis of foods and other products containing xanthan.

## 10.9 Quantitative determination of acidic polysaccharides

Acidic polysaccharides such as pectin and xanthan are of considerable importance as constituents of food and other commercial products. These uses are due to the ability of these polymers to form gels and modify the properties of the products. A reliable quantitative method for the determination of the polymers is needed. Poly(hexamethylenebiguanidinium)chloride is a suitable reagent to test. Acidic polymers undergo cross-linking reactions with this reagent. A method has been developed for the quantitative determination of acidic polysaccharides of industrial value using the hexamethylene polymer (67).

---

**Protocol 18.** Determination of pectin and xanthan

1. Add to 1.0 ml duplicate samples of the acidic polysaccharide, 2.0 ml of hexamethylene polymer (0.3% solution).

2. Stir continuously for 5 min.

3. Remove the precipitate by centrifuging at 3000 r.p.m. for 5 min.

4. Dilute supernatant 100-fold and read UV absorption at 235 nm.

5. Construct a standard curve by measuring absorption as above for varying concentrations of acidic polysaccharides.

---

## 10.10 Electron microscopic observation of waxy maize starch ($\alpha(1 \rightarrow 4)$ $(1 \rightarrow 6)$ glucan)

The fine-structure of polysaccharides and the location of polysaccharides in cellular subparticles can be examined by electron and scanning electron microscopic techniques. Results of such studies on amylopectin of waxy maize have been published (68). These results have been interpreted as evidence for a new type of structure for the amylopectin molecule and the arrangement of these molecules in the starch granule. This interpretation is consistent with the growth rings of granules observable by the light microscope. These results have been obtained with amylopectin from waxy maize starch and with amylodextrin, a crystalline degradation product from amylopectin. The electron micrograph shown in frame A of *Figure 13* shows that acid treated

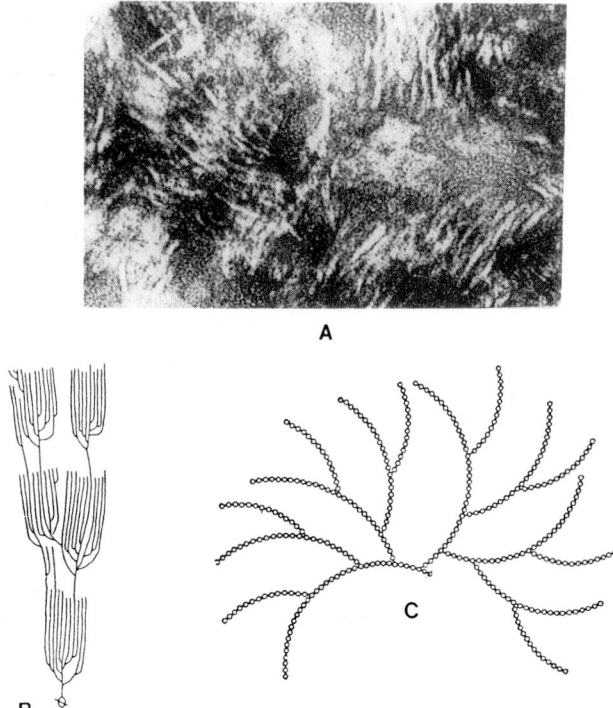

**Figure 13**. Transmission electron micrograph of waxy maize starch after acid treatment and dispersion in water: A = negative stain, B = cluster model for waxy maize amylopectin, C = branch model for amylopectin and glycogen (67).

granules contain amorphous regions and crystalline regions. Similar results are obtainable with the amylodextrin samples. Observations of this type have led to the proposal of a cluster model for the amylopectin molcule as diagrammed in frame B of *Figure 13*. Additional evidence and interpretations for the cluster model structure are presented in a recent review on amylopectin structure (69). This type of structure is quite different from the branched or tree model diagrammed in frame C which has been the accepted model for representing the structure of amylopectin and glycogen (70). No evidence is available at present that glycogen structure conforms to the cluster model. In such a model for glycogen the protein primer will need to be positioned in the model (71).

## 10.11 Monoclonal antibodies and the identification of *Brucella* antigens

Strains of *Brucella*, a disease producing bactrium contain lipopolysaccharide antigens in the cell walls. The polysaccharide moiety of the antigens of strains

A and M are homopolymers of 4,6-dideoxy-4-formamido-α-D-mannopyrnose residues. The carbohydrate residues of A antigen are joined by α(1 → 2) glycosidic linkages and the residues of the M antigens are joined mostly by this type of linkage but a few residues are joined by the α-(1 → 3) linkages. Both antigens consist of repeating pentasaccharide units shown in Formula 10 (72). Monoconal antibodies have been produced against *Brucella* A and M antigens by the cell fusion techniques. These antibodies were used in binding studies with the *Brucella* A and M antigens designed to elucidate the epitopes (immunodeterminents) of the antigens. It was demonstrated that the epitope of antigen A is a pentasaccharide unit of five contiguous residues of the above mannose derivative joined by α-D-(1 → 2) linkages. The epitope of M antigen is also composed of a pentasaccharide containing one α-D-(1 → 3) linkage as well as four α-D-(1 → 2) linkages. The binding of the antigens to the monoclonal antibodies occurs in characteristic fashion indicating a simultaneous expression (73) of both epitopes within a single antigen molecule. It has been reported previously that the structural unit of the A antigen epitope is also present in the M antigen. It now appears that the M epitope is also present as a minor constituent of the A antigen. The monoclonal antibodies may be suitable reagents for the identification of *Brucella* A and M antigens. A method for diagnosis of diseases due to *Brucella* such as undulant fever and contagious abortion may be developed.

[Formula 10]

# Acknowledgements

The author sincerely thanks Eileen McConnell for the excellent assistance in the preparation for the manuscript, Jean Pazur for many constructive improvements, and Susan Magargee and Linda Collins for technical assistance and providing analytical data. All of these contributions were invaluable.

# References

1. Danishefsky, I., Whistler, R. L., and Bettelheim, F. A. (1970). In *The carbohydrates* (2nd edn) (ed. W. Pigman and D. Horton), Vol. IIA, pp. 375–412. Academic Press Inc., New York.
2. Joint Commission on Biochemical Nomenclature. (1982). *J. Biol. Chem.*, **257**, 3352.
3. Sharon, N. (1980). *Sci. Am.*, **243(5)**, 90.
4. Pazur, J. H. (1991). In *Advances in carbohydrate analysis* (ed. C. A. White), Vol. 1, pp. 1–62. JAI Press Ltd., London.
5. Block, R. J., Durrum, E. L., and Zweig, G. (1958). *Paper chromatography and paper electrophoresis* (2nd edn). Academic Press, New York.
6. Dutton, G. G. S. (1973). In *Advances in carbohydrate chemistry and biochemistry* (ed. R. S. Tipson and D. Horton), Vol. 28, pp. 11–160. Academic Press, New York.
7. McGinnis, G. D., Laver, M. L., and Biermann, C. J. (1989) In *Analysis of carbohydrates by GLC and MS* (ed. C. J. Biermann and G. D. McGinnis), pp. 19–26. CRC Press Inc., Boca Raton, Florida.
8. Kennedy, J. F. and Fox, J. E. (1980). In *Methods in carbohydrate chemistry* (ed. R. L. Whistler and J. N. BeMiller), Vol. 8, pp. 3–12. Academic Press, New York.
9. Dionex. (1989). Technical Note 20, Dionex Corp, Sunnyvale, CA.
10. Pazur, J. H., Dropkin, D. J., Dreher, K. L., Forsberg, L. S., and Lowman, C. S. (1976). *Arch. Biochem. Biophys.*, **176**, 257.
11. Taylor, R. L. and Conrad, H. E. (1972). *Biochemistry*, **11**, 1383.
12. Pazur, J. H. and Kleppe, K. (1964). *Biochemistry*, **3**, 578.
13. Pazur, J. H., Knull, H. R., and Chevalier, G. E. (1977). *J. Carbohydr. Nucleos. Nucleot.*, **4**, 129.
14. Gerwig, G. J., Kamerling, J. P., and Vliegenthart, J. F. G. (1978). *Carbohydr. Res.*, **62**, 349.
15. Leontein, K. and Lonngren, J. (1993) In *Methods in carbohydrate chemistry* (ed. J. N. BeMiller, R. L. Whistler, and D. H. Shaw), Vol. 9, pp. 87–9. John Wiley & Sons Inc., New York.
16. Fox, J. D. and Robyt, J. F. (1992). *Carbohydr. Res.*, **227**, 163.
17. Harding, S. E., Varum, K. M., Stokke, B. T., and Smidsrod, O. (1991). In *Advances in carbohydrate analysis* (ed. C. A. White), Vol. 1, pp. 63–144. JAI Press Ltd., London.
18. Pazur, J. H., Kleppe, K., and Anderson, J. S. (1962). *Biochim. Biophys. Acta*, **65**, 369.
19. Martin, R. G. and Ames, B. N. (1961). *J. Biol. Chem.*, **236**, 1372.

20. Fox, J. D. and Robyt, J. F. (1991). *Anal. Biochem.*, **195**, 93.
21. Isbell, H. S. (1973). In *Methods in carbohydrate chemistry* (ed. R. L. Whistler), Vol. 5, pp. 249–50. Academic Press, New York.
22. Whistler, R. L. and Anisuzzaman, A. K. M. (1980). In *Methods in carbohydrate chemistry* (ed. R. L. Whistler and J. N. BeMiller), Vol. 8, pp. 45–53. Academic Press. New York.
23. Kennedy, J. F., Stevenson, D. L., and White, C. A. (1988). *Starch/Stärke*, **40**, 396.
24. Haworth, W. N. (1915). *J. Chem. Soc.* (London), **107**, 8.
25. Hakomori, S.-I. (1964). *J. Biochem.* (Tokyo), **55**, 205.
26. Bjorndal, H., Hellerqvist, C. G., Lindberg, B., and Svensson, S. (1970). *Angew. Chem. Internat. Edit.*, **9**, 610.
27. Jansson, P. E., Kenne, L., Liedgren, H., Lindberg, B., and Lonngren, J. (1976). *Chem. Commun., Univ. Stockholm*, **8**, 1.
28. Pazur, J. H., Cepure, A., Kane, J. A., and Hellerquist, C. G. (1973). *J. Biol. Chem.*, **248**, 279.
29. Hay, G. W., Lewis, B. A., and Smith, F. (1973). In *Methods in carbohydrate chemistry* (ed. R. L. Whistler), Vol. 5, pp. 357–61. Academic Press, New York.
30. Pazur, J. H. and Forsberg, L. S. (1977). *Carbohydr. Res.*, **58**, 222.
31. O'Colla, P. S. (1965). In *Methods in carbohydrate chemistry* (ed. R. L. Whistler, and J. N. BeMiller), Vol. 5, pp. 382–92. Academic Press, New York.
32. Lindberg, B., Lonngren, J., and Svensson, S. (1975). In *Advances in carbohydrate chemistry and biochemistry* (ed. R. S. Tipson and D. Horton), Vol. 31, pp. 185–240. Academic Press, New York.
33. Stewart, T. S., Mendershausen, P. B., and Ballou, C. E. (1968). *Biochemistry*, **7**, 1843.
34. Laine, R. A., Esselman, W. J., and Sweeley, C. C. (1972). In *Methods in enzymology* (ed. V. Ginsburg), Vol. 28, pp. 159–77. Academic Press, New York.
35. Wolfrom, M. L., Thompson, M., and Timberlake, C. E. (1963). *Cereal Chem.*, **40**, 82.
36. Takasaki, S. and Kobata, A. (1986). *Biochemistry*, **25**, 5709.
37. BeMiller, J. N. (1965). In *Starch chemistry and technology* (ed. R. L. Whistler and E. F. Paschall), Vol. I, pp. 521–31. Academic Press, New York.
38. Pazur, J. H. and Forsberg, L. S. (1978). *Carbohydr. Res.*, **60**, 167.
39. Pazur, J. H. and Forsberg, L. S. (1980). *Carbohydr. Res.*, **83**, 406.
40. Kenne, L. and Lindberg, B. (1980). In *Methods in carbohydrate chemistry* (ed. R. L. Whistler and J. N. BeMiller), Vol. 8, pp. 295–9. Academic Press, New York.
41. Jennings, H. J. and Smith, I. C. P. (1980). In *Methods in carbohydrate chemistry* (ed. R. L. Whistler and J. N. BeMiller), Vol. 8, pp. 97–105. Academic Press, New York.
42. Gorin, P. A. J. (1981). In *Advances in carbohydrate chemistry and biochemistry* (ed. R. S. Tipson and D. Horton), Vol. 38, pp. 13–97. Academic Press, New York.
43. Pazur, J. H., Dropkin, D. J., and Forsberg, L. S. (1978). *Carbohydr. Res.*, **66**, 155.
44. Heidelberger, M. and Avery, O. T. (1924). *J. Exp. Med.*, **40**, 301.
45. Lancefield, R. (1940–1941). *Harvey Lectures Ser.*, **36**, 251.

46. Heidelberger, M. (1973). In *Research in immunochemistry and immunology* (ed. J. B. C. Kwapinsky and E. Day), Vol. 3, pp. 1–40. University Park Press, Baltimore, London, Tokyo.
47. Pazur, J. H., Dreher, K. L., and Forsberg, L. S. (1978). *J. Biol. Chem.*, **253**, 1832.
48. Pazur, J. H. (1981). In *Advances in carbohydrate chemistry and biochemistry* (ed. R. S. Tipson and D. Horton), Vol. 39, pp. 405–47. Academic Press, New York.
49. Pazur, J. H., Liu, B., Li, N. Q., and Lee, Y. C. (1990). *J. Prot. Chem.*, **9**, 143.
50. Eisen, H. N. and Reilly, E. B. (1985). *Annu. Rev. Immunol.*, **3**, 337.
51. Pazur, J. H., Tay, M. E., Pazur, B. A., and Miskiel, F. J. (1987). *J. Prot. Chem.*, **6**, 387.
52. Miskiel, F. J. and Pazur, J. H. (1991). *Carbohydr. Polymers*, **16**, 17.
53. Stephen, A. M. (1983). In *The polysaccharides* (2nd edn) (ed. G. O. Aspinall), Vol. 2, pp. 97–193. Academic Press, New York.
54. Pazur, J. H., Miskiel, F. J., Witham, T. F., and Marchetti, N. (1991). *Carbohydr. Res.*, **214**, 1.
55. Pazur, J. H. and Kelly, S. A. (1984). *J. Immunol. Methods*, **75**, 107.
56. Vodonik, S. A. and Gray, G. R. (1988). *Carbohydr. Res.*, **175**, 93.
57. Garegg, P. J., Lindberg, B., Konradsson, P., and Kvarnstrom, I. (1988). *Carbohydr. Res.*, **176**, 145.
58. Adeyeye, A., Jansson, P. E., Lindberg, B., and Henrichsen, J. (1988). *Carbohydr. Res.*, **180**, 295.
59. Slodki, M. E., England, R. E., Platiner, R. D., and Dick, W. E. (1986). *Carbohydr. Res.*, **156**, 199.
60. Marra, M., Palmeri, A., Ballio, A., Segre, A., and Slodki, M. E. (1990). *Carbohydr. Res.*, **197**, 338.
61. O'Neill, M. A., Robinson, P. D., Chou, K. J., Darvill, A. G., and Albersheim, P. (1992). *Carbohydr. Res.* **226**, 131.
62. Aspinall, G. O., Khondo, L., and Puvanesarajah, V. (1989). *Carbohydr. Res.*, **188**, 113.
63. Jager, G., Lay, H., Lehmann, J., and Ziser, L. (1991). *Carbohydr. Res.*, **217**, 99.
64. Daffe, M., Brennan, P. J., and McNeil, M. (1990). *J. Biol. Chem.*, **265**, 6734.
65. Kelco Division of Merck & Company. (1988). In Xanthan Gum, Kelco Division, San Diego CA.
66. Pazur, J. H., Marchetti, N., and Miskiel, F. J. (1993). *FEBS J.*, **7**, 1259.
67. Kennedy, J. F., Melo, E. H. M., Crescenzi, V., Dentini, M., and Martricardi, P. (1992). *Carbohydr. Polymers*, **17**, 199.
68. Yamaguchi, M., Kainuma, K., and French, D. (1979). *J. Ultrastruct. Res.*, **69**, 249.
69. Manners, D. (1989). *Carbohydr. Polymers*, **11**, 87.
70. Manners, D. (1991). *Carbohydr. Polymers*, **16**, 37.
71. Lomako, J., Lomako, W. M., and Whelan, W. J. (1990). *FEBS Lett.*, **268**, 8.
72. Bundle, D. R., Cherwonogrodzky, J. W., Gidney, M. A. J., Meckle, P. J., Perry, M. B., and Peters, T. (1989). *Infect. Immun.*, **57**, 2829.
73. Yancopoulos, G. D. and Alt, F. W. (1986). *Annu. Rev. Immunol.*, **4**, 339.

# Proteoglycans

S. L. CARNEY

## 1. Introduction

Proteoglycans, by definition, are conjugates of protein and glycosaminoglycan which are found largely in connective tissue. Their major role is to provide swelling pressure in these tissues by virtue of a Donnan osmotic effect due to the high fixed negative charge found on these molecules. Proteoglycans by their very nature are related to glycoproteins, but are generally differentiated from this class of molecule because of their distinct glycosaminoglycan component. In glycoproteins, generally the sugar moiety composes only a small percentage of the molecular mass, the sugars are relatively small, generally uncharged (with the exception of sialic acid residues), and frequently highly branched. In contrast, glycosaminoglycans from proteoglycans are large, linear molecules which are highly charged, containing both carboxylic acid and sulfate ester groups, and comprise generally a significant proportion of the total molecular mass.

In the seven years since the publication of the first edition of *Carbohydrate analysis—a practical approach*, there has been significant progress in the molecular biology as related to the components of connective tissues. In particular, molecular biology has allowed biochemists to name and identify proteoglycans, and as a result has reduced confusion in the area. The data have shown that proteoglycans can be categorized into a relatively small number of structural classes, and that diversity arises from alternative splicing and post-translational modification.

Although molecular biological techniques have allowed a more precise definition of the protein structure of proteoglycans, the field has continued to expand since the majority of the mass of proteoglycan molecules is produced by post-translational modification of the protein core. Recent work has indicated that in addition to the physicochemical properties of the proteoglycans, glycosaminoglycans may have specific biological functions conferred upon them by specific sequences within the carbohydrate chain. The control of the sulfation of glycosaminoglycan chains is still a question of extreme interest and it will be hoped that in the next decade similar advances will be made in this area as have been made in molecular biology in the past decade.

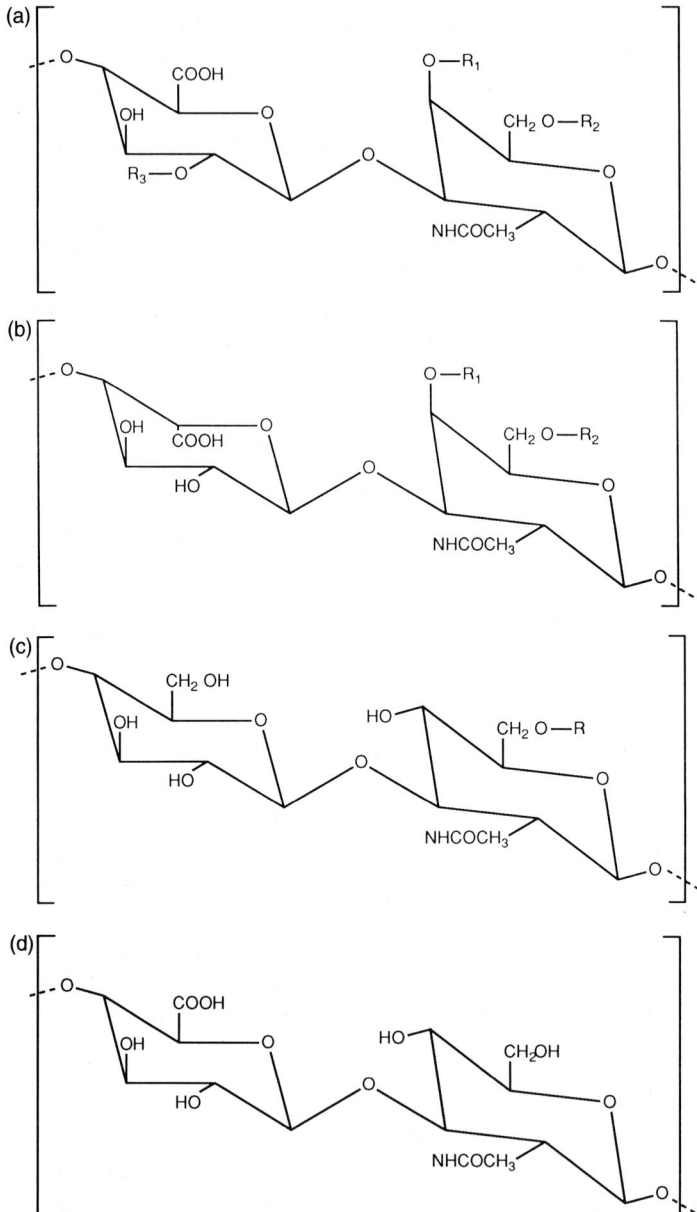

**Figure 1.** (a) The repeating disaccharides of chondroitin. $R_1$, $R_2$, and $R_3$ may be H or $SO_3$. None or all of these R groups may be $SO_3$, but in nature most commonly either $R_1$ or $R_2$ will be $SO_3$ giving rise to chondroitin 4- and 6-sulfate respectively. (b) The repeating disaccharide of dermatan sulfate, $R_1$ or $R_2$ may be either H or $SO_3$. It is more common for $R_1$ to be $SO_3$. (c) The repeating disaccharide structure of keratan sulfate. R may be H or $SO_3$. (d) The repeating structure of hyaluronic acid.

The glycosaminoglycans found covalently attached to proteoglycans are heteropolymeric molecules consisting of repeating disaccharides, which are substituted in various positions of (generally) the constituent hexosamines. Glycosaminoglycans fall into two main categories: the polyuronides, which include chondroitin sulfate and heparan sulfate. They consist primarily or in part, of glucuronic acids in their repeating disaccharide structure. Dermatan sulfate is also a polyuronide which is composed of repeating disaccharides containing iduronic acid. The other category of glycosaminoglycans are the polylactosamines, the main example of which is keratan sulfate. In common with the polyuronides, keratan sulfate is sulfated on the hexosamine residues of its repeat disaccharide units. *Figure 1* shows the structure of the repeating disaccharides of the major classes of glycosaminoglycans.

It is clear that the potential diversity of proteoglycans is huge. It would therefore be of little use to outline specific procedures for any particular proteoglycan. This chapter will outline general methods used for the extraction, purification, and analysis of proteoglycans (generally as applied to aggrecan), and will indicate appropriate modifications required for particular proteoglycans. Since the purification and analysis of proteoglycans has long been achieved by the properties and structure of the component glycosamino-glycans, such a general approach to the isolation of these diverse molecules is altogether appropriate.

# 2. Proteoglycan component structure

## 2.1 Connective tissue glycosaminoglycans

There are five main classes of glycosaminoglycan: the chondroitin sulfates, dermatan sulfate, keratan sulfate, heparan sulfate (and heparin), and hyaluronic acid. All with the exception of hyaluronic acid are sulfated and found covalently linked to protein. Thus, hyaluronic acid does not fall into the category of proteoglycans, however its highly specific and structurally significant interaction with certain proteoglycans (particularly aggrecan) warrants it a place in this section.

### 2.1.1 Chondroitin sulfate

Chondroitin sulfate is found ubiquitously throughout connective tissues. It is composed of repeating disaccharides of glucuronic acid linked $\beta(1 \rightarrow 3)$ to 2-acetamido-2-deoxy-D-galactose (N-acetylgalactosamine). The galactose may be unsulfated or sulfated at C-4 or C-6 or bis-sulfated. In unusual cases the glucuronic acid may be sulfated at C-2 or in very rare cases at C-3. Thus a very diverse range of structures from unsulfated to trisulfated can occur. To the author's knowledge, a tetrasulfated disaccharide of chondroitin has not been reported. Any of these disaccharides can be linked to one another via a $\beta(1 \rightarrow 4)$ linkage between N-acetylgalactosamine and glucuronic acid.

Although the molecular mass of chondroitin sulfate varies from source to source, it can range from 2.4–50 kd with a mean of approximately 20 kd. The structure of the repeating disaccharide of chondroitin sulfate is shown in *Figure 1*.

### 2.1.2 Keratan sulfate

Keratan sulfate is found in a variety of connective tissues, but is most concentrated in cartilage, intervertebral disc, and the cornea. There are two distinct classes of keratan sulfate, both of which share the same repeating disaccharide structure but vary in their linkage to protein. The corneal form of keratan sulfate (KS I) is linked to protein by an alkali-stable bond, similar to the linkage for *N*-linked oligosaccharides. The skeletal form of keratan sulfate however is linked to protein by alkali-labile bonds which are *O*-glycosidic in nature involving *N*-acetylgalactosamine and serine or threonine in the protein core. The repeating unit of keratan sulfate is a disaccharide of galactose linked $\beta(1 \rightarrow 4)$ to *N*-acetylglucosamine. The *N*-acetylglucosamine residue is generally sulfated at C-6, although oversulfation in keratan sulfate is commonly observed. The repeating disaccharides are linked *N*-acetyl-glucosamine $\beta(1 \rightarrow 3)$ galactose to one another.

### 2.1.3 Dermatan sulfate

Dermatan sulfate is an isomer of chondroitin sulfate, with C-5 of the uronic acid epimerized yielding an iduronate residue. The repeat disaccharide of this glycosaminoglycan is therefore iduronic acid linked $\alpha(1 \rightarrow 3)$ to *N*-acetyl-galactosamine. The linkage between the uronic acid and hexosamine is designated $\alpha$ rather than $\beta$ by convention even though the conformation around the C-1 of the uronic acid and the C-3 of the hexosamine is identical in chondroitin and dermatan sulfate, since the iduronate residue is of the L configuration rather than D for the glucuronate. The sulfate esters of dermatan sulfate are generally substituted on C-4 of the hexosamine. Dermatan sulfate is on average larger than chondroitin sulfate.

### 2.1.4 Heparin

Heparin has a much more complex structure than the glycosaminoglycans described previously. The glycosaminoglycan chain contains both iduronate and glucuronate residues and many of the hexosamine residues contain *N*-sulfate groups in addition to the more common *O*-sulfate found in all of the other glycosaminoglycans. Compared with the other glycosaminoglycans, heparin is much more highly sulfated, containing up to three sulfate residues per disaccharide (generally between 1.6 and 3). The repeating disaccharide unit of heparin consists of D-glucuronic acid (or L-iduronic acid) linked $\beta( \rightarrow 4)$ (or $\alpha(1 \rightarrow 4)$) to glucosamine. The disaccharides are linked to one another in the following way: glucosamine-$\beta(1 \rightarrow 4)$-[uronic acid $\alpha/\beta(1 \rightarrow 4)$-glucosamine]$_n$-$\beta(1 \rightarrow 4)$-uronic acid. The amino group on the glucosamine may be either sulfated or acetylated, and the C-6 on the glucosamine is generally sulfated. Extra sulfate groups may be substituted at the C-3 position. The uronic acid may be sulfated at C-2.

### 2.1.5 Heparan sulfate

Heparan sulfate is very closely related to heparin, containing the same repeating disaccharide units, linked in the same manner as in heparin. The major differences between heparin and heparan sulfate is that heparan sulfate contains higher proportions of glucuronic acid, is less *O*-sulfated and more *N*-acetylated, and less sulfated overall than heparin containing on average between 0.4 and 2 sulfate residues per disaccharide. As a result, many of the disaccharide residues may be unsulfated.

### 2.1.6 Hyaluronic acid

Although in the strictest sence, hyaluronic acid is not found as a proteoglycan, it is involved in non-covalent interactions with specific proteoglycan domains that are essential for their biological function. Since the functions of hyaluronic acid and proteoglycan are closely intertwined, it seems only appropriate that this biopolymer be included in this chapter. Hyaluronic acid, in common with the chondroitin sulfates, is a linear heteropolymer of lengths up to 1.6 μm. The repeating disaccharide unit of hyaluronic acid consists of glucuronic acid linked $\beta(1 \rightarrow 4)$ to *N*-acetylglucosamine. The repeating disaccharides are linked to one another in the following manner: *N*-acetylglucosamine $\beta(1 \rightarrow 3)$[glucuronic acid $\beta(1 \rightarrow 4)N$-acetylglucosamine]$_n$-$\beta(1 \rightarrow 3)$ glucuronic acid.

### 2.1.7 Linkage to protein

The polyuronides are all attached to protein via a common linkage region whose structure was largely determined by the work of Rodén (1). The linkage consists of a trisaccharide, galactose-galactose-xylose. The xylose is linked by a glycosidic bond to serine residues in the protein core. The structure of this linkage is outlined in *Figure 14*. The linkage region of keratan sulfates is more complex, involving alkali-stable (*N*-glycosylamine) bonds in KS I between *N*-acetylglucosamine and asparagine in the protein core. Alkali-labile bonds characterize the cartilage keratan sulfates (KS II) between the terminal *N*-acetylgalactosamine and serine or threonine in the protein core.

## 3. Proteoglycan extraction

Certain proteoglycan types undergo interactions in connective tissue rendering them too large to be easily extracted using non-chaotropic extractants. Such proteoglycan types include aggrecan and versican which both can form multimolecular aggregates with hyaluronic acid. Such interactions can be broken using chaotropic agents like 4 M guanidine.HCL or 7 M urea. Such protocols are referred to as dissociative extraction techniques. These extractants have obtained almost universal acceptance for proteoglycan extraction since they are capable of extraction in high yield and upon dialysis to normal physiological conditions, reaggregation will occur spontaneously almost to completion (2). At a pH of about 6 and in the presence of protease inhibitors, proteoglycan degradation during extraction is not detectable. Protease inhibitors generally used in extraction include: 100 mM 6-aminocaproic acid (which inhibits proteases with activities similar to cathepsin D), 5 mM benzamidine.HCl (to inhibit enzymes with specificity similar to trypsin), 10 mM EDTA (to inhibit metalloproteases), 10 mM $N$-ethylmaleimide (which inhibits thiol proteases and also helps to prevent disulphide exchange, a reaction which is common in denaturing solvents), 1 mM PMSF (a general inhibitor of serine proteases). Other protease inhibitors may be added should they be required and agents such as leupeptin, pepstatin, and soybean trypsin inhibitor are often included in protease inhibitor 'cocktails'.

---

**Protocol 1.** Preparation of proteoglycans

*Equipment and reagents*

- Suitable sized glassware
- Magnetic (or pneumatic) stirrer
- High speed centrifuge and rotors
- Cryostat (desirable but not essential)
- Rocking table (or similar apparatus) for cell extractions
- Reagent A for dissociative extraction: 4 M guanidine.HCl (382.12 g/litre)/50 mM sodium acetate (4.1 g/litre)/100 mM 6-aminocaproic acid (13.12 g/litre)/10 mM EDTA disodium salt (3.722 g/litre)/5 mM benzamidine.HCL (0.784 g/litre) dissolved in distilled water and the pH adjusted to 5.8
- Reagent B for associative extraction: 0.5 M guanidine.HCl (47.765 g/litre)/50mM sodium acetate (4.1 g/litre)/100 mM 6-aminocaproic acid (13.12 g/litre)/10 mM EDTA disodium salt (3.722 g/litre)/10 mM $N$-ethylmaleimide

(1.25 g/litre)/5 mM benzamidine.HCl (0.784 g/litre) dissolved in distilled water and the pH adjusted to 5.8
- Reagent C for detergent extraction: 4 M guanidine.HCl (382.12 g/litre)/1% (w/v) CHAPS (10 g/litre)/50 mM sodium acetate (4.1 g/litre)/100 mM 6-aminocaproic acid (13.12 g/litre)/10 mM EDTA disodium salt (3.722 g/litre)/10 mM $N$-ethylmaleimide (1.25 g/litre)/ 5 mM benzamidine.HCl (0.784 g/litre) dissolved in distilled water and the pH adjusted to 5.8
- Reagent D for enzymic liberation of residual proteoglycans: 0.5 M Tris (60 g/litre)/20 mM EDTA (9.31 g/litre)/20 mM cysteine.HCl (0.77 g/litre) dissolved in distilled water, adjusted to a pH of 6.8 with 1 M HCl. A suspension of papain (5 ml/litre; available from Sigma Chemicals Ltd.) was added to the solution to give a final papain concentration of 13 mg% (w/v).

A. *Basic method for extraction of tissues*

**1.** Dissect the tissue to be extracted free from any contaminating related tissues where appropriate, then maintain the excised tissue on ice to prevent degradation and loss of water.

2. Slice the tissue as finely as possible using sharp scalpels, or preferably microtome, to 20 μm using a cryostat. This may be achieved by freezing the tissue on a cryostat chuck using water and following sectioning, the water may be removed by centrifugation.

3. Once sectioned, place the tissue slices into chilled extractant (either reagent A, B, or C). As a general rule, dense connective tissues such as cartilage should be extracted with 10 vol. of extractant per gram of tissue, i.e. 1 litre per 100 g of tissue. For more areolar connective tissues such as aorta, corneal stroma, basement membranes, lung, mesenchyme, etc., one need only use 5 vol. of extractant per gram of tissue.

4. Extract the tissue with constant stirring for periods of up to 48 h at 4 °C.

5. After extraction, centrifuge the tissue at 20 000 $g$ for 20 min. Decant the supernatant away from the extracted tissue. As an alternative to centrifugation, the extract may be separated from the tissue by filtration through glass wool.

B. *Basic method for dissociative extraction of cell cultures*

1. Remove the cell maintenance medium and add extractant (reagent A) for 24 h at 4 °C.

2. Remove the extract and then further extract the cell layer with extractant containing detergent (reagent C) for 24 h at 4 °C. All extraction steps should be performed with agitation if possible using a suitable rocking table.

Extracts may be subsequently dialysed or desalted into more physiological buffers should this be required. This is of particular interest and significance in the case of aggregating proteoglycans from cartilage (aggrecan), where the function of the proteoglycan may be related to its ability to form large multimolecular aggregates at physiological ionic strength and pH. It must be strongly stated at this stage that for optimal extraction of proteoglycans, particularly from dense connective tissues such as cartilage, the preparation of the tissue prior to extraction is of great importance. Bayliss (3) has shown that the microtomy of cartilage to slices of 20 μm or less considerably increases the efficiency of extraction.

In general, it is difficult to extract proteoglycans from dense connective tissues such as cartilage using buffers that are incapable of dissociating proteoglycan aggregates (e.g. 0.5 M guanidinium chloride). In most cases one can extract only a small proportion of the total proteoglycans using these *associative* (since they permit *association* of aggrecan and hyaluronic acid) buffers. In other, more areolar connective tissues, such as rat chondrosarcoma, a much more significant proportion of the total proteoglycan can be extracted without the need to resort to dissociative solvents such as 4 M guanidinium chloride or 7 M urea.

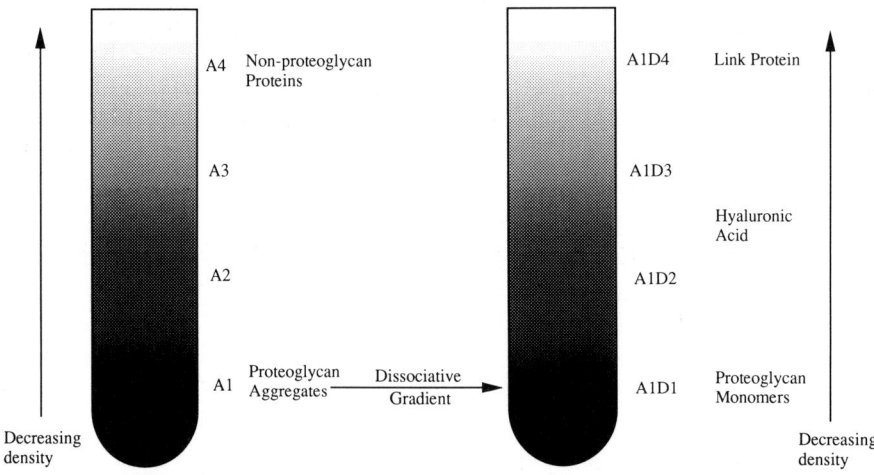

**Figure 2.** A diagrammatic representation of the nomenclature of the fractions obtained in associative and dissociative density gradient centrifugation. The diagram shows the expected distribution of various connective tissue components within the gradient.

**Protocol 2.** Basic method for associative extraction of tissues and cell cultures

See *Protocol 1* for equipment and reagents.

1. The tissue or cell culture to be extracted would be treated in exactly the ways outlined in *Protocol 1A* and *B*.

2. Extract the tissues for 48 h at 4 °C in 10 vol. of reagent B in the case of dense connective tissues, and in 5 vol. in the case of looser, more areolar connective tissues.

3. Following extraction, the tissue can be separated from the extract by filtration or centrifugation as outlined in *Protocol 1A*.

4. Cell cultures may be treated in the following manner: using reagent B, extract the cell layer for 24 h at 4 °C with constant agitation. The use of a rocking or vibrating table is recommended for this procedure.

No matter how diligently one may extract proteoglycans from tissues, it is very unlikely that one may be able to remove all of the proteoglycans. In many respects it is frequently more interesting to examine the glycosaminoglycans which remain within the tissue as they frequently differ from those extracted from the tissue. As yet, it is difficult to obtain total removal of proteoglycans from tissue after dissociative extraction, although good results have been obtained using collagenase. It is difficult to give precise conditions for the removal of proteoglycans using collagenase and one would have to examine the literature to obtain conditions for specific extractions. A more widespread method for the removal of 'residual' proteoglycans is the enzymic liberation of glycosaminoglycans by the use of the proteolytic enzyme, papain.

**Protocol 3.** Enzymic liberation of residual glycosaminoglycans

See *Protocol 1* for equipment and reagents.

1. If the cartilage to be digested has been extracted previously with dissociative buffers such as 4 M guanidinium chloride, it is necessary to wash out any remaining guanidinium with several changes of distilled water to prevent inhibition of the enzyme.

2. Add the papain solution (reagent D) to the residual cartilage. Generally 2 ml of reagent D is sufficient for up to 200 mg (wet weight) of cartilage.

3. The vials containing the cartilage and reagent D should be tightly stoppered and incubated at 60–65 °C for 16 h.

4. Following digestion, the reaction may be terminated by heating the solution at 100 °C for 5 min. Alternatively, stop the reaction by adding iodoacetamide (400 mM) to a final concentration of 40 mM (i.e. 222 µl/2 ml of digest). It is not however, always necessary to inactivate the enzyme, since subsequent analyses may not be interfered by the presence of an active enzyme.

5. Should the subsequent analysis require an active protein entity, for example immunoassay, then the inactivation steps should be followed.

There are other methods of liberating the residual glycosaminoglycans, for example by alkaline β-elimination as outlined in the first edition of this work. In general, however this method has been largely superceded by papain digestion, except in particular cases, such as when one would like to introduce a radiolabel into the chains liberated.

# 4. Proteoglycan purification

## 4.1 Purification by centrifugal techniques

When subjected to high centrifugal fields, caesium salts (particularly CsCl, CsBr, and $Cs_2SO_4$ for the separation of biological molecules) spontaneously form concentration gradients. When extracts of connective tissues are centrifuged in caesium salts (most commonly CsCl), the components of the extract migrate in the gradient according to their buoyant density and band there. The buoyant density of a protein or proteoglycan is in essence related to the amount of $Cs^+$ ions that can be bound by the molecule in question. Therefore, proteoglycans with their high concentrations of carboxyl and sulfate ester groups bind considerably more caesium than proteins and therefore have a much higher buoyant density. As a result, when subjected to isopycnic CsCl density gradient centrifugation, proteoglycans are found in the bottom (most dense) fractions, whereas proteins occupy the top (least dense) fractions of the gradient. The technique was first used to purify glycoproteins by Franek and Dunstone (4) and was adapted by Sajdera and Hascall (5) for proteoglycans. The technique has achieved almost universal acceptance as a technique for the purification of proteoglycans from complex mixtures of biological molecules, since it has the following advantages over many other procedures:

- it has a very high capacity
- it does not require gradient-formers
- proteoglycans may be purified in an aggregating (native) form
- aggregating proteoglycans may be subsequently fractionated into their components by the use of gradients run in dissociating conditions
- the separation is not dependent upon size, therefore small proteoglycans can be separated from high molecular weight contaminants which would be difficult using other techniques

---

**Protocol 4.** Associative density gradient centrifugation

This technique will separate proteoglycans in a native state from contaminating connective tissue components. It is most commonly used to separate the cartilage proteoglycan, aggrecan, in an aggregated form.

---

## Equipment and reagents

- Ultracentrifuge and suitable rotors
- Means of fractionating the gradients i.e. peristaltic pump in addition to a suitable displacement or gravity fractionation device
- Methanol/solid $CO_2$ freezing-bath and hacksaw
- Reagent A: 0.5 M guanidinium chloride (47.765 g/litre)/50 mM sodium acetate (4.1 g/litre of the anhydrous salt)/100 mM aminocaproic acid (13.12 g/litre)/10 mM EDTA (3.724 g/litre)/5 mM benzamidine.HCl (0.784 g/litre)/10 mM N-ethylmaleimide (1.252 g/litre)/1 μg/ml soybean trypsin inhibitor (1 mg/litre) pH 5.8. Just before use, pepstatin (172 mg) and PMSF (176 mg) were dissolved in methanol (1 ml) and added to 1 litre of solution to give a final concentration of 0.4 mM and 1 mM respectively. Aminocaproic acid, EDTA,

benzamidine, N-ethylmaleimide, soybean trypsin inhibitor, pepstatin, and PMSF are included as protease inhibitors.
- Reagent B: as reagent A but containing 4 M of guanidinium chloride (382.12 g/litre)
- Reagent C: 8 M guanidinium chloride (764.24 g/litre)/50 mM sodium acetate (4.1 g/litre of the anhydrous salt) containing the same protease inhibitors as in reagent A
- Reagent D: 500 mM sodium acetate (41 g/litre of the anhydrous salt) containing the same protease inhibitors as in reagent A
- Reagent E: 50% sucrose (500 g/litre)/0.2 M $CaCl_2$ (29.4 g/litre)/1 mM Tris–HCl (0.12 g/litre) pH 7.5
- Reagent F: as reagent E but without sucrose

## Method

1. The connective tissue extract must be dialysed against reagent A to allow reaggregation of the components. Alternatively, other physiological solutions may be used but practically there are certain advantages in working with guanidinium solutions, especially if link protein is to be examined at a later date.

2. The connective tissue extract should ideally contain less than 6 mg/ml of uronic acid for optimum separation. The system will tolerate higher levels of uronic acid if required but this may result in loss of proteoglycan into fractions of lower buoyant density than would be expected and will result in a decreased overall yield of material.

3. Following dialysis, adjust the proteoglycan solution to the appropriate density by the addition of solid CsCl. For large proteoglycans such as aggrecan a suitable starting density is 1.5 g/ml. This density can be achieved by adding 3.4 g of solid CsCl to every 4 ml of proteoglycan solution. The final volume will be 125% of the starting volume.

4. Centrifuge the solution at 100 000 g for 48 h at 5–10 °C.

The centrifugation time is directly determined by path length, as a result, vertical rotors (which have a very short path length) reach equilibrium more quickly than fixed angle or swing-out rotors of comparable volume. As a general rule, analytical vertical rotors (8 or 12 × 15 ml or similar) need to be centrifuged for 8–16 hours. Preparative rotors (8 × 35 ml or similar) need to be run for about 16 hours or more. The disadvantage of vertical rotors is that gradients are less steep than fixed angle rotors because of the short path through which the gradient can form, hence resolution is reduced. The major advantage of vertical rotors is, however, in the bulk preparation of proteoglycans because of their large volume and rapid separation. One would generally however choose a fixed angle or swing-out rotor for finer more analytical work, or when the volumes of proteoglycan solutions were relatively small (5–100 ml). *Note*: The initial density ($\varrho_i$) used in this protocol is lower than stated by many other authors. This protocol has the following advantages:

- it ensures that CsCl does not precipitate at the bottom of the gradient
- it serves to compact proteoglycans of the aggrecan type at the densest part of the gradient

Other proteoglycans, especially those that are less substituted with glycosaminoglycan chains have a lower buoyant density than aggrecan and hence would require centrifugation at a lower starting density. It is not possible to give a 'rule of thumb' by which the starting density required for preparation of various proteoglycans may be derived, however generally, the lower the glycosaminoglycan substitution the lower the required starting density.

Following centrifugation the gradients require fractionation to obtain the proteoglycan sample. There are four basic methods by which gradients may be fractionated:

- by upward displacement
- by piercing and draining from the bottom of the tube
- by aspiration from the top of the gradient
- freezing and sawing the gradient into equal fractions by volume

As a general rule, when preparing bulk amounts of proteoglycans, upward displacement and downward draining of the tubes is not particularly satisfactory, since the proteoglycan forms a dense gel which cannot be easily pumped or drained. Aspiration of fractions from the top of the gradient is an adequate method of fractionation in this case. Freezing of the gradients rapidly in a methanol–solid $CO_2$ bath allows them to be sawn and hence fractionated. This method is perhaps the quickest and easiest procedure for preparing large amounts of proteoglycans. When fractionating analytical amounts of proteoglycans, any of the methods mentioned are suitable and selection of any of the procedures is largely a matter of personal choice and often dictated by the availability of appropriate equipment.

---

**Protocol 5.** Dissociative density gradient centrifugation

This method has been particularly useful in the past in the study of the aggregating structure of aggrecan, allowing the separation and purification of the aggregate components.

See *Protocol 4* for equipment and reagents.

1. Following associative density gradient centrifugation, mix the A1 fraction (see *Figures 2* and *3*) with an equal volume of reagent C.

2. Add CsCl to the resulting solution according to *Table 1*. As previously stated, it is not possible to give precise values for the starting densities required for every proteoglycan, although in this case it is probably unlikely that one need to use starting densities outside of the range of 1.5–1.6 g/ml.

3. Centrifuge the solution at 100 000 *g* for 48 h at 5–10 °C (except for vertical rotors when one should use significantly shorter run times as described previously).

4. Following centrifugation, the gradients may be fractionated as described above.

---

Proteoglycan distribution on associative and subsequent dissociative density gradients are shown in *Figure 3*.

---

**Table 1.** Preparation of dissociative gradients

| Density of starting material (g/ml) (A) | Required density (g/ml) (B) | | | | |
|---|---|---|---|---|---|
| | 1.35 | 1.40 | 1.45 | 1.50 | 1.55 |
| 1.35 | 0.153 | 0.238 | 0.323 | 0.408 | 0.493 |
| 1.40 | 0.100 | 0.187 | 0.274 | 0.361 | 0.448 |
| 1.45 | 0.047 | 0.136 | 0.225 | 0.314 | 0.403 |
| 1.50 | — | 0.085 | 0.176 | 0.267 | 0.358 |
| 1.55 | — | — | 0.127 | 0.220 | 0.313 |
| 1.60 | — | — | 0.078 | 0.173 | 0.268 |
| 1.65 | — | — | 0.029 | 0.126 | 0.223 |
| 1.70 | — | — | — | 0.079 | 0.178 |

Column (A) refers to the density of the A1 fraction which is to be subsequently centrifuged under dissociative conditions. Columns (B) refer to the required starting density for the dissociative density gradient run. The figures in the table refer to the amount of CsCl (in g/ml of solution) to be added to the A1 fraction (after addition of an equal volume of 8 M guanidine.HCl to it) to achieve the required starting density for the dissociative gradient.
   *Note*: The addition of 0.5 g CsCl/ml of solution produced an approximately 10% increase in final volume.

---

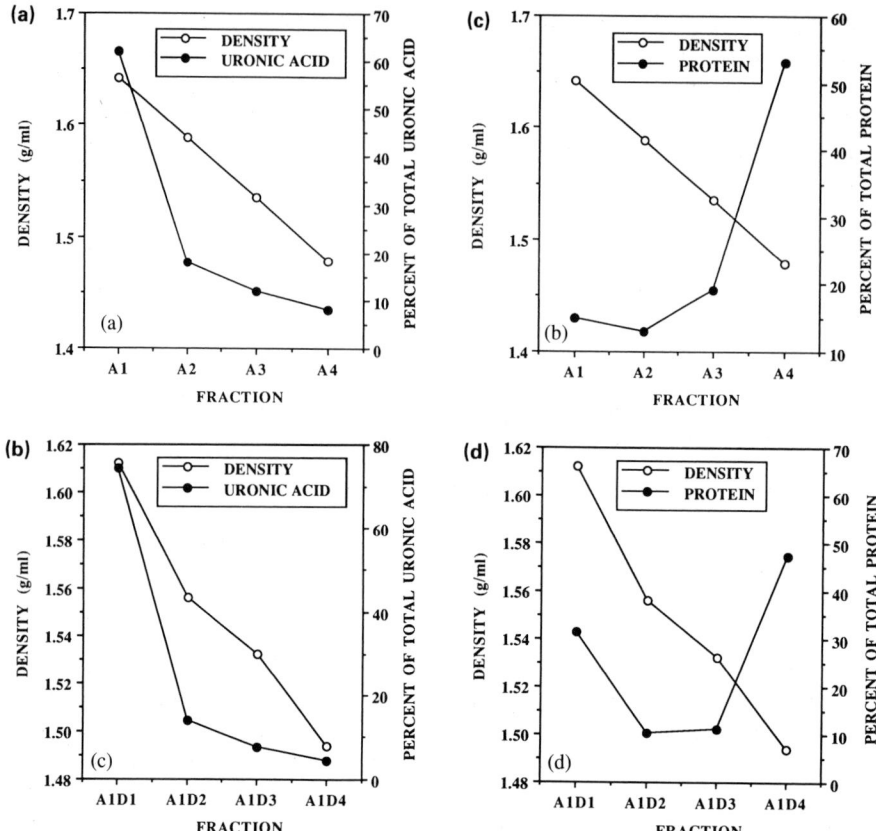

**Figure 3.** (a) The distribution of uronic acid on an associative CsCl density gradient. The starting density of the gradient was 1.53 g/ml. The graph shows the accumulation of uronic acid components in the densest (A1) fraction of the gradient. (c) The distribution of protein in the gradient shown in *Figure 3* (a). In this case it can be seen that the bulk of the protein accumulates in the least dense fraction of the gradient (A4). (b) The A1 fraction from the associative gradient centrifuged under dissociative conditions. The densest fraction (A1D1) contains the bulk of the uronic acid components. (d) The same gradient as in (b) but showing the distribution of protein components. It can be seen that a significant amount of protein is found in the A1D1 fraction (this is due to the protein core of the proteoglycans) and in the A1D4 fraction. This is largely due to link protein from the aggregate structure, but there will also be some contaminating protein that was present in the aggregate structure. The starting density of the gradient was 1.567 g/ml. The material centrifuged was a 4M guanidine extract of beagle articular cartilage.

For proteoglycans which do not participate in aggregation with hyaluronic acid one may perform direct dissociative density gradient centrifugation to effect purification. This technique is also useful should one wish to prepare monomeric proteoglycans direct from extracts of tissues containing aggregating proteoglycans. Add solid CsCl to the dissociatively extracted proteoglycan solution, or alternatively one may dialyse associative extracted material against reagent B before the addition of CsCl. Again, a starting density of about 1.5 g/ml is generally adequate for large proteoglycans that are highly substituted with glycosaminoglycans. This value may be adjusted to as low at 1.35 g/ml when the proteoglycan to be purified has a particularly low buoyant density.

---

**Protocol 6.** Direct dissociative density gradient centrifugation

See *Protocol 4* for equipment and reagents.

1. To achieve a starting density of 1.5 g/ml add 6.52 g of CsCl to 10 ml of dissociative extract which will give a final volume of 11.7 ml.

2. Centrifuge the solution at 100 000 *g* for 48 h at 5–10 °C.

3. Following centrifugation the gradients may be fractionated as described previously.

---

The buoyant density of cartilage proteoglycans in CsCl is so high that it exceeds the maximum density that it is possible to obtain for CsCl in aqueous solution. As a result, the bulk of proteoglycans sediment at the bottom of such CsCl gradients. This is useful for the bulk preparation of proteoglycans, but prevents examination of the polydispersity of such preparations as a result of their varying buoyant density. Several reports have shown that proteoglycans have considerably reduced buoyant density in $Cs_2SO_4$ solutions. It would appear that such changes in buoyant density are due to reductions in the hydration of the proteoglycan. Reductions in buoyant density can be observed in CsCl gradients by increasing the guanidine.HCl concentration due to a cation effect of the $Cl^-$ species. A study of proteoglycan polydispersity using $Cs_2SO_4$ isopycnic centrifugation has been carried out by Bonnet *et al.* (6) from which this method has been taken.

---

**Protocol 7.** Isopycnic ultracentrifugation in $Cs_2SO_4$

See *Protocol 4* for equipment and reagents

1. Extract the proteoglycans by dissociative or associative extraction (*Protocols 4* and *5*) then prepare the required proteoglycan fraction by CsCl density gradient centrifugation.

---

**Protocol 7.** *Continued*

2. Dialyse the proteoglycan preparation into reagent D. The concentration of proteoglycan for these purposes should be about 0.5 mg/ml.

3. Adjust the density of the solution to 1.53 g/ml by the addition of solid $Cs_2SO_4$ (0.635 g/ml).

4. Centrifuge the solution at 100 000 *g* for 48 h at 10–20 °C.

5. Following centrifugation, fractionate the gradients as described (in the section following *Protocol 4*). A typical gradient profile is shown in *Figure 4*.

---

As with all such procedures, conditions for centrifugation and starting density will vary from proteoglycan to proteoglycan. The above set of procedures was derived for bovine nasal cartilage proteoglycan. The conditions for other proteoglycans will have to be determined empirically. The effect of guanidine.HCl on this separation is highly complex and if one wishes to perform such centrifugations under dissociative conditions, it is strongly advised that one reads the paper of Bonnet *et al.* (6) before commencing.

The technique of rate zonal centrifugation in sucrose gradients (7) relies upon the principle that larger macromolecules sediment at faster rates than smaller ones in centrifugal fields. In this respect it is similar to sedimentation velocity centrifugation which was outlined in the first edition of this work. It

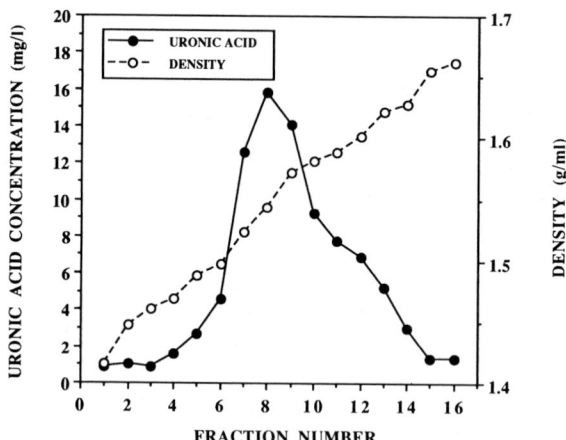

**Figure 4.** A $Cs_2SO_4$ density gradient of bovine nasal cartilage aggregating proteoglycans centrifuged as outlined in *Protocol 7*.

is, however far more akin to the moving band sedimentation velocity technique performed in Vinograd cells, where the sample is applied as a narrow band and one detects the passage of solutes through the cell as distinct zones. In this particular protocol, proteoglycans are centrifuged through a linear sucrose gradient from 10% to 50% (w/v) at high speed. Under these conditions, proteoglycan aggregates can be separated from proteoglycan monomers which migrate far more slowly. In contrast with the other centrifugation methods used for the separation of proteoglycans, this technique does not attain an equilibrium position, but instead the bands continue to migrate toward the bottom of the gradient under the influence of the centrifugal field.

---

**Protocol 8.** Rate zonal centrifugation in sucrose density gradients

See *Protocol 4* for equipment and reagents.

1. To prepare the gradients mix reagents E and F in the following proportions: for 50% sucrose mix 100% E and 0% F, for 40% sucrose mix 80% E and 20% F, for 30% sucrose mix 60% E and 40% F, for 20% sucrose mix 40% E and 60% F, and finally for 10% sucrose mix 20% E and 80% F.

2. Pipette into the bottom of centrifuge tubes of a suitable swing-out rotor (in this case tubes for the 8 × 14 ml MSE rotor) 2.6 ml of the 50% solution.

3. Carefully layer on this a further 2.6 ml of the 40% solution and repeat this for the 30%, 20%, and 10% solutions.

4. Allow these stepped gradients to diffuse together for about 2 h at 4 °C. The resultant gradient will be continuous and linear.

5. The proteoglycan solution(s) to be centrifuged should be dissolved in reagent F to a concentration of about 5 mg/ml.

6. Layer the sample (1 ml) on to the surface of the gradient and centrifuge at 200 000 *g* for 55 min at 20 °C.

7. Following centrifugation the samples may be recovered by any of the basic fractionation methods mentioned above.

Sample results for the separations are shown in *Figure 5*.
*Note*: When using different rotors, it may be necessary to modify the centrifugation conditions. This must be done empirically and depends on the proteoglycan being centrifuged and the dimensions of the rotor.

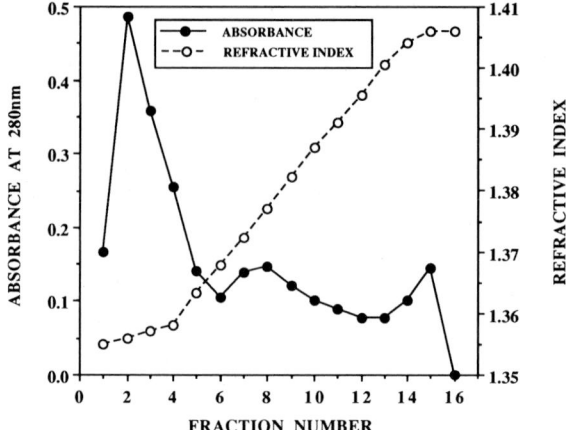

**Figure 5.** Rate zonal centrifugation of beagle articular cartilage proteoglycans in sucrose gradients according to *Protocol 8*. This proteoglycan preparation contained approximately 25% aggregating proteoglycan as assessed by gel filtration chromatography. The peak at the least dense part of the gradient contains the non-aggregating proteoglycans.

## 4.2 Proteoglycan preparation by chromatographic techniques

### 4.2.1 Proteoglycan purification by ion-exchange chromatography

As a result of the high fixed charge conferred on proteoglycans by the carboxyl and sulfate ester groups of the substituent glycosaminoglycans, these molecules bind tightly to anion exchangers such as DEAE (diethylaminoethyl) cellulose. This property can be used to separate such molecules from proteins which generally are considerably less negatively charged. An ion-exchange technique for proteoglycan preparation using DEAE–cellulose as the anion exchanger was developed by Antonopoulos *et al.* (18). In this resin the charged group is the quaternary ammonium of the diethylaminoethyl $[-OCH_2CH_2NH^+(CH_2CH_3)_2]$ substituent on the cellulose matrix. The following procedure is performed in the presence of 7 M urea for two reasons: first, it maintains the proteoglycans in a disaggregated form (in the case of the aggregating proteoglycans) since aggregates do not bind particularly well. Secondly, it prevents the formation of colloidal suspensions which can occur on dialysis to associative conditions.

The urea used in this procedure should be as free from ionic contaminants as possible since at concentrations of 7 M these may interfere with the binding of molecules to the exchanger and affect their subsequent elution. Either one may use ultrapure urea (which is expensive) or alternatively one may remove the ionic contaminants (largely cyanates) by passing through a column of mixed bed ion-exchange resin. The urea may then be freeze-dried and used as required.

Ion-exchange chromatography is generally performed in relatively short, broad columns in contrast with gel permeation chromatograpy (see Section 4.2.2). The bed supports should be selected such that with continued use they will not become clogged. Hence, glass wool and coarse glass sinters are not recommended for long-term use. Generally one need not use columns of greater than 20 cm in length, although column dimensions are dictated by the amount of applied material. One may load as much as 2 mg of proteoglycan per millilitre of DEAE–cellulose. Unlike gel permeation chromatography, separation in ion exchange is dictated not by partition between stationary and mobile phases, but by the ionic strength and/or pH of the mobile phase. Hence ion-exchange separations can be performed at much higher flow rates than one would use for gel permeation. Since maximum flow rate in a column is highly dependant upon cross-sectional area, one would tend to select columns that are as short and wide as practical for ion-exchange separations.

---

**Protocol 9.** Ion-exchange chromatographic purification of proteoglycans

*Equipment and reagents*

- Chromatography column
- Sintered glass funnel
- Source of vacuum
- Large side-arm conical flask
- Peristaltic pump
- Fraction collector
- Gradient former
- On-line UV monitor (optional)

- Reagent A: 7 M urea (420 g/litre)/50 mM Tris–HCl (6.06 g/litre) adjusted to pH 6.5 with HCl
- Reagent B: as reagent A but containing 0.15 M NaCl (8.76 g/litre) pH 6.5
- Reagent C: as reagent A but containing 2 M NaCl (108.88 g/litre) pH 6.5
- Reagent D: 0.5 M HCl (43 ml/litre)

A. *Preparation of chromatographic resin*

1. Add DEAE–cellulose to distilled water (10 g/litre) and allow to swell. This takes about 16–24 h at room temperature.

2. Transfer the slurry to a sintered glass funnel and wash with 50–100 vol. of distilled water, pulling the washings through the resin by vacuum.

3. Resuspend the resin cake in about 10–20 vol. of reagent D in a large measuring cylinder, and allow the suspension to settle for 1–2 h at room temperature.

4. Carefully decant off the supernatant (this removes finings from the exchanger resin) and wash the resin extensively with distilled water until the pH of the washings is constant and approximately equal to the distilled water.

5. Resuspend the resin cake in reagent A (about 10–20 vol.) and allow to stand in a measuring cylinder for 1–2 h at room temperature.

**Protocol 9.** *Continued*

**6.** Decant off the supernatant (which will remove any remaining finings) and dilute the packed slurry 1:1 (v/v) with reagent A.

B. *Chromatographic column packing*

**1.** Degas the DEAE–cellulose slurry under vacuum for about 20 min.

**2.** Whilst preparing the resin, mount the column vertically and flush the afferent tubing with buffer to release any trapped air, then clamp and close the outlet.

**3.** Carefully pour the exchanger slurry into the column down the side of a glass rod, taking care to avoid air bubbles.

**4.** Open the column outlet and pump the eluent at a rate twice that to be used for the separation (up to 40 ml/cm$^2$/h).

**5.** When all of the exchanger has been packed into the column, reduce the flow rate to that to be used for the separation, overlay with 2–3 cm of reagent A, then connect to a reservoir containing reagent A. The column inlets and outlets may now be adjusted such that the operating pressure is about 1 cm per cm height of exchanger (i.e. the difference in height between the inlet from the reservoir and the outlet from the column is the same as the height of exchanger).

**6.** When the pH and conductivity of the buffer in the reservoir and the column eluent are equal, the exchanger is correctly equilibrated and ready for use. At this stage a UV monitor may be put in-line with the column eluent if available (or required).

C. *Impure proteoglycan sample preparation*

**1.** Guanidine.HCl extracts of connective tissue must be adjusted to 7 M urea/50 mM Tris–HCl pH 7.5 before application to the column. This is most conveniently achieved by dialysis of the sample against this buffer (with several changes).

**2.** Alternatively, one may exchange the buffer by desalting on columns of Sephadex G-25 equilibrated and eluted with reagent A. Although this method is efficient, it suffers in that the sample is diluted, however this is not normally a problem in ion-exchange separations since the sample is normally concentrated by binding to the ion-exchange resin.

There are three basic ways by which a sample may be applied to the column: by the use of end-piece adaptors, by application of dense samples under the buffer, and by application on to a drained bed surface. In practice the most common and most convenient method is to apply to a drained gel bed. To achieve this, one must disconnect the buffer inlet and allow the buffer to elute from the column. When the buffer level is just at the level of the top of the exchanger, stop the pump, and gently and evenly layer the sample on to the surface of the gel. Reconnect the pump and start the fraction collector, then allow the sample to drain in to the exchanger. Replace the buffer on top of the surface (again, about 2–3 cm) and reconnect the column to the buffer reservoir.

There are two forms of elution which can be used for the separation of proteoglycans from other connective tissues on ion-exchange resins. First, one may use stepwise elution. This particular stepwise elution (*Protocol 10*) is useful for the large scale preparation of proteoglycans, however should one be interested in the separation of various proteoglycan species or if it is not clear how avidly a particular proteoglycan binds to DEAE, it is often preferable to use an alternate form of column elution, that of gradient elution.

---

**Protocol 10.** Stepwise elution of chromatographic column

See *Protocol 9* for equipment and reagents.

1. After sample application, the column is eluted with reagent A until those proteins which do not bind to the column (or those which bind only weakly) have eluted from the column. Normally two or three bed volumes is sufficient for this purpose, although generally one may apply around five volumes just to be sure.

2. Following this initial elution step, elute the column with reagent B (up to five column volumes). This elutes the non-proteoglycan connective tissue proteins from the column.

3. Finally the column is eluted with reagent C (up to five column volumes) which elutes the proteoglycan.

---

**Protocol 11.** Gradient elution of chromatographic column

See *Protocol 9* for equipment and reagents.

1. After sample application the column is eluted with up to five volumes of reagent A as in *Protocol 10*.

2. Apply a linear gradient from 0–2 M NaCl to the column.

**Protocol 11.** *Continued*

3. The linear gradient is prepared in the following way: using a suitable gradient former, pour 10–25 bed volumes of reagent C into the compartment for the densest component (i.e. that compartment most distant from the outlet). Into the other compartment pour an equal volume of reagent A. Using a magnetic stirrer, ensure the compartment containing reagent A is well mixed. Connect the gradient former to the column then open the connection between the compartments containing reagents A and C.

4. Continue to elute the column at the normal rate until the gradient former is empty, at which time one should continue to elute the column with reagent C (up to five bed volumes) to ensure that all material has been eluted from the column. Should an on-line UV monitor be used, the elution position of the proteoglycans will be clear, if not the fractions collected will have to be assayed by one of the methods outlined in Section 8.

Although this section has described ion-exchange chromatography of proteoglycans using DEAE–cellulose as the anion exchanger, there are now many alternative support material available for such preparations. Of particular interest are DEAE–Sephacel, a microcrystalline bead-formed exchanger whose support matrix is cellulose. As a result of its bead structure, if offers greater resolution than existing DEAE–cellulose exchangers. This may be of importance if one wishes to separate proteoglycans having only narrowly differing elution characteristics on DEAE–cellulose. Furthermore, a DEAE exchanger using a gel permeation matrix as support (DEAE–Sepharose CL-6B) has been developed which has better flow characteristics than any of the ion exchangers previously mentioned. Moreover, such a matrix affords a certain element of gel permeation to the separation, which may be of use when preparing low molecular weight proteoglycans of a size which can be fractionated by Sepharose 6B.

Although this method has concentrated on the separation of intact proteoglycans, it is possible with only minor alteration to prepare chondroitin sulfates by a similar technique. One applies and elutes the glycosaminoglycan in 0.2 M NaCl/50 mM Tris–HCl pH 7.4 essentially as described in *Protocol 10* (it is not necessary to use urea when chromatographing glycosaminoglycan chains). The column may then be eluted with 2 M NaCl/50 mM Tris–HCl pH 7.4 which will elute the glycosaminoglycan chains. Should one wish to separate various glycosaminoglycan chains, one can adapt this method for gradient elution essentially as described previously. For comparison it is of interest to examine the separation of chondroitin sulfate disaccharides in Section 5.3 by the use of HPAE (high performance anion exchange).

## 4.2.2 Proteoglycan purification and analysis by gel permeation chromatography

Although in the previous edition of this work, gel permeation chromatography was placed in the category of analytical techniques, it is perhaps more appropriate to include it in a section on preparation. In many respects most of the gel permeation carried out on proteoglycans is of a semi-preparative or preparative scale.

Gel permeation chromatography of proteoglycans has become the method of choice for the routine laboratory analysis and preparation based upon hydrodynamic size. Such separations may be achieved on large-pore matrices based on either agarose or dextran gels or alternatively on controlled pore glass supports. Although all these matrices have been successfully used to fractionate proteoglycans, the agarose-based gels have become most widely accepted for this purpose. The most commonly used agarose gels are the Sepharoses (produced by Pharmacia, Uppsala, Sweden) and will be described in the following section.

Gel permeation should be performed in long narrow columns, since the resolution of macromolecules increases as a function of the square root of the column length. Excessively broad columns should be avoided since they produce an unwanted dilution of sample and band broadening. Very narrow columns however, are hard to pack and resolution of components may be affected by cohesive wall effects. As a general rule, however, the length of the column should be dictated by the optimal resolution required for the separation, whereas the diameter should be selected according to the quantity of sample to be chromatographed. The outlet deadspace should be as small as possible to prevent the remixing of resolved analytes. A column of dimensions 100 cm × 0.8–1.5 cm internal diameter would be sufficient to resolve up to 400 μg of proteoglycan (expressed as its uronic acid equivalent).

---

**Protocol 12.** Gel permeation chromatographic purification of proteoglycans

*Equipment and reagents*

- Chromatography column
- Peristaltic pump
- Fraction collector
- UV on-line monitor (optional)
- Reagent A: associative buffer—0.5 M sodium acetate (41 g/litre) adjusted to pH 6.8 with acetic acid

- Reagent B: dissociative buffer—2 M guanidine.HCl (190 g/litre)/0.5 M sodium acetate (41 g/litre) adjusted to pH 6.8 with acetic acid
  Before use, filter all buffers through a Millipore filter (0.45 μm, and degas under vacuum with gentle heating for about 20 min.

### A. *Preparation of chromatographic gel matrices*

1. Transfer the Sepharose slurry to a sintered glass funnel and wash with 10–20 vol. of reagent A, which serves to equilibrate the gel and remove

---

**Protocol 12.** *Continued*

merthiolate present in the gel as a preservative. Sepharose is supplied preswolen so no swelling of the matrix is required.

2. Suspend the gel cake in 10 vol. of reagent A, and allow to settle in a suitably sized measuring cylinder.

3. Carefully decant off the supernatant to remove finings should they be present.

4. Resuspend the slurry in reagent A in the ratio of 1 vol. of slurry to 1 vol. of reagent A.

5. Degas the slurry under vacuum with very gentle warming for about 20 min.

*Note*: The gel matrix prepared as described above has been equilibrated for elution under associative conditions. To prepare the gel for elution under dissociative conditions, all equilibration steps should be performed with reagent B instead of reagent A.

B. *Chromatographic column packing*

1. Fill the column with degassed buffer (reagent A or B depending on whether the column is to be eluted under associative or dissociative conditions).

2. Flush any air from the column effluent end-pieces and clamp the column closed.

3. Open the column outlet and add the gel slurry to the column, preferably in one step if the column has an adaptor which allows this, or alternatively add the slurry periodically as the level of the buffer drops during packing.

4. The peristaltic pump should now be switched on and set to approximately twice the speed that will be used for chromatography.

5. When the gel has completely packed down, connect the top of the column to a suitable reservoir and reduce the flow rate to that to be used for separation (about 10–15 ml/cm$^2$/h is suitable for Sepharoses), and if required, plumb in the on-line UV monitor.

6. Allow two bed volumes of buffer to elute before use. It is advisable to 'precondition' the column (i.e. to prevent subsequent non-specific binding of proteoglycans) by applying a proteoglycan sample, about 200–300 μg (as uronic acid) before the application of more precious samples.

The application of samples to gel permeation columns is performed exactly as stated for ion-exchange chromatography in Section 4.2.1. Concomitant with sample application, fraction collection should be commenced.

Sepharose 2B (or CL-2B) is an agarose-based gel permeation matrix having an approximate agarose concentration of 2%. It has a very high exclusion limit, such that monomeric aggrecan will be included on columns packed with this material. Proteoglycans interacting with hyaluronic acid (e.g. aggrecan) will be sufficiently large to be excluded from the column. Thus, chromatography on Sepharose 2B can be used to determine the proportion of proteoglycans which can interact with hyaluronic acid. Eluting proteoglycans under dissociative conditions on these columns can be used to estimate the hydrodynamic size of such preparations. Typical elution profiles of aggrecan eluted under associative and dissociative conditions are given in *Figure 6*.

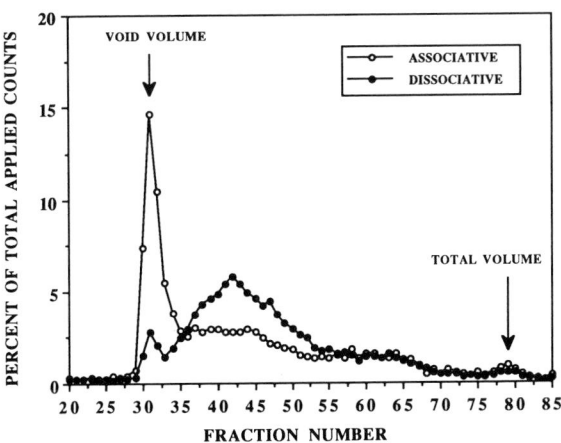

**Figure 6.** Gel permeation chromatography of biosynthetically labelled proteoglycans from 4 M guanidine extracts of guinea-pig articular cartilage. The figure shows the same sample eluted under associative and dissociative conditions as outlined in *Protocol 12*A and B.

Although aggrecan is included on columns of Sepharose 2B eluted with dissociative buffers, they are completely excluded from columns of Sepharose-4B under these conditions. Such columns are useful however in the examination of smaller proteoglycans and may also be used to examine the breakdown products of larger proteoglycans such as aggrecan. Sepharose 6B has an agarose concentration of about 6% and can be used to fractionate proteoglycans too small to be effectively included on either Sepharose 2B or 4B. This matrix is, however, far more commonly used to examine the size distribution of the proteoglycan constituent glycosaminoglycans following their liberation from the core protein by proteolysis or alkaline β-elimination.

When comparing relative proteoglycan sizes, it is common to refer to their '$K_{av}$' on a particular gel matrix. This term is a distribution coefficient and is related to the proportion of the column matrix that is *a*vailable to a particular analyte. The $K_{av}$ may be derived from the following expression:

$$K_{av} = (V_e - V_o)/(V_t - V_o)$$

where $V_e$ is the elution volume of the solute in question, $V_o$ is the void volume of the column (i.e. that volume in which very large molecules which are excluded from the column will elute), and $V_t$ is the total volume of the column (i.e. that volume in which very small molecules which are completely included on the gel will elute).

# 5. Proteoglycan analysis

## 5.1 Analysis of proteoglycans by gel electrophoresis

### 5.1.1 Polyacrylamide–agarose gel electrophoresis of proteoglycans (9, 10)

The efficiency of resolution of proteoglycan species by standard biochemical techniques is impaired due to the high degree of polydispersity of such molecules. The electrophoresis of proteoglycans on polyacrylamide–agarose gels is capable of resolving proteoglycans (especially aggrecan) into a number of discrete bands. The separation is dependent upon factors such as charge density, molecular size, and electroendosmosis, but the precise details of why various species resolve is not altogether clear.

The large hydrodynamic size of proteoglycans generally prevents their penetration into polyacrylamide gels. At concentrations below 2%, acrylamide will not form mechanically stable gels; however, even at this low concentration, the pore size is too small to allow proteoglycans to enter the gel. To increase the pore size, yet maintain rigidity, it is necessary to include agarose at a concentration of about 0.5%. Empirically, it has been determined that optimum proteoglycan resolution is achieved at acrylamide concentrations of 1.2% (w/v) and agarose concentrations of 0.6% (w/v). This electrophoretic technique is useful for the qualitative examination of proteoglycans from different tissues and species, or alterations in proteoglycans produced by ageing or disease (11).

---

**Protocol 13.** Polyacrylamide–agarose gel electrophoretic separation of proteoglycans

*Equipment and reagents*

- Vertical electrophoresis tank
- Vertical gel casting apparatus
- Power pack
- Heating oven
- Magnetic stirrer/hotplate
- Cold room or refrigerator
- Gel drier
- Staining trays
- Reagent A: 40 mM Tris (4.84 g/litre)/1 mM sodium sulfate (0.142 g/litre) adjusted to pH 6.8 (at 4 °C) with acetic acid
- Reagent B: $\beta$-dimethylaminopropionitrile (640 μl in 10 ml distilled water)

- Reagent C: ammonium persulfate (300 mg in 10 ml of distilled water)—this reagent must be prepared freshly
- Reagent D: acrylamide (291 g/litre)/bis-acrylamide (9 g/litre)—this reagent is stable for two or three weeks if kept cool and protected from light
- Reagent E: *n*-butanol saturated with distilled water
- Reagent F: dilute reagent A fourfold and include sucrose (200 g/litre) and bromo-phenol blue (0.01 g/litre)

In this section, all quantities of gel etc., have been calculated for gel plates for casting gels of the following approximate dimensions: 14 cm × 14 cm × 0.2 cm.

A. *Gel casting*

1. Wash the glass gel plates thoroughly, dry, and assemble in the casting apparatus. Mix reagent D (10 ml) with distilled water (18.47 ml).

2. Add TEMED (30 μl) and reagent C (1.5 ml).

3. Ensure complete mixing by swirling in a conical flask, then transfer to the assembled plates to a depth of about 30% of the total plate length. Overlay this acrylamide plug with reagent E.

4. After polymerization is complete (around 30 min) drain off reagent E and wash the gel surface with distilled water. Blot the plates dry, taking care not to touch the gel surface.

5. Whilst the polyacrylamide–agarose gel is being prepared, transfer the gel plates containing the polyacrylamide plug to an oven at about 60 °C to warm.

6. To prepare the separating (polyacrylamide–agarose) gel, mix agarose (0.24 g) with reagent A (22.5 ml) and heat with constant stirring to 100 °C using a magnetic heater/stirrer.

7. Allow the agarose to cool gradually with constant stirring. Whilst the agarose cools, dissolve acrylamide (0.456 g) and bisacrylamide (0.024 g) in reagent A (9.7 ml), then add reagent B (4.8 ml). This

---

**151**

**Protocol 13.** *Continued*

solution should be gently warmed by placing it on the hotplate where the agarose solution is cooling.

8. When the agarose solution has cooled to 52 °C, add reagent C (3 ml) to the acrylamide solution, mix well and add to the agarose solution, and continue mixing for a few seconds.

9. Remove the warmed gel plates from the oven and add the agarose–acrylamide mixture.

10. Insert the well former and transfer to the cold room or refrigerator. Overlay the surface with reagent E to exclude air. The well former must not be removed for at least 1 h, since polymerization in these gels occurs relatively slowly under these conditions.

B. *Sample preparation*

1. Dissolve the proteoglycan sample (which must be monomeric) in reagent F. To detect the proteoglycan by toluidine blue staining, the proteoglycan concentration should not be less than 1 mg/ml.

2. Apply about 5–10 µl to each well.

3. For detection by fluorography apply no less than 500–1000 d.p.m. (as [35]S) to each well.

If one wishes to electrophorese aggregated proteoglycans, the proteo-glycan sample must be dissolved in reagent F containing 7 M urea and left at 4 °C for 16 h. In addition the gel must be equilibrated in buffer A diluted fourfold and containing 4 M urea, by immersing the gel in a gel tank containing this buffer. The urea containing buffer is also used for electrophoresis. All other steps in electrophoresis are as stated in the following sections.

C. *Electrophoresis*

1. Drain all reagent E from the gel surface, then remove the well former, and place the gel into a vertical electrophoresis tank containing a fourfold dilution of reagent A.

2. Apply the proteoglycan sample under the buffer using a Hamilton syringe or an automatic pipette with a gel loading tip.

3. Apply a voltage of 60 V until the dye enters the gel, then increase the voltage to 120 V until the bromophenol blue tracking dye has migrated 3 cm. This generally takes around 2–3 h.

## 5.1.2 Detection of proteoglycans in polyacrylamide–agarose gels

Although proteoglycans may be stained by a number of dyes, toluidine blue staining has long been used for the treatment of agarose–polyacrylamide gels due to its quickness and ease. *Figure 7* shows a typical example of proteoglycans electrophoresed on such gels, stained with toluidine blue.

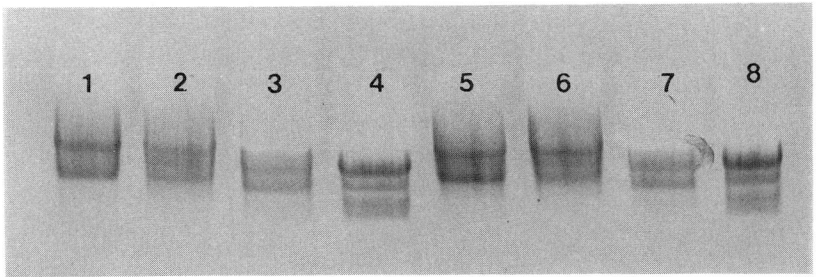

**Figure 7**. Polyacrylamide–agarose gel electrophoresis of proteoglycans from beagle articular cartilage. Tracks 1, 2, 5, and 6 are guanidine extracts of cartilage labelled *n vitro* with $^{35}SO_4$ and cultured for 48 hours. Tracks 3, 4, 7, and 8 are the proteoglycans released from these explants into the tissue culture medium during the culture period. Tracks 1, 3, 5, and 7 were from normal animals, tracks 2, 4, 6, and 8 were from animals that had had an experimental osteoarthritis for six months. Electrophoresis was as outlined in *Protocol 13* and the gel was stained with toluidine blue (*Protocol 14*A). One can see from tracks 3, 4, 7, and 8 that this methodology can be used to monitor degradation of proteoglycans and also to show differences in the patterns of breakdown in health and disease.

---

### Protocol 14. Detection of proteoglycans in polyacrylamide–agarose gels

#### Equipment and reagents

- See *Protocol 13*
- Reagent G: toluidine blue (2 g/litre) in 0.1 M acetic acid
- Reagent H: 3% acetic acid (30 ml/litre) in distilled water
- Reagent I: sodium acetate (1.23 g/litre) in absolute ethanol
- Reagent J: diphenyloxazole (4 g/litre) in reagent I
- Reagent K: phosphate-buffered saline (commercially available tablet preparations are adequate for this purpose)
- Reagent L: powdered milk (50 g/litre) in reagent K
- Reagent M: powdered milk (20 g/litre) in reagent K

- Reagent N: diaminobenzidine (50 mg) in reagent K (100 ml) containing 3 ml of cobalt chloride (10 g/litre)/ammonium nickel sulfate (10 g/litre)—take care with preparation and use of this reagent since it is thought that diaminobenzidine may be carcinogenic
- Reagent O: as reagent N but containing 100 μl of $H_2O_2$ (5 μl $H_2O_2$(60% v/v) in 300 μl distilled water) per 100 ml of reagent N
- Wet or semi-dry blotting apparatus
- Rocking platform
- Heat sealing device and appropriate plastic sheeting (a home freezer bag sealer is generally adequate)
- Staining trays

**Protocol 14.** *Continued*

A. *Detection by toluidine blue*

1. Immerse the gel slab in reagent G for 20 min.

2. Remove the dye and destain the gel in reagent H for 90 min with several changes of reagent.

3. Continue to destain with distilled water until clear. Proteoglycans appear as purple bands against a faint blue background.

B. *Detection by fluorography*

1. Fix the gel slab by immersion in reagent J (100 ml) for 16 h. This step may be performed on unstained gels or on gels that have been stained with toluidine blue as outlined in part A.

2. Remove reagent J and rinse the gel for 1 h in distilled water with several changes (or alternatively against running water). This step precipitates the diphenyloxazole within the gel and will make it appear white and opaque.

3. Dry the gel down on to Whatman's No. 1 filter paper or equivalent using a commercial gel drier.

4. In the dark-room, put the gel into contact with X-ray film that has been preflashed to an optical density of 0.2 A at 520 nm (see ref. 12 for full details). Ensure good contact between gel and film and clamp or staple in place if necessary.

5. Seal the gel–film sandwich into a light-tight box and store at −70 °C for the required length of time.
   As a rule of thumb, [$^{35}$S]proteoglycans containing 500 d.p.m. require four to five days development time.

6. Remove the film and develop according to the manufacturers instructions.

C. *Immunodetection following Western blotting*

1. Following electrophoresis, remove the gel and electrophoretically blot on to a piece of Nylon 66 membrane. As there are so many blotting apparati on the market it is impossible to give precise instructions as to how to perform this step, so one should be guided by the manufacturers instructions for use with standard proteins.

2. The electrophoretic transfer buffer should be a fourfold dilution of reagent A.

3. Following transfer, block any non-specific binding sites by incubation in reagent L for 1 h at 37 °C or overnight at 4 °C.

4. Remove the blocking buffer and incubate the membrane in reagent L

containing 0.025 U/ml of chondroitinase ABC for 1 h at 37 °C. Digestion by the enzyme chondroitinase ABC is not always necessary, and for certain antibodies, their epitope is destroyed by this treatment. However, in most cases enzyme digestion is required to prevent steric inhibition of antibody–antigen binding. Unless one is specifically indicated not to perform digestion, this step should always be included.

5. The membrane must now be washed six times, 5 min per wash with reagent K, then sealed into a plastic bag to which is added 10 ml reagent M containing antibody at a suitable dilution.

6. Incubate for 90 min at room temperature to allow complete antigen–antibody interaction. It is not possible to give precise details of the antibody dilution required, since they will vary for different antibodies and indeed from batch to batch. As a general rule, most of the antibodies used for determination of glycosaminoglycan structures need to be diluted in the order of 1:1000 to 1:10 000. Sealing the antibody solution into plastic bags allows one to incubate the blot in smaller volumes and hence save precious antibody.

7. Wash the membrane six times, 5 min each wash with reagent K. Following washing, incubate the membrane with reagent M containing an appropriate dilution of anti-Ig-horseradish peroxidase conjugate for 1 h at room temperature. This conjugate is widely commercially available and in our laboratory we use a preparation from Sigma which is a mixture of an anti-mouse IgG, an anti-mouse IgA, and an anti-mouse IgM which are conjugated to horseradish peroxidase. We use this conjugate at a dilution of 1:1000 in reagent M.

8. Wash the membrane six times, 5 min each wash in reagent K, then the proteoglycans may be detected by incubation for 15 min at room temperature with reagent N.

9. Reagent N must then be removed and discarded.

10. Without washing the membrane, incubate in reagent O until the bands have developed.

11. After complete development, thoroughly wash the membrane with distilled water, and dry in a stream of warm air.

An example of proteoglycans visualized by fluorography is given in *Figure 8*.

The recent availability of mono- and polyclonal antibodies against various components of aggrecan structure has been of great importance. It has shown that in disease states there are subtle yet significant changes in proteoglycan structure that could not have been determined by other means. The

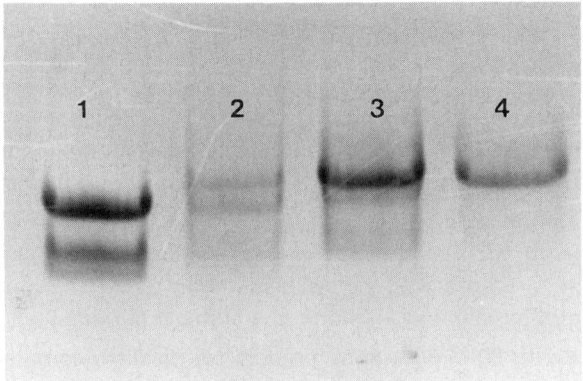

**Figure 8.** A fluorogram of $^{35}$S-labelled proteoglycans prepared as outlined in *Figure 7*. Tracks 3 and 4 are guanidine extracts from arthritic and normal cartilage respectively. Tracks 1 and 2 are proteoglycan breakdown products from the maintenance medium of arthritic and normal cartilage respectively. Again differences in breakdown patterns can be demonstrated by this technique. The fluorography was performed as outlined in *Protocol 14*B.

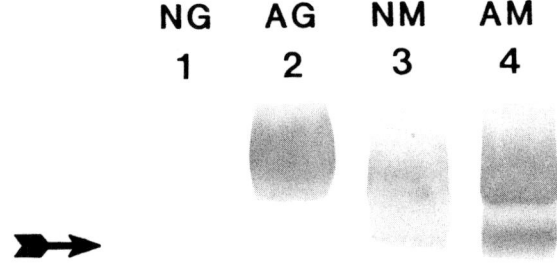

**Figure 9.** Immunodetection of proteoglycans using the monoclonal antibody 7-D-4 (kindly provided by Professor Bruce Caterson). The tracks marked 1–4 contain the same samples marked 4–1 respectively in *Figure 7*. As can be seen, the epitope recognized by 7-D-4 is only present to any great degree in samples derived from osteoarthritic tissue. This has produced a great deal of interest in that it may provide the potential means by which osteoarthritis may be monitored by biochemical means, which is not possible at present. Immunodetection was performed as outlined in *Protocol 14*C.

development of Western blotting techniques has allowed investigations into the breakdown of proteoglycans which would have been difficult, if not impossible by other means. An example of an immunolocalized Western blot is given in *Figure 9*.

## 5.2 Analysis of glycosaminoglycans by gel electrophoresis

### 5.2.1 Analytical vertical slab gel electrophoresis of glycosaminoglycans

It has become clear in the past few years that the sulfation of chondroitin is not a random event and that sequence structure may be contained within the chain. This has been shown to be the case for the heparin pentasaccharide that is responsible for anti-thrombin binding. Antibodies produced by Bruce Caterson have been used to show the appearance of specific chondroitin sulfate substructures that can be expressed by cells during differentiation and in disease. Analysis of the oligosaccharides produced by digestion of glycosaminoglycans by hydrolases or eliminates may therefore be of importance in elucidating specific sequence structure within linear glycan chains. An example of glycosaminoglycan separations is given in *Figure 10*.

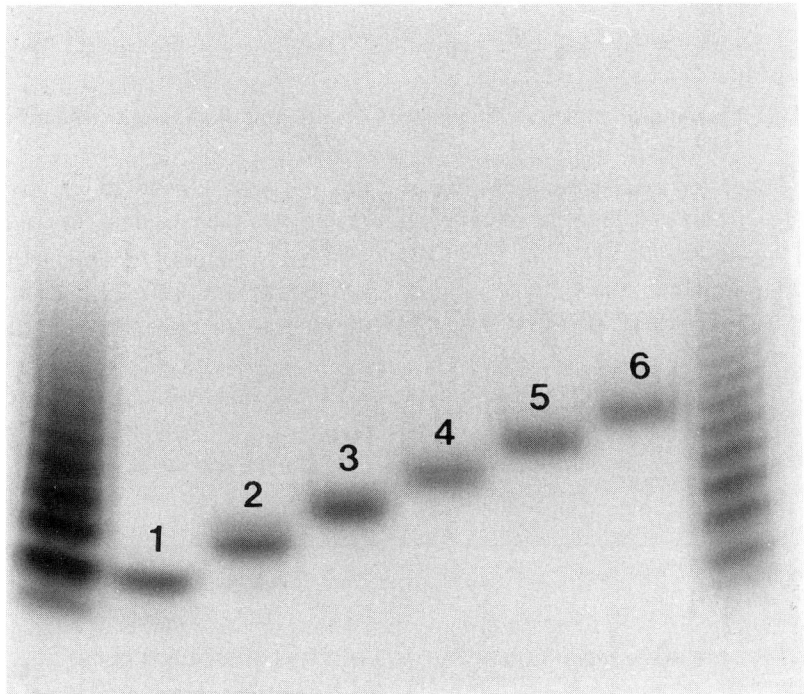

**Figure 10.** Polyacrylamide gel electrophoresis (20% acrylamide) of rat xiphysternal chondroitin sulfate oligosaccharides, depolymerized using bovine testicular hyaluronidase as outlined in *Protocol 20*. The tracks numbered 1–6 are aliquots of those fractions obtained from preparative gel electrophoresis in *Figure 13*. Hence track 1 contains an aliquot of pool 1 from the preparative column. This gel shows the relative purity of oligosaccharides prepared from only a single run on the Bio-Rad PrepGel. Electrophoresis was performed as outlined in *Protocol 15*. The two outer tracks (unlabelled) represent the unfractionated depolymerized chondroitin sulfate sample.

---

**Protocol 15.** Polyacrylamide gel electrophoresis of chondroitin sulfate oligosaccharides

*Equipment and reagents*

- Reagent A: 900 mM Tris (108.9 g/litre)/ 900 mM boric acid (55.65 g/litre)/24 mM EDTA (7.014 g/litre) pH 8.3
- Reagent B: acrylamide (291 g/litre)/ bisacrylamide (9 g/litre) in distilled water
- Reagent C: ammonium persulfate (150 mg in 10 ml distilled water)

- Reagent D: alcian blue 8GX (5 g/litre) in acetic acid (20 ml/litre)
- Reagent E: acetic acid (20 ml/litre)
- Vertical gel electrophoresis tank
- Electrophoresis casting apparatus and glass plates
- Staining trays

The method outlined is essentially that outlined by Cowman *et al.* (13). The following instructions are for the preparation of 20% (w/v) gels with 3% cross/linking and of dimensions 15 × 15 × 0.2 cm.

*Method*

1. Mix reagent A (3 ml) and reagent B (20 ml) in a suitable conical flask with side arm.

2. To this solution add distilled water (5.47 ml), and degas under vacuum for 10–20 min.

3. Following degassing, add reagent C (1.5 ml) and TEMED (30 μl), mix well, and pour into gel plates assembled in a suitable casting frame.

4. Samples were dissolved in a tenfold dilution of reagent A containing bromophenol blue and phenol red (both 0.001% w/v) as tracker dyes. Sucrose was added to a final concentration of 5% (w/v) to facilitate loading.

5. Run the gel at 13.33 V/cm until the phenol red tracking dye has migrated 12 cm.

6. Stain the gel for 45 min with reagent D, then destain with reagent E until clear.

## 5.2.2 Preparative gel electrophoresis of glycosaminoglycans

Optimization of the gel concentration is achieved by casting analytical gels (as in *Protocol 15*) and measuring migration parameters as in *Protocol 16*. Examples of gel optimization are given in *Figures 11* and *12*. An example of a preparative gel run is given in *Figure 13*.

**EFFECT OF ACRYLAMIDE CONCENTRATION ON THE MIGRATION OF SHARK CHONDROITIN SULPHATE OLIGOMERS**

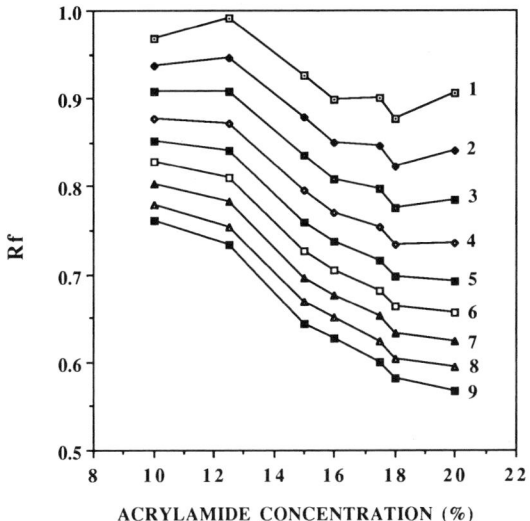

**Figure 11.** Electrophoresis of the unfractionated, depolymerized sample used in *Figure 10* on polyacrylamide gels of varying concentrations. The graph represents a plot of the $R_f$ of the various bands plotted against acrylamide concentration. The curves are numbered in order, where 1 would be the fastest migrating to 9 the slowest migrating oligosaccharide. Conditions were as stated in *Protocol 16*A.

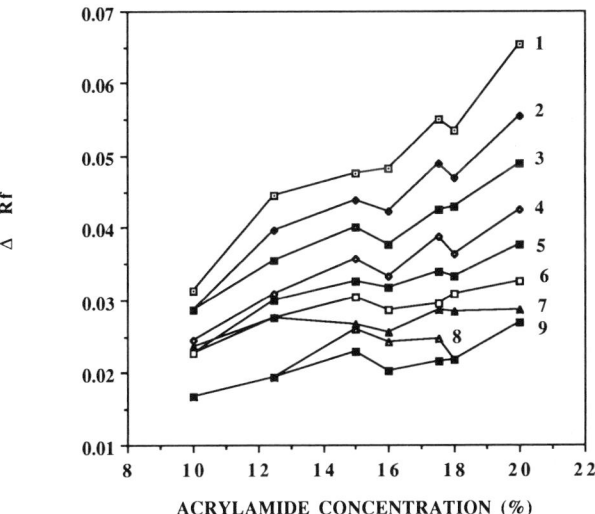

**Figure 12.** A plot of the difference in $R_f$ ($\Delta R_f$) between an oligosaccharide and its nearest neighbour. This value was plotted for each oligosaccharide as a function of polyacrylamide concentration. There is an inflection point seen at about 16% which would be the optimal concentration for separation. Details of this procedure are given in *Protocol 16*A.

159

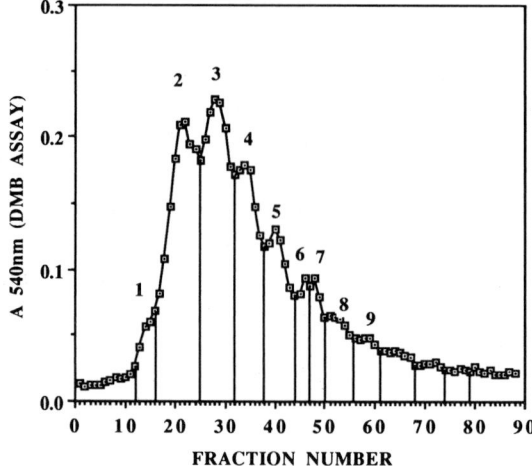

**Figure 13.** An example of a preparative gel run using a 20% acrylamide gel and a sample of rat xiphysternal chondroitin sulfate depolymerized with bovine testicular hyaluronidase as in *Figure 10*. Detection of the oligosaccharides was by a dye binding assay (*Protocol 24*) and the peaks numbered 1–9 were collected and aliquots of each applied to an analytical gel (*Figure 10*) to assess purity. Electrophoresis of the preparative gel was as outlined in *Protocol 16*B.

---

**Protocol 16.** Preparative gel electrophoresis of chondroitin sulfate oligosaccharides

*Equipment and reagents*

- See *Protocol 15*
- Preparative gel electrophoresis system (e.g. Bio-Rad model 491 Prep Cell)

A. *Optimization of gel concentration*

1. Cast gels of the following acrylamide concentrations: 10, 12, 15, 16, 17.5, 18, and 20% (w/v) by varying the proportion of reagent B in the mixture.

2. Apply a sample of hyaluronidase-depolymerized chondroitin sulfate to each of the gels, and electrophorese as described previously.

3. Determine the relative mobility ($R_f$) of each oligosaccharide compared with the phenol red tracking dye following staining of the gel with reagent D (as described in *Protocol 15*).

4. From these measurements, calculate and plot the difference in $R_f$ ($\Delta R_f$) of each oligosaccharide from its nearest neighbour. An inflection point can be seen at which optimal separation could be obtained using preparative gel electrophoresis.

**160**

B. *Running preparative gels*

1. Cast gels using the Bio-Rad model 491 Prep Cell, according to manufacturers recommendations and at a concentration determined from part A. The gel composition, running buffers, and sample preparation are exactly as outlined in *Protocols 15* and *16*A.

2. Gels are typically 10–12 cm in length and are run at approximately 25 V/cm. A typical gel run would take approximately 5 h to complete.

3. Collect fractions (2 min fractions, approximately 2 ml) from the preparative gel when the phenol red tracker dye has migrated to within 3 cm of the end of the gel rod. Fractions are collected and assayed using the dimethylmethylene blue GAG-dye binding assay of Farndale *et al.* (the method is outlined in Section 8.2, *Protocol 24*) modified for microtitre plate analysis.

## 5.2.3 Capillary electrophoresis of oligosaccharides and disaccharides of chondroitin sulfate and hyaluronan (14)

The high resolution separation techniques available to the analytical chemist in the form of HPLC have until recently been relatively unused by the connective tissue biochemist. There are many reasons for this, all of which are related to the physical properties of the glycosaminoglycans and their component di- and oligosaccharides. Chondroitin sulfate is not generally amenable to HPLC techniques because of its large hydrodynamic size, its high water solubility, the presence of many charge groups, and its high polarity. The final three factors however are the very reasons why these molecules are particularly suitable for separation and analysis by capillary electrophoresis. Capillary electrophoresis has become popular recently, due in part to the availability of affordable (approximately equivalent price to a standard HPLC system), reliable commercial systems. Although it is now attracting great interest, the technique is far from new and the application to separation in microbore capillaries dates back to 1979 when Mikkers *et al.* (15) reported on its use. The technique has incredible resolving power (considerably greater than HPLC) and is largely dependent on two factors for separation. First, the technique separates molecules according to their charge : mass ratio and presumably, dipole moment. Secondly, the technique can resolve uncharged molecules as a result of electroendosmosis which is a significant player in separations at neutral or alkaline pH. Moreover, the introduction of micellar-forming reagents such as SDS gives another dimension to separation due to analytes relative solubility in such micelles. In general this technique is referred to as micellar electrokinetic chromatography (MEKC). In the next section several capillary electrophoresis techniques will

161

be given as will their application to the examination of connective tissue carbohydrates.

The depolymerization of proteoglycans using enzymes specific for the degradation of particular glycosaminoglycans has become a popular method of analysis in recent years (16, 17). The enzymes chondroitinase ABC (chondroitin ABC lyase) and chondroitinase AC (chondroitin AC lyase) have been extensively used for the examination of chondroitin structure. Chondroitinase ABC will cleave hyaluronic acid, chondroitin 4- and 6-sulfate, as well as non-sulfated chondroitin and dermatan sulfate. Chondroitinase AC will not however, degrade dermatan sulfate. The chondroitin lyases are eliminases and as a result, introduce a double bond between carbons-4 and 5 of the glucuronic acid residue. The conjugated system of double bonds created by this elimination produces a chromophore at 232 nm in the resultant disaccharides or residual 'stub' as shown in *Figure 14*. The introduction of an unsaturated terminal glucuronate into the linkage region stub renders the structure much more antigenic and this has been the strategy for the production of a wide range of monoclonal antibodies directed against specific chondroitin sulfate structures (18).

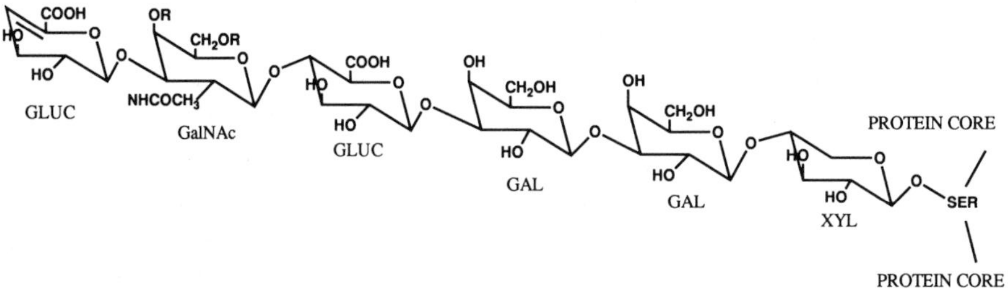

**Figure 14.** The structure of the linkage region 'stub' of chondroitin sulfate retained on the protein core after depolymerization with chondroitin ABC lyase.

---

**Protocol 17.** Capillary electrophoresis of glycosaminoglycan oligosaccharides

*Equipment and reagents*

- Capillary electrophoresis system
- Water-bath
- High speed bench (Eppendorf) centrifuge
- Reagent A: 200 mM *ortho*-phosphoric acid (12.48 ml/litre) pH 3.0
- Reagent B: 40 mM disodium hydrogen phosphate (5.678 g/litre)/40 mM SDS (11.552 g/litre)/10 mM sodium tetraborate (3.814 g/litre) pH 9.0

- Reagent C: 0.1 M NaOH (volumetric solution)
- Reagent D: 1 M NaOH (volumetric solution)
- Reagent E: 0.1 M Tris (12.1 g/litre)/150 mM sodium acetate (1.23 g/litre) adjusted to pH 8.0 by the addition of acetic acid
- Reagent F: exactly as reagent E but adjusted to pH 7.4

Before use, the capillaries are preconditioned by treatment with reagent D for about 1 h, followed by a significant wash in the buffer to be used. Lengths of times required for this washing step are unclear but at least another hour should be used and some suggested as long as 3–4 h may be required. This procedure is only required for new capillaries, however, or should one be drastically changing buffer conditions.

### A. *Preparation of chondroitin sulfate disaccharides*

1. Dissolve the proteoglycan or glycosaminoglycan in reagent E if performing digestion with chondroitinase ABC, and in reagent F if performing digestion with chondroitinase AC, to a concentration of about 200 μg/ml.

2. Add to the solution 0.05 U/ml of the required enzyme and incubate at 37 °C for 40 min.

3. Following digestion the reaction may be terminated by the addition of an equal volume of 8 M guanidine.HCl or more conveniently, by brief boiling (about 2 min). If you are unsure as to whether the reaction has reached completion the absorbance of the reaction mixture may be monitored at 232 nm. When the absorbance remains constant, the reaction has ended.

   When the reaction is complete, the disaccharides may be analysed direct, although it is often preferable to perform the following clean up step, which will remove any unwanted protein from your sample.

4. Chill the digested sample on ice, then add 4–5 vol. of ice-cold absolute ethanol, and maintain the sample on ice for about 30 min.

5. The solution may now be centrifuged at high speed to pellet out the precipitate. The supernatant contains the digested disaccharides of chondroitin and may now be removed from the precipitate and dried down, either by rotary evaporation or by drying in a stream of dry gas (preferably nitrogen).

6. Redissolve the disaccharide samples in aqueous buffers.

### B. *Electrophoresis at acid pH*

1. The conditions are as follows: 2 min wash with reagent C, 5 min rinse with reagent A.

2. Introduce the samples by vacuum (typically 1–2 sec, approximately 4–8 nl) of solutions containing in the order of 0.1 mg/ml.

3. Perform electrophoresis at 40 °C and −15 kV in reagent A.

4. Monitor the analytes at 232 nm.

### C. *Electrophoresis at alkaline pH*

1. The conditions are as follows: 2 min wash with reagent C, 5 min rinse with reagent B.

**Protocol 17.** *Continued*

2. Introduce the samples by vacuum (typically 1–2 sec) essentially as stated in part B.

3. Perform electrophoresis at 40 °C and at +15 kV in reagent B.

4. Monitor the analytes at either 200 or 232 nm.

The polarity of the electrodes is reversed in *Protocols 17*B and C for the following reasons. At low pH, the sulfate groups on chondroitin will still be negatively charged and will therefore migrate towards the anode since at this pH, the electroendosmotic flow is negligible. As a result, the positive electrode should be on the detector side and this is referred to as a negative potential, hence '−15 kV'. In contrast, at pH 9, the sulfate and carboxylate groups will both be charged and as before, will migrate towards the cathode. There is, however, at this pH a considerable electroendosmotic flow in the direction from anode to cathode. In *Protocol 17*C, the overall migration is controlled by the electroendosmosis. Although their migration due to electrophoresis is toward the anode, the overall migration is toward the

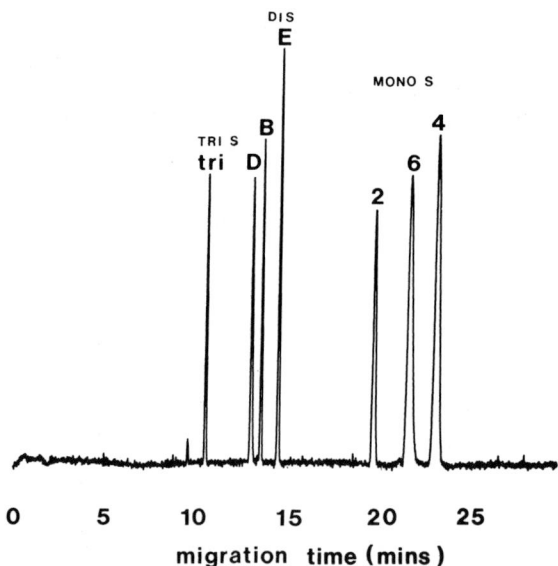

**Figure 15.** Separation at acid pH of the disaccharides of chondroitin sulfate as outlined in *Protocol 17*B. The peak marked TRI S refers to the trisulfated chondroitin sulfate disaccharide where $R_1$, $R_2$ and $R_3$ shown in *Figure 1*(a) are all $SO_3$. The peaks marked DI S refers to the disulfated disaccharides of chondroitin. Peak D, $R_1$ is H, and $R_2$ and $R_3$ are $SO_3$. Peak B, $R_1$ is $SO_3$, $R_2$ is H, and $R_3$ is $SO_3$. Peak E, $R_1$ and $R_2$ are $SO_3$, and $R_3$ is H. The peaks marked MONO S refer to the monosulfated chondroitin disaccharides, where for the peak marked 2, $R_1$ and $R_2$ are H, and $R_3$ is $SO_3$. Peak 6, $R_1$ is H, $R_2$ is $SO_3$, and $R_3$ is H. Peak 4, $R_1$ is $SO_3$, $R_2$ and $R_3$ are H. The capillary was monitored at 232 nm.

cathode due to the bulk flow as a result of electroendosmosis. Thus the detector should be on the cathode side and this is referred to as the 'normal' mode and potentials in this format are positive, hence '+15 kV'.

Examples of separations of di- and oligosaccharides are given in *Figures 15*, *16*, and *17*.

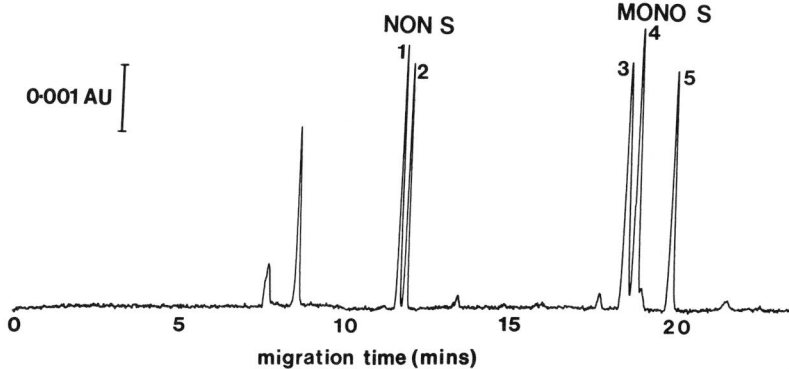

**Figure 16.** Electrophoresis at alkaline pH of chondroitin sulfate disaccharides as outlined in *Protocol 17*C. The peaks marked NON S refer to the non-sulfated disaccharides of hyaluronic acid (peak 1) and chondroitin (peak 2). The peaks marked MONO S refer to the monosulfated disaccharides of chondroitin. Peak 3 corresponds to peak 6 (Δ di 6S) on *Figure 15*, peak 4 corresponds to peak 4 (Δdi 4S) on *Figure 15*, and 5 corresponds to 2 (Δdi UA2S) on *Figure 15*. The capillary was monitored at 232 nm.

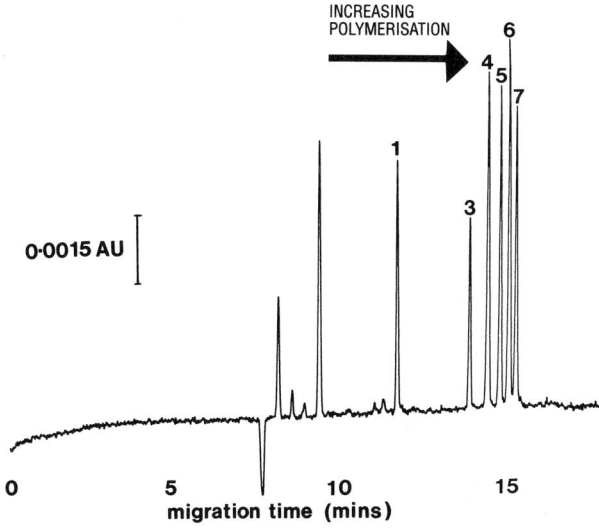

**Figure 17.** The electrophoresis of depolymerized hyaluronic acid at alkaline pH as outlined in *Protocol 17*C. The peaks marked 1–7 refer to the degree of polymerization of the hyaluronic acid where 1 refers to the disaccharide, 3 to the hexasaccharide, 4 to the octasaccharide, and so on. The capillary was monitored at 200 nm.

## 5.3 Examination of chondroitin sulfate disaccharides by HPAE

As previously stated, HPLC techniques have not made as significant an impact into the connective tissue world as perhaps other fields of biochemical investigation. Recently, new developments in high performance anion-exchange chromatography (HPAE) have shown very promising results. The principles of HPAE are essentially as outlined for soft gel anion exchange described in Section 4.2.1. The real advance in this area has come from Dionex who have developed the pellicular anion exchangers which are particularly good for analysis of sugars and liquid chromatography systems that are PEEK lined to prevent corrosion at high pH. In essence, the principle of separation is that at a high enough pH, the hydroxyl groups on neutral sugars become negatively charged and therefore can be separated on anion exchangers. Chondroitin sulfates, by virtue of their carboxylate and sulfate ester groups bind even more avidly than neutral sugars and need gradient elution to remove them. The method outlined here is an adaptation of some work published by Midura and colleagues (19, 20) using TFA gradients and a more efficient PAC 100 column than the PAC 1 used in the published work.

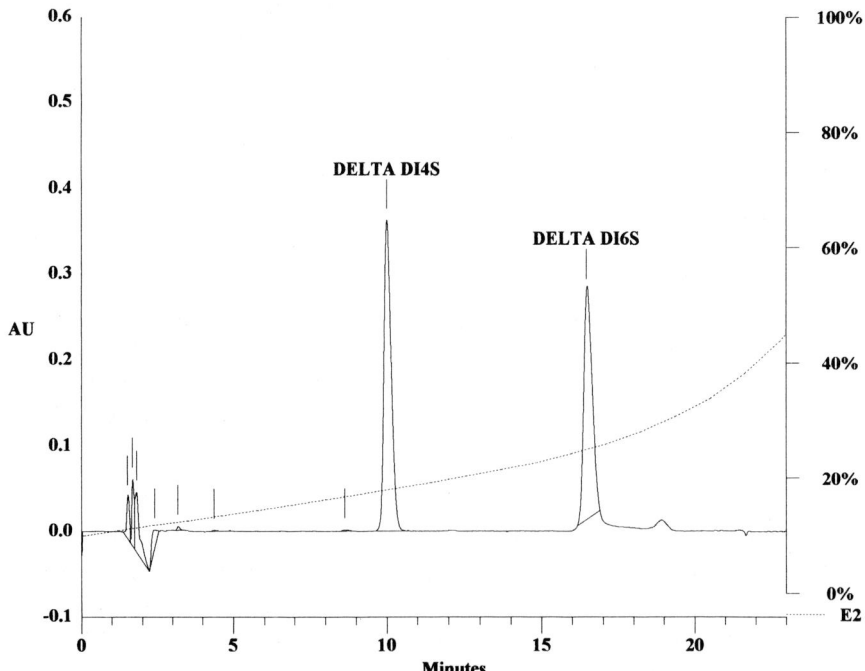

**Figure 18.** The chromatography of the reduced disaccharides of chondroitin 4-sulfate (DELTA DI 4S) and chondroitin 6-sulfate (DELTA DI 6S) on a CarboPac PA-100 column as outlined in Section 5.3. The absorbance of the eluate was monitored at 232 nm and the axis E2 refers to the percentage of the TFA in the eluate (i.e. the percentage of reagent B). The overlayed dotted line shows the gradient profile.

The quality of deionized water used in this procedure is of great importance, its resistance should be greater than 10 MΩ/cm. Before use, all water should be sparged with helium, then the other components (e.g. NaOH, TFA) added and sparging continued. This procedure prevents the formation of carbonates which will precipitate out and damage components of the system, particularly seals and pump heads.

The major drawback to this technique is the destruction of chondroitin sulfates at high pH by 'chain-peeling' reactions probably involving epoxide formation or cyclic sulfate intermediates. To prevent this occurring, the reducing terminal needs to be reduced by reaction with sodium borohydride to the galactosaminitol. This protects the sulfate ester group and prevents its elimination. Moreover, the reduction step allows the radiolabelling of the terminal aldehyde by the use of tritiated borohydride.

---

**Protocol 18.** High performance anion-exchange chromatography of glycosaminoglycan disaccharides

*Equipment and reagents*

- Dionex DX 300 series chromatography station
- Dionex CarboPac PA-100 column
- Reagent A: 100 mM NaOH (5.68 ml of a 46–48% volumetric solution/litre)
- Reagent B: as reagent A but containing 500 mM trifluoroacetic acid (38.6 ml/litre)
- Reagent C: deionized water
- Reagent D: 100 μM NaOH (1 ml of reagent A/litre)/1 M NaBH$_4$ (37.83 g/litre)
- Reagent E: 1 M acetic acid (57.5 ml/litre)

A. *Glycosaminoglycan disaccharide sample preparation*

1. To a solution containing glycosaminoglycan add reagent D to produce a final concentration of borohydride of 200 mM (i.e. 1 vol. of reagent D to 4 vol. of sample).

2. Incubate at room temperature of 1 h, then end the reaction by the gentle addition of reagent E to a final concentration of 300 mM (i.e. three parts reagent E to seven parts reaction mixture), maintaining the mixture on ice. Be careful to add the acetic acid gently as the liberation of hydrogen from the reaction mixture may be violent.

B. *Chromatography glycosaminoglycan disaccharide samples*

1. Apply the sample to the column by use of a 25 μl sample loop.

2. Monitor the eluate by absorbance at 232 nm, by pulsed amperometric detection, or by monitoring fractions for radioactivity if the sample has been labelled with sodium borotritide. The flow rate should be in the region of 1 ml/min.

3. A hyperbolic gradient of 10–45% reagent B was applied as shown in *Figure 18*. The elution positions of Δdi4S and 6S are indicated on the figure.

---

The advantages of this system over capillary electrophoresis are that sample recovery is easier and the amount of material that can be recovered is in the tens of micrograms per application. The main drawback however, is that the recovered material contains by necessity a terminal hexosaminitol residue.

# 6. Preparation of protein domains of aggregated proteoglycans (aggrecan)

Hyaluronic acid binding region (HABR; G1 domain) and link protein are important components of the aggrecan aggregate structure (*Protocol 25*A). As will be shown later, the G1 domain can be put to practical use in the analysis of proteoglycan structures. The conservation of sequence across a wide range of species and in other proteoglycan types suggests that they play a key role in the structural integrity of many connective tissues, but particularly articular connective tissue.

The G1 domain is a globular region of the aggrecan core protein which specifically interacts with hyaluronic acid to allow the formation of ternary aggregates along with link protein. It has a high affinity for hyaluronic acid $(K_d \sim 2 \times 10^{-8}M)$. Although this binding is strong, it is reversible and the interaction can be competitively inhibited by incubation with hyaluronate oligosaccharides of decasaccharide or larger. When G1 domain is found in a ternary complex with hyaluronic acid and link protein, the strength of interaction is so great, that to all intents and purposes, it is not reversible and cannot be competitively inhibited by oligosaccharides of hyaluronic acid. These data allow the development of methods by which hyaluronic acid may be assayed and also functional assays for link protein be developed.

The presence of LP in the ternary complex protects the HABR from limited proteolytic digestion (21); however, the rest of the proteoglycan molecule is highly degraded. The large molecular size of the HA-LP-G1 complex allows it to be easily separated from the degraded proteoglycan fragments by preparative gel chromatography under associative conditions. The LP and G1 domain may then be separated from one another by chromatography under dissociative conditions.

---

**Protocol 19.** Preparation of proteoglycan G1 domain and link protein

*Equipment and reagents*

- Apparatus outlined in *Protocols 1, 4,* and *12*
- Freeze-drier
- Columns of Sepharose CL-6B and Sephacryl S-300
- Reagent A: 0.1 M Tris (12.1 g/litre)/50 mM sodium acetate (4.1 g/litre) pH 7.3
- Reagent B: 4 M guanidine.HCl (380 g/litre)/ 50 mM sodium acetate (4.1 g/litre)/1 mM EDTA (0.372g/litre) pH 5.8

Filter both reagents through a Millipore filter *(0.45 μm) and thoroughly degas before use.*

---

*Method*

1. Prepare proteoglycan aggregates exactly as outlined in *Protocols 1*A and *4*, and one of the four basic fractionation methods discussed.

2. Dissolve the proteoglycan aggregates in reagent A to a concentration of about 10–20 mg/ml.

3. Add chondroitinase ABC (3.5 U/g proteoglycan) and incubate at 37 °C for 50 min. This procedure will remove the bulk of the chondroitin sulfate chains leaving only linkage region 'stubs' attached to the protein core. The removal of chondroitin sulfate will facilitate digestion with trypsin and subsequent separation.

4. To the chondroitinase digested proteoglycan, add trypsin (diphenyl-carbamoyl treated; 2 mg/g proteoglycan) and digest for 6 h at 37 °C.

5. Following trypsin digestion, freeze the solution, and partially freeze-dry to about one third of its volume.

6. Chromatograph aliquots of this concentrated solution on a column of Sepharose CL-6B equilibrated and eluted under associative conditions (*Protocol 12*A and B. A protein rich peak (peak A; *Figure 19*) is excluded from the CL-6B column. This peak contains the HA-G1-LP complex.

7. Collect the peak, dialyse against distilled water, and freeze-dry. Dissolve the freeze-dried material in reagent B (this buffer will dissociate the complex into its component parts) and chromatograph on a column of Sephacryl S-300 equilibrated and eluted with reagent B.

Typical chromatograms are given in *Figure 19* and *20*. Link protein and G1 domain are only partially separated from one another using this procedure, and the pooled peaks containing G1 domain and link protein will need to be re-chromatographed to homogeneity.

## 7. Functional assays for the degree of link stabilization of proteoglycan aggregates

These assays are based on the observation that the interaction of aggrecan with hyaluronan (HA) is strong but freely reversible under normal physiological conditions. The ternary complex between aggrecan, link protein, and HA, however, is effectively an irreversible interaction, requiring chaotropic agents to dissociate it. It is therefore possible, by including a large molar excess of oligosaccharides of HA varying in length from 10 to about 30 monosaccharides, to compete the aggrecans that are not 'link stabilized' from their high molecular weight HA. Link stabilized aggrecans will not be competed off the

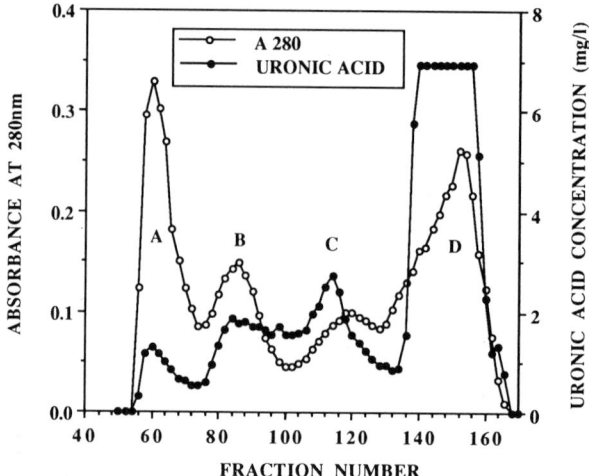

**Figure 19.** Chromatography of chondroitinase ABC/trypsin digested bovine nasal proteoglycan aggregates on Sepharose CL-6B eluted under associative conditions. Peak A contains the protein-rich G1-HA-LP complex, peak B contains the keratan sulfate-rich region, peak C contains CS 'stub'-peptides, and peak D contains chondroitin sulfate disaccharides and small peptides. Uronic acid was determined by the method outlined in *Protocol 23*.

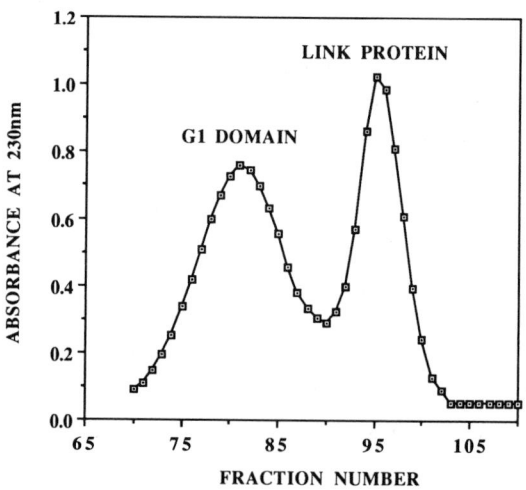

**Figure 20.** Chromatography on dissociative Sephacryl S-300 of peak A, *Figure 19*, after dialysis against water, freeze-drying, and resuspension in dissociative buffer as outlined in *Protocol 19*. The profile shows the elution positions of the G1 domain and link protein. The column effluent was monitored at 230 nm.

HA by excesses of HA oligosaccharides. This interaction can be examined in two ways, classically by gel filtration chromatography, but more recently and perhaps more conveniently by agarose–polyacrylamide gel electrophoresis.

## 7.1 Determination of link stabilization by column chromatography

---

**Protocol 20.** Preparation of oligosaccharides of hyaluronan (22)

*Equipment and reagents*

- Apparatus for column chromatography as outlined in *Protocol 12*
- Rotary evaporator
- Reagent A: 0.15 M NaCl (88.8 g/litre)/0.1 M sodium acetate (8.2 g/litre) pH 5.0
- Reagent B: 0.25 M pyridinium acetate (20.147 ml/litre of distilled water) adjusted to pH 6.5 with glacial acetic acid
- Reagent C: associative buffer—0.5 M sodium acetate (41 g/litre) adjusted to pH 6.8 with acetic acid

*Method*

1. Dissolve the hyaluronan (25 mg) in reagent A (2 ml), add bovine testicular hyaluronidase (50 µg), and incubate at 37 °C for 3 h.
2. Terminate the reaction, by heating the digest for 5 min at 100 °C, then cool to room temperature.
3. Chromatograph the digest on a suitable column of Sephadex G-50 superfine equilibrated and eluted in reagent B.
4. Collect and retain those fractions with $K_{av}$ between 0.3 and 0.45, pool, and rotary evaporate to dryness.
5. Dissolve in water (up to 10 ml) and repeat the rotary evaporation.
6. Continue to dissolve and evaporate until no trace of pyridine can be detected. When the digest is pyridine free, dissolve in a suitable volume of distilled water.

---

**Protocol 21.** Determination of link stability by chromatography

*See Protocol 20* for equipment and reagents.

1. Dialyse the proteoglycan test sample into reagent C, then add 10 mg of HA oligosaccharides (*Protocol 20*) per milligram of proteoglycan (both measured as uronic acid equivalents).

**Protocol 21.** *Continued*

2. Incubate the mixture at 4 °C for 16 h.

3. Chromatograph the sample on a column of Sepharose 2B as outlined in *Protocol 12*A and B.

4. Assay the eluate fractions for proteoglycan by dye binding or uronic acid if analytical amounts of proteoglycan are being used. If you are using [35]S-labelled proteoglycans, one may simply measure the radioactivity in each sample.

In the case of biosynthetically labelled samples, there is often insufficient material to detect by standard analytical procedures, it is appropriate therefore, to merely add a massive molar excess of oligosaccharides. Excessive quantities of oligosaccharide will not have any effect on the determination of link stability. A sample determination is shown in *Figure 21*. The degree of link stabilization can be determined from the following relationship: % link stabilization = (area of excluded peak) × 100/(area of excluded peak) + (area of included peak).

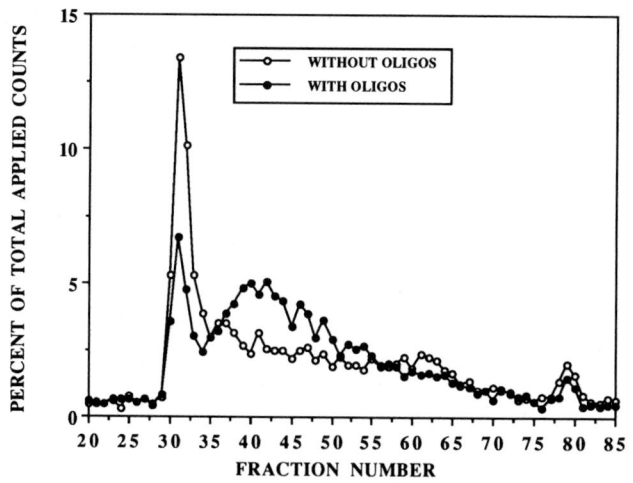

**Figure 21.** Determination of link stability by chromatography. The figure shows the elution profiles of guinea-pig proteoglycans on a Sepharose CL-2B column eluted under associative conditions. The two traces show identical samples, one which was not treated with oligosaccharides of hyaluronic acid (without oligos), and one that has been treated with oligosaccharides of hyaluronic acid (with oligos). It can be seen that there is a reduction in the proportion of material excluded from the column when treated with oligosaccharides. This corresponds to the non-link stabilized proteoglycans. The degree of link stabilization can be calculated as stated in Section 7.1.

## 7.2 Determination of link stability by agarose–polyacrylamide gel electrophoresis

---

**Protocol 22.** Determination of link stability by electrophoresis

See *Protocol 13* for equipment and reagents.

1. Prepare oligosaccharides of hyaluronan as described in *Protocol 20*.

2. Samples should be dialysed into reagent A (*Protocol 13*) diluted fourfold with distilled water.

3. Add 10 mg of hyaluronan oligosaccharides to every milligram of proteoglycan (as uronic acid equivalents) and incubate for 16 h at 4 °C.

4. After incubation, add a volume of a twofold dilution of reagent A containing sucrose (200 g/litre) and bromophenol blue (0.01 g/litre) to every volume of sample, hence diluting both 1:1.

5. Electrophorese the samples exactly as outlined in *Protocol 13*A.

6. Proteoglycans may be visualized by the methods outlined in *Protocol 14*A. Quantitation may be achieved by scanning densitometry of the dried gels (for toluidine blue) or of the X-ray films (for radiolabelled material). The determination of link stability is exactly as described in *Protocol 21*.

---

# 8. Proteoglycan constituent analytical techniques

Since the last edition of *Carbohydrate analysis: a practical approach* there have been significant changes in the way in which proteoglycans are analysed. Although automated analysis by dedicated autoanalysers are still popular, more and more laboratories have developed assays using microtitre plate readers. In this section I will concentrate on some of the more important assays which can and have been adapted for microtitre plate analysis. There are several advantages in this approach in that many laboratories will already possess plate readers for ELISA-based assays. The methods are economic, in that they require relatively little reagent. The results are digital and do not require peak height determination as is frequently required using auto-analysis equipment. Finally, they are generally quick procedures, allowing many samples to be analysed. The manual and automated assays for the proteoglycan components have been dealt with in great depth in the first edition of this work and will not be described here.

In contrast with the first edition, there will be no methods given for the immunoassay of proteoglycan components. This is for several reasons: first, the number of antibodies raised against proteoglycan components has increased exponentiallty and it would be impossible to give a general method which one might apply to each. Secondly, the area of immunoassay alone could quite comfortably fill a work as large as the whole of this edition. Finally, as most antibodies are in effect a 'cottage industry', assays are developed in those laboratories raising the antibodies. It seems unlikely that a laboratory capable of raising antibodies to proteoglycan components would need to consult a laboratory manual such as this to develop an assay method.

## 8.1 Uronic acid assay using microtitre plate format

This assay is an adaptation of several earlier methods, particularly that of Bitter and Muir (23) and Dische (24). It involves the reaction of carbazole with the unstable acid hydrolysed dehydrated derivatives of hexuronic acid.

---

**Protocol 23.** Determination of uronic acid by carbazole assay

*Equipment and reagents*

- Microtitre plate shaker
- Oven
- −70 °C freezer
- Multichannel pipettes
- Reagent A: sodium tetraborate decahydrate (9.56 g/litre) in concentrated sulfuric acid. It is difficult to dissolve borate in sulfuric acid and warming produces only small increases in the rate of dissolution (and is potentially hazardous). It is

therefore desirable to allow the borate to dissolve overnight. High grade sulfuric acid should be used for the assay as the presence of heavy metal ions causes anomalous colour formation. Anecdotally we have found that bubbling dry nitrogen through the reagent for a few hours before use reduces the likelihood of unwanted artefacts in the assay procedure
- Reagent B: reagent A containing carbazole (0.48 g/litre)

*Note*: Concentrated sulfuric acid is a very hazardous reagent. All care and precautions should be taken when handling this material. Gloves, protective glasses, and lab coats should be worn at all times. Although the method has been designed to prevent wherever possible the exposure of the researcher to strong acid one can never underestimate the risk of this reagent.

*Method*

1. To each well of a 96 well microtitre plate add reagent B (150 µl), seal

---

the plate, and store at −70 °C. Plates can be stored for long periods of time at this temperature without obvious deterioration.

2. When the acid (reagent B) has frozen, the plate is ready for use.

3. Remove the plate and keep it cool by placing it on dry ice.

4. Gently layer the samples (30 µl) and standards on to the surface of the reagent in the wells (up to 2 µg uronic acid in 30 µl is sufficient).

5. When the samples have been added, remove the plate from the dry ice, and seal it with a self adhesive plastic sealing sheet.

6. Gently shake the plate on a microtitre plate shaker for 15 min.

7. Heat the plate at 80 °C for about 25 min, or until colour development is complete.

8. Read the plate at 500 nm on a microtitre plate reader.

Cooling the acid in this assay has two main advantages: first, it allows the prepared plates to be stored safely since the acid is frozen. Adding sample to frozen acid prevents excessive heat production due to mixing since as the plate warms up, there is a very gradual mixing and any heat generated can be dissipated into the heat sink provided by the rest of the frozen acid. It is important that during colour development the temperature of the heating oven does not exceed 80 °C since in our experience, most microtitre plates begin to deform above this temperature and as a result will not fit into the plate reader. Placing the microtitre plates into a glass dish in the oven will reduce risk of accidental spillage, should the temperature be set, accidentally, above 80 °C.

## 8.2 Dye binding assay for glycosaminoglycan using 1,9-dimethylmethylene blue

There are various procedures in the literature for the determination of glycosaminoglycans using dyes such as dimethylmethylene blue and alcian blue. The basis of the assay is that the binding of dye to the polyanion causes a metachromatic shift in the absorbance spectrum of the dye. The degree of metachromatic shift is proportional to the amount of polyanion and can be measured spectrophotometrically. This assay is particularly suited to adaptation for a microtitre plate format and has proved to be very useful for routine and rapid determination of glycosaminoglycans.

---

**Protocol 24.** Determination of glycosaminoglycan by a microtitre plate adaptation of dimethylmethylene blue assay of Farndale *et al.* (25)

*Equipment and reagents*

- Microtitre plate reader
- Spectrophotometer capable of measuring in the visible region
- Multichannel pipettes
- Reagent A: 0.2% sodium formate (2 g/litre) adjusted to pH 3.5 with formic acid
- Reagent B: 1,9-dimethylmethylene blue (16 mg) dissolved in absolute ethanol (50 ml) and made up to a final volume of 100 ml with reagent A. This stock dye solution is relatively stable at room temperature if stored in a brown glass container
- Reagent C: dilute reagent B (stock dye solution) 1:9 with reagent A. This is the working dye solution which should have an optical density between 0.37 and 0.43 at 525 nm. If the solution falls outside of these values adjust the solution with formate buffer (if the OD > 0.43) or with reagent B (if the OD < 0.37)

*Method*

1. Add to each well of a 96 well microtitre plate either samples or standards (25 µl) of concentrations between 1 and 50 µg/ml chondroitin sulfate (or equivalent).

2. When all samples have been added, quickly pipette the working dye solution (200 µl) prepared as outlined above into each well.

3. Mix rapidly and thoroughly, then read the absorbance at 540 nm in a suitable microtitre plate. If a plate reader is available that is capable of measuring differential optical densities this should be used in preference to a single wavelength machine since it improves the reliability, selectivity, and reproducibility of the assay. In this case the reference wavelength should be set to 570 nm and the test wavelength to 540 nm.

## 8.3 Competitive inhibition assay for hyaluronan using proteoglycan G1 domain

This method is essentially that outlined by Fosang *et al.* (26). The basis of the assay is the competition of binding of biotinylated proteoglycan G1 domain to an immobile microtitre plate coat of hyaluronan by samples of hyaluronan in solution.

**Protocol 25.** Competitive inhibition assay for hyaluronan

*Equipment and reagents*

- Microtitre plate reader
- Commercially available biotinylation kit
- Multichannel pipette
- Reagent A: hyaluronan solution (200 μg/ml) in distilled water
- Reagent B: 0.2 M anhydrous sodium carbonate (21.2 g/litre) adjusted to pH 99.2 with acetic acid
- Reagent C: 0.1% bovine serum albumin (1 g/litre in phosphate-buffered saline (commercial PBS buffer tablets are adequate for all of the reagents listed below) below)

- Reagent D: 5% bovine serum albumin (50 g/litre) in phosphate-buffered saline
- Reagent E: 0.02 M sodium azide (1.3 g/litre) in phosphate-buffered saline
- Reagent F: 0.05% (v/v) TWEEN 20 (0.5 ml/litre) in phosphate-buffered saline
- Reagent G: 1% bovine serum albumin (10 g/litre) in phosphate-buffered saline
- Reagent H: 1% sodium dodecyl sulfate (10 g/litre) in water

### A. *Coating of plates with hyaluronan*

1. Mix equal volumes of reagents A and B, and pipette the resultant solution into each well of a 96 well microtitre plate (200 μl/well).

2. Incubate the plates for 16 h at 4 °C, remove the hyaluronan coating solution, then wash the plates six times with reagent C.

3. Block non-specific binding sites by the addition of reagent D (200 μl) to each well. Incubate the plate at 4 °C for 6 h.

4. Discard the contents of the wells and wash twice with reagent E.

5. Following washing, the plates may be stored by filling the wells with reagent E, and maintaining at 4 °C. The plates are stable for at least a few months if stored under these conditions.

### B. *Preparation of biotinylated G1 domain*

This procedure must be begun in advance, as it takes a considerable time to prepare the G1 domain.

1. Prepare G1 domain as outlined in *Protocol 19*.

2. Biotinylate the purified protein. This is most conveniently achieved using commercially available biotinylation kits (e.g. from Amersham International).

   In our experience, the biotinylation procedure does not affect the binding of G1 domain to hyaluronan, and as a result, it is not necessary to biotinylate the G1 domain as a complex with hyaluronan. This is therefore a much easier preparation than the biotinylation of a G1-HA complex, because one does not have to subsequently separate the G1 domain from the hyaluronan before use.

**Protocol 25.** *Continued*

C. *The assay of hyaluronan*

1. Take the coated plate, discard the contents of the wells, and wash five times with reagent F.

2. Into the top two rows of the plate, pipette HA standard solutions prepared by doubling dilution ranging from 10 μg/ml down to 5 ng/ml (100 μl/well). The standard solutions should be dissolved in reagent G.

3. Unknown HA samples may now be pipetted into the remainder of the wells (100 μl/well).

4. When all samples have been added, add biotinylated G1 domain (100 μl) to each well, and incubate for 16 h at 4 °C. The G1 domain should be diluted in reagent G, however, the actual dilution required must be determined by trial and error. In our experience, using the biotinylation kit available from Amersham, a dilution of the final biotinylated product of 1:5000 gives optimal results.

5. After 16 h at 4 °C, the G1, immobilized, and mobile hyaluronan have achieved equilibrium. The contents of the plate should be discarded and the plate washed five times with reagent F.

6. To each well add avidin horseradish peroxidase (100 μl of a 1:5000 dilution in reagent G) and incubate for 1 h at 37 °C. Avidin horseradish peroxidase is available from a number of manufacturers, the preparation used in our laboratory was supplied by Vector laboratories.

7. After incubation with the peroxidase conjugate, wash the plate five times with reagent F, then add ABTS (100 μl) to each well, and shake at room temperature. The ABTS used in our laboratory was obtained from Kirkegaard and Perry Laboratories.

8. The colour reaction must be stopped by the addition of reagent H (100 μl) when the optical density of the strongest coloured standard (i.e. standards of HA containing 5 ng/ml) approaches 1.6 at 410 nm.

9. The plates may now be read on a suitable microtitre plate reader at 410 nm.

As stated in the previous section, should a plate reader capable of differential optical density measurements be available this is preferable to single wavelength measurement. In this case the samples should be measured with the reference wavelength set to 490 nm and the test wavelength to 410 nm. An example of a typical competition inhibition curve is given in *Figure 22*.

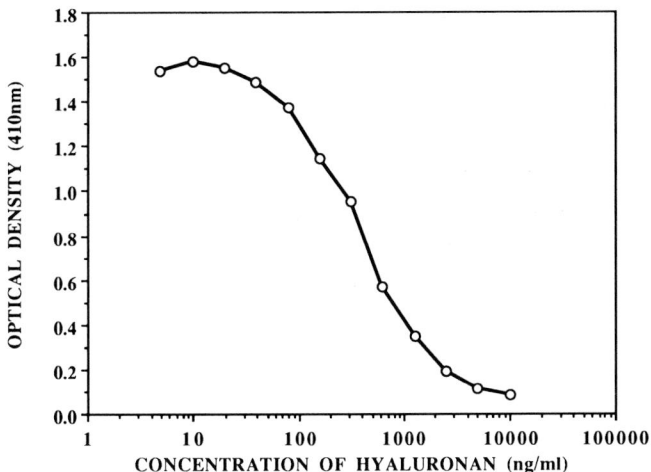

**Figure 22.** Determination of hyaluronic acid by a competition inhibition assay as outlined in *Protocol 25*.

# Acknowledgements

This work was supported by Lilly Research Laboratories. I extend my thanks to all of my colleagues at Lilly Research Centre Limited, in particular Richard Broadmore and Helen Penticost for their efforts, and to my wife Eileen who has introduced me to an altogether alternative interpretation of separation technology terminology.

# References

1. Rodén, L. (1980). In *Biochemistry of glycoproteins and proteoglycans* (ed. E. Lennarz), pp. 267–371. Plenum, New York.
2. Hascall, V. C. and Sajdera, S. W. (1969). *J. Biol. Chem.*, **244**, 2384.
3. Bayliss, M. T., Venn, M., Maroudas, A., and Ali, S. Y. (1983). *Biochem. J.*, **209**, 387.
4. Franek, M. D. and Dunstone, J. R. (1966). *Biochim. Biophys. Acta*, **338**, 108.
5. Sajdera, S. W. and Hascall, V. C. (1969). *J. Biol. Chem.*, **244**, 77.
6. Bonnet, F., Périn, J.-P., and Jollès, P. (1980). *Biochim. Biophys. Acta*, **623**, 57.
7. Heinegård, D., Wieslander, J., Sheehan, J., Paulsson, M., and Sommarin, Y. (1985). *Biochem. J.*, **225**, 95.
8. Antonopoulos, C. A., Axelsson, I., Heinegård, D., and Gardell, S. (1974). *Biochim. Biophys. Acta*, **338**, 108.
9. McDevitt, C. A. and Muir, H. (1971). *Anal. Biochem.*, **44**, 612.
10. Carney, S. L., Bayliss, M. T., Collier, J. M., and Muir, H. (1986). *Anal. Biochem.*, **156**, 38.

11. Carney, S. L., Billingham, M. E. J., Caterson, B., Ratcliffe, A., Bayliss, M. T., Hardingham, T. E., and Muir, H. (1992). *Matrix*, **12**, 137.
12. Laskey, R. A. and Mills, A. D. (1975). *Eur. J. Biochem.*, **56**, 335.
13. Cowman, M. K., Slahetka, M. F., Hittner, D. M., Kim, J., Forino, M., and Gadelrab, G. (1984). *Biochem. J.*, **221**, 707.
14. Carney, S. L. and Osborne, D. J. (1991). *Anal. Biochem.*, **195**, 132.
15. Mikkers, F. E. P., Everaerts, J. M., and Verheggen, T. P. E. M. (1979). *J. Chromatogr.*, **169**, 11.
16. Saito, H., Yamagata, T., and Suzuki, S. (1968). *J. Biol. Chem.*, **243**, 1536.
17. Oike, Y., Kimata, K., Shinomura, T., Nakazawa, K., and Suzuki, S. (1980). *Biochem. J.*, **191**, 193.
18. Christner, J. E., Caterson, B., and Baker, J. R. (1980). *J. Biol. Chem.*, **255**, 7102.
19. Shibata, S., Midura, R. J., and Hascall, V. C. (1992). *J. Biol. Chem.*, **267**, 6548.
20. Deutsch, A. J., Midura, R. J., and Plaas, A. H. K. (1993). In *Transactions of the 39th annual meeting of the American orthopaedic research society*, p. 94. San Francisco, California.
21. Heinegård, D. and Hascall, V. C. (1974). *J. Biol. Chem.*, **249**, 4250.
22. Hascall, V. C. and Heinegård, D. (1974). *J. Biol. Chem.*, **249**, 4242.
23. Bitter, T. and Muir, H. (1962). *Anal. Biochem.*, **4**, 330.
24. Dische, Z. (1947). *J. Biol. Chem.*, **167**, 189.
25. Farndale, R. W., Sayers, C. A., and Barrett, A. J. (1982). *Connect. Tiss. Res.*, **184**, 177.
26. Fosang, A. J., Hey, N. J., Carney, S. L., and Hardingham, T. E. (1990). *Matrix*, **10**, 306.

# 5

# Glycoproteins

JEAN MONTREUIL, STEPHANE BOUQUELET, HENRI DEBRAY, JEROME LEMOINE, JEAN-CLAUDE MICHALSKI, GENEVIEVE SPIK, and GERARD STRECKER

## 1. Introduction

Glycoproteins (for recent reviews, see refs 1–4), which together with glycolipids constitute the class of glycoconjugates, result from the covalent association of carbohydrate moieties (glycans) with proteins. Glycosylation of proteins represents one of the most important post-translational events because of the universality of the phenomenon. Most proteins are glycosylated, the glycoproteins being widely distributed in animals, plants, microorganisms, and viruses. In addition, the development of recombinant proteins of therapeutic interest represents a formidable challenge since the structure of recombinant glycoproteins must conform with that of the natural ones (5). It has been established that glycans perform important biological roles including:

(a) the protection of the peptide chain against proteolytic attack;
(b) the induction and maintenance of the protein conformation in a biologically active form;
(c) the decrease of the immunogenicity of proteins;
(d) the recognition and association with viruses, enzymes, and lectins;
(e) the control of the life span of circulating glycoproteins and cells;
(f) intercellular recognition and adhesion;
(g) cell contact inhibition.

In addition, we know that the glycan structure of the cell membrane glycoproteins is profoundly altered in cancer cells. This molecular transformation may be related to the appearance of cell surface neoantigens and could be a factor of immune response and metastatic diffusion.

Knowledge of the primary structure and conformation of glycans is clearly necessary in order to understand the mechanisms of glycan metabolism and to lay the foundations of the molecular biology and biotechnology of glycoprotins.

## 1.1 Types of glycan–protein linkages

Glycans are conjugated to peptide chains through two types of primary covalent linkages (*N*-glycosyl and *O*-glycosyl) leading to the definition of three classes of glycoproteins (*N*-glycosylproteins, *O*-glycosylproteins, and *N*, *O*-glycosylproteins). The most distributed *N*-glycosidic bond presently found in glycoproteins is *N*-acetylglucosaminyl-asparagine (GlcNAc($\beta$1-*N*)Asn). In contrast, the *O*-glycosidic bond presents a wide variety of linkages, the most common being the following:

(a) Mucin-type: the alkali-labile linkage between *N*-acetyl-D-galactosamine and L-serine or L-threonine (GalNAc($\alpha$1 → 3)Ser or GalNAc($\alpha$1 → 3) Thr) found in very numerous glycoproteins which are said to be of the 'mucin-type'.

(b) Proteoglycan-type: the alkali-labile linkage between D-xylose and L-serine (Xyl($\beta$1 → 3)Ser) involved in the acidic mucopolysaccharide–protein bond of proteoglycans.

(c) Collagen-type: the alkali-stable linkage between D-galactose and 5-hydroxy-D-lysine (Gal($\beta$1 → 5)OH-Lys) characterized in collagens.

(d) Extensin-type: the alkali-stable linkage between L-arabinofuranose and 4-hydroxyl-L-proline (1-Ara*f*($\beta$1 → 4)OH-Pro) identified in plant glycoproteins.

## 1.2 Primary structure of glycoprotein glycans

Glycan structures are not randomly constructed. They may be divided into families with similar structures and common oligosaccharide sequences, whether they originate from animals, plants, micro-organisms, or viruses. Consequently a number of classes and concepts are firmly established.

(a) *Concept of the common 'inner-core'*: the carbohydrate moiety of *N*- and *O*-glycosylprotein is derived from the substitution of oligosaccharide structures common to all glycans of a given class of glycoproteins. These non-specific and invariant structures are conjugated to the peptide chain and constitute the most internal part of glycans (the *inner-core, Figure 1*).

(b) *Concept of the antenna*: on the basis of their morphology, their flexibility, and their property of being recognition signals, the term *antenna* has been proposed for the outer variable arms substituting the inner-cores.

(c) *Microheterogeneity of glycans*: in addition to genetic variants expressed as variations in their polypeptide chains, almost all glycoproteins reveal another form of polymorphism associated with their carbohydrate residues. A given glycan located at a given amino acid in a glycoprotein, often presents a structural heterogeneity which is produced by partial substitution of sugar residues on a similar core structure. This type of diversity is termed 'microheterogeneity' or 'peripheral heterogeneity' because it often involves the number and position of the most externally situated monosaccharides in the glycan. The microheterogeneity is related to variations in the level of sialylation or to more profound modifications like the number of antennae in *N*-glycosylproteins. In all cases, the polymorphism of glycoprotein glycans still poses one of the most formidable problems encountered in the primary structure determination of glycans.

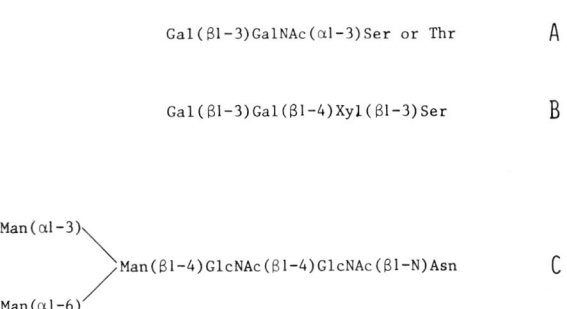

Gal(β1–3)GalNAc(α1–3)Ser or Thr     A

Gal(β1–3)Gal(β1–4)Xyl(β1–3)Ser     B

Man(α1–3)
           Man(β1–4)GlcNAc(β1–4)GlcNAc(β1–N)Asn     C
Man(α1–6)

**Figure 1.** Oligosaccharide inner-cores of glycoprotein glycans. Core A exists in the *O*-glycosylproteins of the mucin-type. Core B is the terminal sequence of almost all of the glycosaminoglycans. Core C is common to all *N*-glycosylproteins.

## 1.2.1 Primary structure of *O*-glycosylprotein glycans

(a) *Glycans conjugated through a GalNAc (α1 → 3)Ser or Thr linkage*; this group mainly comprises mucin structures. These glycans are often designated as mucin-like structures, even if they are present in plasma cell membranes and in glycoproteins from biological fluids (*Figure 2*).

(b) *Glycans conjugated through a Xyl (β1 → 3)Ser linkage*: this very homogeneous family of glycans consists of the acid mucopolysaccharides (glycosaminoglycans). They are generally linear polymers made up of disaccharide repeating units (*Figure 3*) and *O*-glycosidically linked through the trisaccharide inner-core B of *Figure 1* to the peptide chain in the proteoglycans.

## 1.2.2 Primary structure of *N*-glycosylprotein glycans

The *N*-glycosylproteins are divided into three families according to the nature of the carbohydrate moiety linked to the pentasaccharidic inner-core C (*Figure 1*). In the first family, the glycans contain mannose and *N*-acetylglucosamine only. They are called glycans of the oligomannosidic or

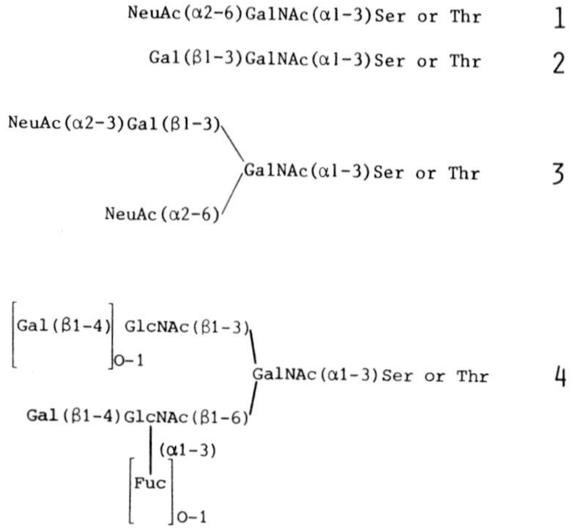

NeuAc(α2–6)GalNAc(α1–3)Ser or Thr     1

Gal(β1–3)GalNAc(α1–3)Ser or Thr     2

NeuAc(α2–3)Gal(β1–3)
                          GalNAc(α1–3)Ser or Thr     3
NeuAc(α2–6)

$$\left[ Gal(β1–4) \right] GlcNAc(β1–3)$$
$$\quad\quad\quad\quad\quad _{0–1}$$
                          GalNAc(α1–3)Ser or Thr     4

Gal(β1–4)GlcNAc(β1–6)
           (α1–3)
$$\left[ Fuc \right]_{0–1}$$

**Figure 2.** Some examples of structures of glycans *O*-glycosidically conjugated to protein through the linkage GalNAc(α1 → 3)Ser or Thr. 1: Submaxillary mucins, human erythrocytes. 2: Anti-freeze glycoproteins from antarctic fish, human chorionic gonado-tropin *β*-subunit, human serum IgA, epiglycanin of TA3-Ha cells, T-reactive erythrocytes, rat brain glycoproteins. 3: Human glycophorin and gonadotropin *β*-subunit, fetuin, bovine kappa-casein, lymphocyte plasma membrane. 4: Bronchial mucin of patients suffering from cystic fibrosis.

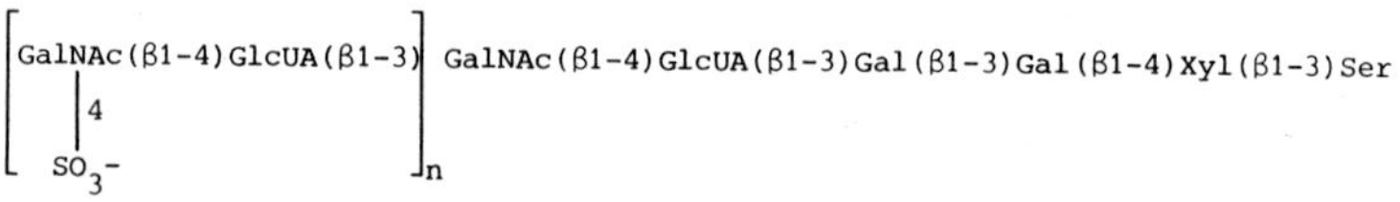

$$\left[ \begin{array}{c} GalNAc(β1–4)GlcUA(β1–3) \\ \quad\quad 4 \\ SO_3^- \end{array} \right]_n \; GalNAc(β1–4)GlcUA(β1–3)Gal(β1–3)Gal(β1–4)Xyl(β1–3)Ser$$

**Figure 3.** Primary structure of seryl-chondroitin 4-sulfate.

high mannose-type (*Figure 4*). In the second family, the sugar composition of glycans is more complex. These glycans contain galactose, fucose, and sialic acids in addition to mannose and *N*-acetylglucosamine. They derive funda-mentally from the addition to the pentasaccharide inner-core C of a variable number of *N*-acetyllactosamine residues (Gal(β1 → 4)GlcNAc). These structures have been called glycans of the *N*-acetyllactosaminic or complex type (*Figure 5*). In some glycans, galactose residues are replaced by *N*-acetylgalactosamine residues and sialic acid residues by sulfate groups. Glycans containing oligo- or poly(*N*-acetyllactosamine) sequences have been found and named poly(glycosyl)-peptides or poly(*N*-acetyl-lactosamine)-peptides (*Figure 6*).

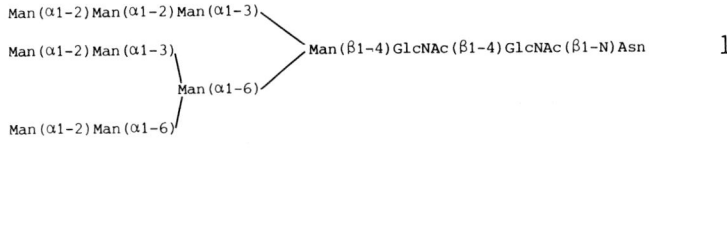

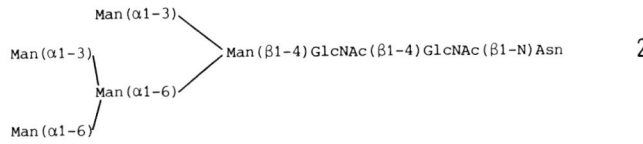

**Figure 4.** Structure of glycans of the oligomannosidic-type. 1: (conventionally assigned as $M_9$) is present in calf thyroglobulin unit A, human IgD and myeloma IgM, Chinese hamster ovary cell glycoproteins, bovine lactotransferrin, soybean agglutinin, Newcastle virus, and scorpion haemocyanin. 2: (conventionally assigned as $M_5$) has been found in Taka-amylase, hen ovalbumin, and human myeloma IgM. For reviews, see refs 1–4.

In the third family, the glycans contain both structures of the oligo-mannosidic and *N*-acetyllactosaminic types. They belong to the oligo-mannosido-*N*-acetyllactosaminic or hybrid type (*Figure 7*).

In numerous membranes, complexes of glycolipids and glycoproteins, called glycosylphosphatidyl inositol (GPI) anchors (*Figure 8*), are increasingly characterized (for review see ref. 6).

### 1.2.3 General considerations about the methodology of glycan primary structures

The problem of the determination of glycan primary structure can be considered as virtually solved due to the development of efficient methods for the isolation of glycoprotein variants and of their glycan residues, chiefly by using HPLC and affinity chromatography on immobilized lectins, and also the progress in chemical, enzymatic, and physical methodologies. In this connection, the collaboration between our laboratory and the one of J. F. G. Vliegenthart led to the introduction of a very efficient and sensitive method for determining the complete primary structure of glycans by associating the permethylation procedure with 360–600 MHz $^1$H-NMR spectroscopy (7, 8). This technique, which requires 25–100 µg of total sugars, is of a general application and is fully described in Section 9 of this chapter. Recent developments in mass spectrometry, particularly the 'fast atom bombardment' method (FAB) (9), and 'matrix assisted laser desorption–time of flight mass spectrometry' (MALD–TOF) (10), have resulted in considerable

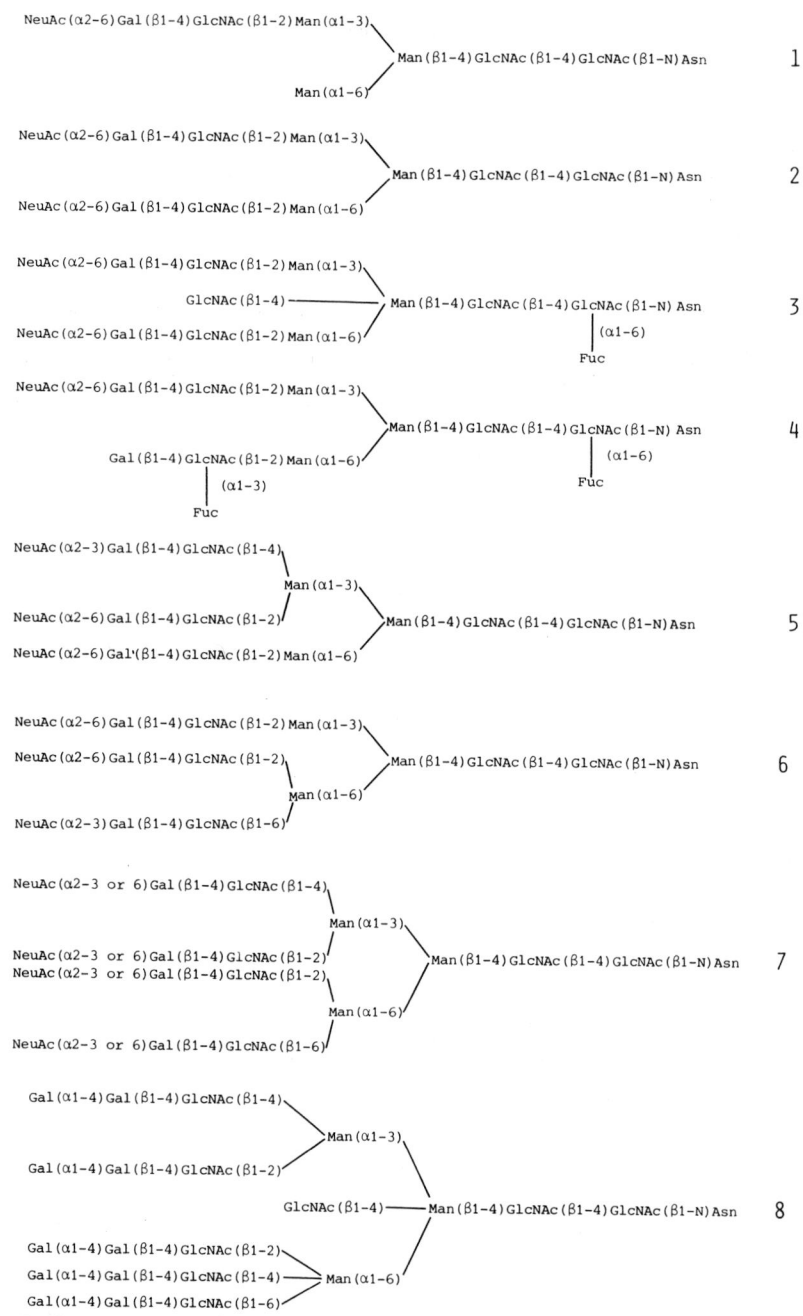

**Figure 5.** Structure of glycans of the *N*-acetyllactosamine-type. 1: Monoantennary glycan of the secretory component from human milk and human chorionic gonadotropin. 2: Biantennary glycan of human serum transferrin. 3: Monofucosylated and 'bisected' (presence of a bisecting *N*-acetylglucosamine residue) biantennary glycan of human IgG. 4: Difucosylated biantennary glycan of human lactotransferrin. 5, 6: Triantennary glycans of human serum transferrin. 7: Tetra-antennary glycan of human α₁-acid glycoprotein. 8: Penta-antennary glycan of turtle-dove ovomucoid. For reviews, see refs 1–4.

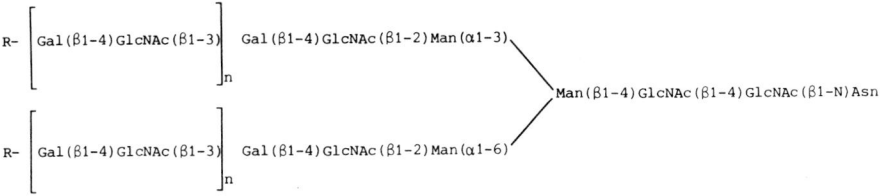

**Figure 6.** General basic structure of the biantennary glycans of the poly(glycosyl)-peptides in which n varies from 1 to 15. In human erythrocyte membrane, R = NeuAc($\alpha$2 $\rightarrow$ 3)Gal($\beta$1 $\rightarrow$ 4)GlcNAc($\beta$1 $\rightarrow$ 3), NeuAc($\alpha$2 $\rightarrow$ 6)Gal($\beta$1 $\rightarrow$ 4)GlcNAc($\beta$1 $\rightarrow$ 3), Fuc ($\alpha$1 $\rightarrow$ 2)Gal($\beta$1 $\rightarrow$ 4)GlcNAc($\beta$1 $\rightarrow$ 3) (Blood group O), or GalNAc($\alpha$1 $\rightarrow$ 3) [Fuc($\alpha$1 $\rightarrow$ 2)]Gal($\beta$1 $\rightarrow$ 4)GlcNAc($\beta$1 $\rightarrow$ 3) (Blood group A), or Gal($\alpha$1 $\rightarrow$ 3) [Fuc($\alpha$1 $\rightarrow$ 2)] Gal($\beta$1 $\rightarrow$ 4)GlcNAc($\beta$1 $\rightarrow$ 3) (Blood group B).

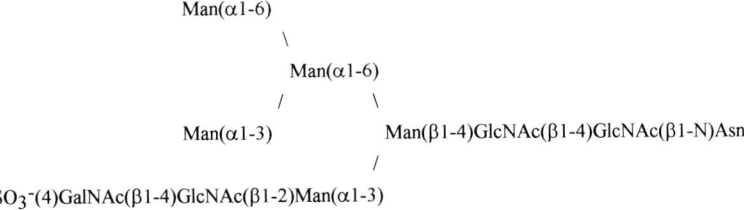

**Figure 7.** Hybrid-type glycan of a pituitary hormone.

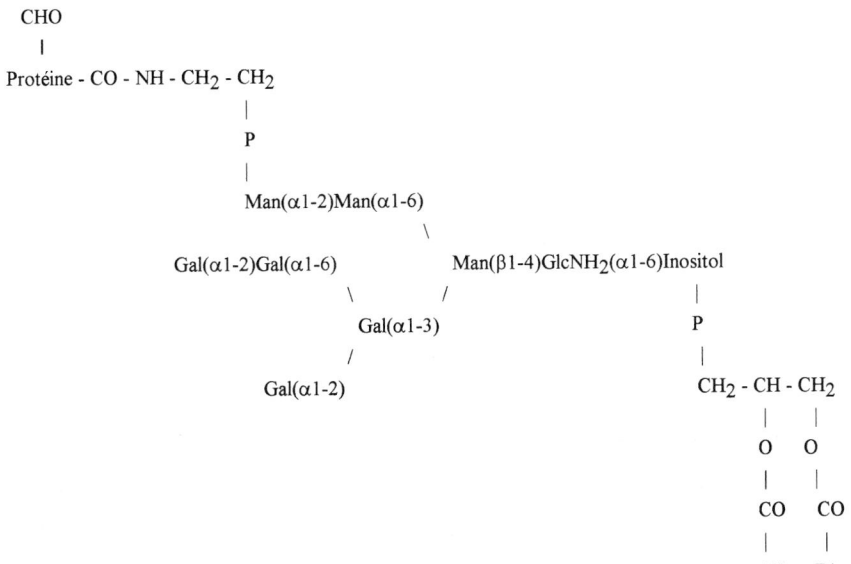

**Figure 8.** Structure of the variant surface glycoprotein of *Trypanosoma brucei*. P: Phosphoryl group; FA: fatty acid ($C_{14}$ to $C_{24}$) generally saturated.

**187**

progress. The description of the excellent, but not yet widely applied method of Kobata *et al.*, which associates with permethylation procedure and the stepwise degradation of glycans by glucosidases, has also been omitted due to lack of space (for details, see ref. 11).

In the following paragraphs, we aim to describe the methods we currently use in our laboratory and have tested over numerous years for determining the primary structure of the glycoprotein glycans. Many of them have been performed by my co-workers, each of whom will describe his own experience in glycan primary structure determination involving

(a) the isolation of glycoproteins, glycopeptides, and glycans,

(b) the determination of their composition with respect to simple sugars,

(c) the determination of the sequential arrangement of monosaccharide units, as well as the linkage between these units and the anomeric configuration of glycosidic bonds,

(d) the determination of the nature of the glycan–peptide linkage.

# 2. Chemical cleavage of *O*- and *N*-glycosidic linkages of glycans

The chemical liberation of the carbohydrate moieties of glycoproteins or glycopeptides can be achieved in different ways. The action of alkali in the presence of sodium or potassium borohydride liberates, by a '*β*-elimination' mechanism, the reduced glycans which were *O*-glycosidically linked to serine and threonine. *N*-Glycosidically linked glycans can be removed by hydrazinolysis or by alkaline cleavage with total or partial *N*-deacetylation. The liberated glycans must be re-*N*-acetylated before further fractionation or investigation.

## 2.1 Alkaline cleavage of *O*-glycosidic linkages

*O*-Glycosidic linkages between glycans and the *β*-hydroxyamino acids serine and threonine are easily split by dilute alkali solution (0.05–0.1 M NaOH or KOH) in mild conditions (40–45 °C for 0.5–6 days) by *β*-elimination leading to the liberation of the glycans. In order to prevent the destruction or modification of the latter by a 'peeling reaction', the reaction should be performed in the presence of a reducing agent (0.8–2 M NaBH$_4$). Under these conditions, *N*-glycosidic linkages and *O*-glycosidic linkages with hydroxyproline, tyrosine, and hydroxylysine are not cleaved.

Concomitantly, under non-reducing conditions, serine and threonine are transformed into dehydro-amino acids (2-aminopropenoic acid and 2-amino-2-butenoic acid, respectively) which strongly absorb UV light at 240 nm. This property is used in order to detect and follow a $\beta$-elimination reaction in non-reductive medium.

The $\beta$-elimination does not occur when the carboxyl group of serine and threonine is free.

---

**Protocol 1.** Reductive cleavage of *O*-glycosidic linkages (12)

1. Adjust the solution of glycoprotein (10–20 mg/ml) to pH 10 with dilute NaOH and add an equal volume of cold freshly prepared 0.1 M NaOH in 2 M NaBH$_4$.[a]

2. Incubate under reflux at 45 °C for 16 h.

3. Neutralize the cooled solution by adding 50% acetic acid (until pH 6) and subject the material to gel filtration on a Sephadex G-50 fine or Bio-Gel P-4 column (75 cm × 2 cm i.d.).

4. Liberated oligosaccharide-alditols can be characterized by thin-layer chromatography (Merck SiO$_2$-plates, solvent systems: *n*-butanol/acetic acid/water 20:10:15 (by vol.), or *n*-butanol/ethanol/acetic acid/pyridine/water 1:10:0.3:1:3 (by vol.)) and isolated by HPLC (see Section 3.2). *N*-Glycosylpeptides are efficiently fractionated by affinity chromatography on immobilized lectins (see Section 4.4).

---

[a] Under these conditions, the *O*-glycosidically linked glycans released as oligosaccharide-alditols are generally well separated from the *N*-glyco-oligopeptides (two to four amino acid residues), if present, formed by alkaline degradation of the peptide chain when both types of linkages co-exist in the studied glycoproteins.

---

## 2.2 Alkaline cleavage of *N*-glycosidic linkages

Although *N*-glycosidic linkages are stable in the mild alkaline conditions of $\beta$-elimination, they are quantitatively cleaved under drastic alkaline conditions (1 M NaOH or saturated barium hydroxide at 100 °C for 6–12 hours). Under these conditions,

(a) alkaline attack must be performed in a reducing medium (1–2 M NaBH$_4$) in order to prevent the 'peeling reaction' of liberated *N*-glycosidically linked glycans but without avoiding the profound degradation of *O*-glycosidically linked glycans.

(b) *N*-Acetylglucosamine residues are deacetylated so that, in a second step, liberated glycans have to be re-*N*-acetylated with acetic anhydride (sometimes, radiolabelled).

---

**Protocol 2.** Alkaline cleavage of *N*-glycosidic linkages (13)

1. Dissolve the *N*-glycosylprotein (50 mg/ml) in a 1 M NaOH–1 M NaBH$_4$ solution.

2. Heat under reflux for 6 h at 100 °C.

3. Neutralize the mixture, maintained in an ice-bath, with 50% acetic acid to a pH value of 6.0.

4. Purify the material on a Bio-Gel P-2 column (40 cm × 2 cm i.d.), pool the carbohydrate containing fractions, and lyophilize.

5. In order to re-*N*-acetylate, dissolve the material in saturated NaHCO$_3$ (1 ml/mg of glycan), add five aliquots of 10 µl of acetic anhydride at 5 min intervals, desalt the reaction mixture on a Dowex 50 × 8 (25–50 mesh; H$^+$) column (5 cm × 1 cm i.d.), and lyophilize the effluent.

6. Fractionate the glycan-alditols either by HPLC (see Section 3.2) or by affinity chromatography on immobilized lectins (see Section 4).

---

## 2.3 Hydrazinolysis of *N*-glycosidic linkages

Hydrazine cleaves the *N*-glycosidic linkages of *N*-glycosylpeptides and *N*-glycosylproteins. It liberates the *N*-deacetylated glycans as their hydrazones. Thus the procedure for obtaining the carbohydrate moieties is carried out in three steps: hydrazinolysis, re-*N*-acetylation, and reduction in order to stabilize the molecule.

---

**Protocol 3.** Hydrazinolysis of *N*-glycosidic linkages (14)

1. In a screw cap tube with a Teflon-silicone disc seal, introduce the dried glycopeptide or glycoprotein plus the minimum amount of anhydrous hydrazine (Pierce Chemical Co., Rockford, IL, USA) necessary in order to cover the sample.

2. Heat at 100 °C for 12–14 h. Do not stir using a Vortex, which deposits the material on the wall of the tube.

3. Eliminate the excess of hydrazine by repeated evaporation in presence of toluene (1 vol.) under a stream of nitrogen. Remove the last traces of hydrazine in a vacuum desiccator over H$_2$SO$_4$.

4. Purify the material on a Bio-Gel P-2 column (40 cm × 2 cm i.d), pool the carbohydrate fractions, and lyophilize.

5. Re-*N*-acetylation is carried out as described in *Protocol 2*, step 5.

---

6. Dissolve the oligosaccharide fraction in 0.05 M NaOH (1 ml/mg), reduce with NaBH$_4$ (5 mg/mg of sugar) for 16 h at 20 °C, and stop the reaction by adding Dowex 50 × 8 (25–50 mesh; H$^+$).

7. Filter, wash the filter with water, and concentrate the filtrate *in vacuo*.

8. Eliminate boric acid by repeated co-distillation *in vacuo* with methanol.

9. Purify the material by gel filtration on a Bio-Gel P-2 column (25 cm × 2 cm i.d.).

10. Glycans-alditols are fractionated as described in Section 3.2.2 (b).

*Notes*

(a) Radiolabels may be introduced by re-*N*-acetylation with radiolabelled acetic anhydride, or reduction with sodium borotritide. This procedure is currently applied in our laboratory for fractionating the cell membrane glycoprotein glycans by affinity chromatography on immobilized lectins (see Section 4.3).

(b) The *N*-acetylglucosamine residue involved in the glycosylamine linkage is converted into the hydrazone derivative. This leads to the formation of by-products in the course of the following reactions:

    *i.* The re-*N*-acetylation step gives rise to 1-acetylhydrazo-*N*-acetylglucosamine, 60% of which is converted, during the treatment with the cationic resin, into *N*-acetylglucosamine which will be transformed during the reduction into *N*-acetylglucosaminitol, while the remaining 40% of the hydrazone derivative is reduced into 1-*N*-acetylhydrazino-*N*-acetylglucosaminitol.

    *ii.* The latter will be converted, in case of permethylation, into a mixture of permethylated epimeric 1-deoxyalditols.

    *iii.* Other minor products, such as 1-deoxy-*N*-acetylglucosaminitol and 1-di-*N*-acetylhydrazino-*N*-acetylglucosaminitol are also present as final products of reduction and are responsible of the complex pattern of *N*-acetyl signals observed by NMR spectroscopy.

(c) Incomplete *N*-deacetylation of C-3 substituted *N*-acetylglucosamine has been reported and must be taken into consideration when quantitative re-*N*-acetylation with radiolabelled acetic anhydride is performed.

(d) The behaviour toward hydrazine of *O*-glycosidically linked glycans has not yet been elucidated. However, some results show that they are profoundly degraded until an alkali-stable linkage stops the 'peeling' reaction.

## 2.4 Production of glycans from *O, N*-glycosylproteins

When glycosylproteins possess both *O*- and *N*-glycosidically linked glycans, we use two procedures:

- successive use of reductive *β*-elimination and alkaline cleavage
- combining reductive *β*-elimination with hydrazinolysis

---

**Protocol 4.** Successive *β*-elimination and alkaline cleavage

1. Incubate the glycoprotein or glycopeptide under reflux at 45 °C for 16 h in 0.05 M NaOH/1 M NaBH₄ as described in *Protocol 1.*[a]

2. Add and equal volume of 2 M NaOH/1 M NaBH₄ and heat again under reflux for 6 h at 100 °C.

3. Neutralize the solution, purify the carbohydrate material as described in *Protocol 1*, and re-*N*-acetylate and fractionate according to *Protocol 2*.

[a] Only the *O*-glycosidically linked glycans are liberated, in the first step, by reductive–mild alkaline treatment as oligosaccharide-alditols, which are stable in the drastic conditions applied in a second step in order to split the *N*-glycosidic linkages.

---

**Protocol 5.** Successive *β*-elimination and hydrazinolysis

1. Incubate the glycoprotein or glycopeptide under reflux at 45 °C for 16 h in 0.05 M NaOH/1 M NaBH₄ as described in *Protocol 1.*[a].

2. Neutralize the solution and purify the carbohydrate material as described in *Protocol 1*.

3. Collect the *N*-glycosylpeptides and oligosaccharide-alditols and lyophilize.

4. Submit the *N*-glycosylpeptides to the hydrazinolysis procedure (see *Protocol 3*.

5. Fractionate the glycan-alditols as described in *Protocol 2*. If the oligosaccharide-alditols and *N*-glycosylpeptides are not well fractionated by gel filtration, submit both to the hydrazinolysis.

[a] *O*-Glycosidically linked glycans are first liberated as oligosaccharide-alditols by *β*-elimination and separated by gel filtration from the *N*-glycosylpeptides which are later hydrazinolysed separately.

---

# 3. Glycopeptide and glycan isolation

The preparation of glycoproteins is not described in this chapter because these compounds cannot be isolated by a uniform procedure, due to the variability in amount and properties of the carbohydrate moieties they carry. In general, the techniques applied are the classical methods for isolating proteins. However, an additional and efficient procedure has been developed in the last few years which is based on the use of lectins and which takes advantage of the presence of the carbohydrate part (see Section 4.3).

In contrast, fractionation and isolation of glycans and glycopeptides is relatively easy, although the mixtures are often complex due to micro-heterogeneity.

## 3.1 Isolation of glycopeptides

Glycopeptides and glyco-amino acids are obtained by the action of various proteases, the most efficient one being Pronase from *Streptomyces griseus*.

---

**Protocol 6.** Proteolysis of glycoproteins

1. To 100 mg of glycoprotein dissolved in 10 ml of 0.02 M calcium acetate in 0.04 M Tris pH 8.0, add 2 mg of Pronase E (Merck, Darmstadt, FRG).

2. Incubate the mixture at 40 °C for 24 h. After 4, 8, and 20 h, add the same amount of Pronase.

3. After 24 h, adjust the proteolytic digest to pH 4.5 with dilute acetic acid and concentrate to 1 ml.

4. Add 10 ml of cold (−10 °C) ethanol. Leave the mixture at room temperature for 2 h and then at 2 °C overnight.

5. Centrifuge and discard the precipitate.

---

**Protocol 7.** Purification of glycopeptides

A. *Purification by ion-exchange chromatography*

1. Add an equal volume of a 10% (w/v) aqueous solution of trichloro-acetic acid to the solution obtained in *Protocol 6*, step 5.

2. Centrifuge and desalt the supernatant by passing through coupled columns of Dowex 50 × 8 (25–50 mesh, H⁺) and Dowex 1 × 8 (25–50 mesh, HCOO⁻).

3. Wash the columns with water and concentrate the eluates under vacuum.

**Protocol 7.** *Continued*

B. *Purification by gel filtration*

1. Remove the large peptides by a gel filtration on Sephadex G-25 or Bio-Gel P-2 eluting with water or 5% (v/v) acetic acid.

---

**Protocol 8.** Fractionation of glycopeptides

A. *Gel filtration chromatography*

1. Equilibrate two coupled columns (140 cm × 1.8 cm i.d.) of Bio-Gel P-4 in 1% (v/v) acetic acid at room temperature.

2. Elute the glycopeptides at a flow rate of 5 ml/h and collect fractions of 1 ml.

3. Analyse each fraction by paper electrophoresis (see below) or by TLC on silica-gel G-60 using *n*-butanol/ethanol/water/acetic acid/pyridine (10:100:30:3:10 by vol.) as solvent.

4. After 5 h, stain the silica gel plates with a reagent containing 0.2% (w/v) orcinol in 20% $H_2SO_4$ (v/v) and heat at 110 °C.

B. *Ion-exchange chromatography*

1. Anion-exchange chromatography of acidic glycopeptides: perform the ion-exchange chromatography of glycopeptides (40 mg) on a Dowex 1 × 2 (200–400 mesh) column (36 cm × 2.5 cm i.d.), equilibrated with 2 mM pyridinium acetate, pH 5.

2. Elute the neutral glycopeptides with the starting buffer, mono-sialylated glycopeptides with 50 mM pyridinium acetate (pH 5), disialylated glycopeptides with 500 mM pyridinium acetate (pH 5).

3. Monitor the glycopeptides with the phenol–sulfuric reagent described in Chapter 1, *Protocol 2*.

4. Concentrate each fraction under vacuum, lyophilize, and purify by gel filtration on a Bio-Gel P-2 column.

5. Cation-exchange chromatography of neutral glycopeptides: fractionate the glycopeptides (40 mg) on a Dowex 50 × 2 (200–400 mesh) column (42 cm × 1.6 cm i.d.) equilibrated with 2 mM pyridinium acetate, pH 5.

6. Elute the glycopeptides, first with the starting buffer, and then with a linear gradient from 2 mM to 500 mM of pyridinium acetate, pH 5.

7. Monitor and purify the fractions as described.

C. *HPLC*

1. Glycopeptides can be fractionated by high performance liquid chromatography performed on an anion-exchange 10 μm Micro-Pack

AX-10 column (30 × 0.4 cm) using a gradient of 500 mM potassium dihydrogen phosphate (adjusted to pH 4.0 with phosphoric acid) as follows: elution with distilled water for 15 min, linear gradient to 2.5% $KH_2PO_4$ (500 mM, adjusted to pH 4.0 with phosphoric acid) for 5 min, isocratic elution for 5 min, linear gradient to 5% $KH_2PO_4$ (500 mM; pH 4.0) for 10 min, and isocratic elution for 20 min (15). Flow rate: 1 ml/min.

2. Fractionation of glycopeptides may be performed by hydrophobic bonding and reverse-phase chromatography. In particular, the glycopeptides are fractionated on a C18 Bondapack column (0.39 × 30 cm) (Waters) and eluted with a linear gradient distilled water/acetonitrile (UV grade) 80%. Flow rate: between 0.75 and 2.5 ml/min at ambient temperature. Chromatographic separations are monitored by UV absorption at 214 nm.

### D. *Affinity chromatography on immobilized lectin*

Affinity chromatography on immobilized lectins (see Section 4.4) constitutes one of the best tools for isolating glycopeptides since the peptide moiety does not interfere in the binding to the lectin.

### E. *Preparative electrophoresis*

1. High voltage electrophoresis. Perform high voltage electrophoresis on Whatman No. 1 paper at pH 6.4 (pyridinium acetate), for 2 h, at 4000 V.

2. Low voltage electrophoresis. Carry out overnight the low voltage electrophoresis on Whatman No. 3 paper at pH 2.4 (1 M acetic acid) or at pH 3.9 (pyridine/acetic acid/water, 15:50:1935 by vol.).

3. Detection of glycopeptides in the fractions can be carried out by depositing a drop of each of the fractions (on sheets of Whatman paper) and by applying to the dried spots a colour generating reaction specific to either the peptides or sugar moiety.

    (a) To detect peptides dip the Whatman paper sheet in a ninhydrin reagent prepared as follows: to a solution of 1 g of cadmium acetate in a mixture of glacial acetic acid (50 ml) and water (100 ml), add a solution of 10 g of ninhydrin in acetone and make up to 1 litre with more acetone. Develop the violet colour by heating the paper at 150 °C.

    (b) To detect carbohydrate dip the sheet of Whatman paper in a solution of 0.1 M periodic acid in acetone. Dry at room temperature and dip the sheet in a 0.15% (w/v) of benzidine in acetone. Sugars develop a yellow colour on a green background.

## 3.2 Isolation of glycans by HPLC

Glycans can be fractionated and isolated by applying the classical methods of electrophoresis and of adsorption, partition and ion-exchange chromatography. However, these procedures have recently been gradually replaced by the more sensitive and rapid methods of HPLC and of affinity chromatography on immobilized lectins (see Section 4). Therefore, we shall restrict ourselves to the description of the procedures of glycan isolation by HPLC.

HPLC is widely applied to the separation and preparation of oligosaccharides liberated from glycopeptides by alkaline cleavage (see Sections 2.1 and 2.2), hydrazinolysis (see Section 2.3), or hydrolysis by endoglycosidases (see Section 8.3). We currently use in our laboratory the following HPLC techniques:

- anion-exchange chromatography of sialyloligosaccharides
- partition chromatography of neutral and acidic oligoscaccharides on primary amine-bonded silica or alkyl diol-bonded silica
- reverse-phase chromatography of neutral oligosaccharides on $C_2$ and $C_{18}$-bonded silica
- high pH anion-exchange chromatography of oligosaccharides (HPAE–Dionex)
- separation of derivatized oligosaccharides (pyridylamino derivatives)

### 3.2.1 Anion-exchange chromatography of acidic (sialylated or sulfated) oligosaccharides

Sialyloligosaccharides can be fractionated by HPLC or by medium pressure anion-exchange chromatography.

*i. Preparative anion-exchange HPLC of sialyloligosaccharides on Micropak AX-10 anion-exchange column*

Perform HPLC of sialyloligosaccharides liberated by reductive alkaline cleavage of *O*- and *N*-glycosylpeptides (see Sections 2.1 and 2.2) or by hydrazinolysis of *N*-glycosylpeptides (see Section 2.3), on a 10 μm Micropak AX-10 column (50 cm × 0.8 cm i.d.; Varian Associates, Walnut Creek, CA, USA) with a Spectra-Physics liquid chromatograph (Model 8700) equipped with Model 8400 variable wavelength detector connected to a Model 4100 computing integrator (Spectra-Physics Inc., San Jose, CA, USA).

---

**Protocol 9.** Preparative anion-exchange HPLC (16)

**1.** Dissolve 20 mg of the oligosaccharides in 60 μl of distilled water and submit to HPLC using a gradient of 0.5 M $KH_2PO_4$ adjusted to pH 4.0 with phosphoric acid under the following conditions: stepwise elution

---

with distilled water for 5 min, linear gradient to 2.5% (v/v) 0.5 M KH$_2$PO$_4$ for 15 min, isocratic elution for 25 min, linear gradient to 5% (v/v) 0.5 M KH$_2$PO$_4$ for 35 min, isocratic elution for 45 min. The flow rate throughout should be maintained at 2 ml/min. Oligosaccharides are detected at 200 nm with detector sensitivity 0.32 and integrator attenuation 16 using an integrator chart speed of 0.5 cm/min.

2. Purify each fraction (4 ml) by gel filtration on Bio-Gel P-2 (200–400 mesh) column (30 cm × 1.9 cm i.d.), the elution being carried out with water at a flow rate of 12 ml/h.

3. Detect the sugars present in each fraction with orcinol–sulfuric acid reagent (see Chapter 2, *Protocol 1*), on silica gel plates (precoated silica gel-60; Merck, Darmstadt, FRG).

4. Detect the phosphate salts eluted from the Bio-Gel column by precipitation with silver nitrate (see *Figure 9* for an example).

## ii. Preparative medium pressure anion-exchange chromatography on Mono Q® of acidic oligosaccharides

Acidic oligosaccharides can be fractionated according to their charge (or sialic acid content) by using medium pressure chromatography on the anion-exchange Mono Q® (17).

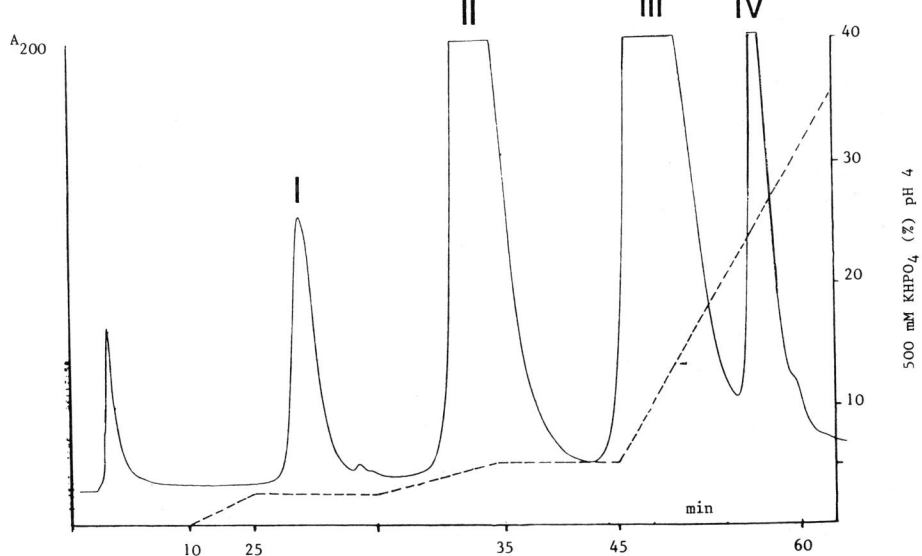

**Figure 9.** HPLC on 10 μm Micropak AX-10 column of sialoglycans liberated by hydrazinolysis of α$_1$-acid glycoprotein. I, II, III, IV: mono-, di-, tri-, and tetrasialylated glycans. The recovery was 91%.

(a) Perform anion-exchange chromatography using a traditional gradient system equipped HPLC apparatus or on a Pharmacia Fast protein liquid chromatography apparatus (FPLC). Carry out the separation at room temperature on an analytical prepacked Mono Q HR 5/5 column (50 × 5 mm) (Pharmacia).

(b) Wash the column with 2 ml of water. Inject the sample dissolved in 50 μl of water. Realize the charge separation of acidic oligosaccharides with a linear gradient from 0–100 mM NaCl in water (10 ml) at a flow rate of 2 ml/min. The eluate is monitored at 214 nm. Wash the column with 2 ml of 100 mM NaCl.

Monosialyl oligosaccharides are eluted at a 15 mM NaCl concentration, disialyl at 40 mM, trisialyl at 50 mM, and tetrasialyl at 70 mM.

### 3.2.2 Partition HPLC of neutral and acidic oligosaccharides on primary amine-bonded silica (18)

i. *Partition HPLC on bonded primary amine packings of neutral oligosaccharides obtained by hydrazinolysis of N-glycosylpeptides*

Carry out analysis and preparative chromatography of glycans liberated by hydrazinolysis (see Section 2.3) of non-sialylated *N*-glycosylpeptides with a Spectra-Physics apparatus Model 8700, equipped with an UV 8400 variable wavelength detector connected to a 4100 computing integrator (Spectra-Physics Inc., San Jose, CA, USA). Perform the chromatography on a 5 μm Amino AS-5A (Brownlee Labs, Santa Clara, CA, USA) column (25 cm × 4 mm i.d.) equilibrated with acetonitrile/water (65:35, v/v). Inject a solution of 5 mg of oligosaccharides dissolved in 30 μl of acetonitrile/water (50:50, v/v) and apply the following stepwise elution conditions: linear gradient from acetonitrile/water (65:35, v/v) to acetonitrile/water (60:40, v/v) for 30 min, followed by isocratic conditions for 30 min, and then a linear gradient to acetonitrile/water (50:50, v/v) for 30 min. The flow rate should be 1 ml/min. Oligosaccharides are detected at 200 nm with detector sensitivity 0.32 and integrator attenuation 50.

The separation obtained within 90 min of mixture of 17 reduced oligosaccharides liberated from hen ovomucoid by hydrazinolysis is shown in *Figure 10.*

ii. *Partition HPLC on bonded primary amine packings of sialyloligosaccharides obtained by alkaline borohydride treatment of O-glycosylpeptides*

Analyse the oligosaccharide-alditols obtained by β-elimination from *O*-glycosylpeptides (see Section 2.1) with the same Spectra-Physics material as

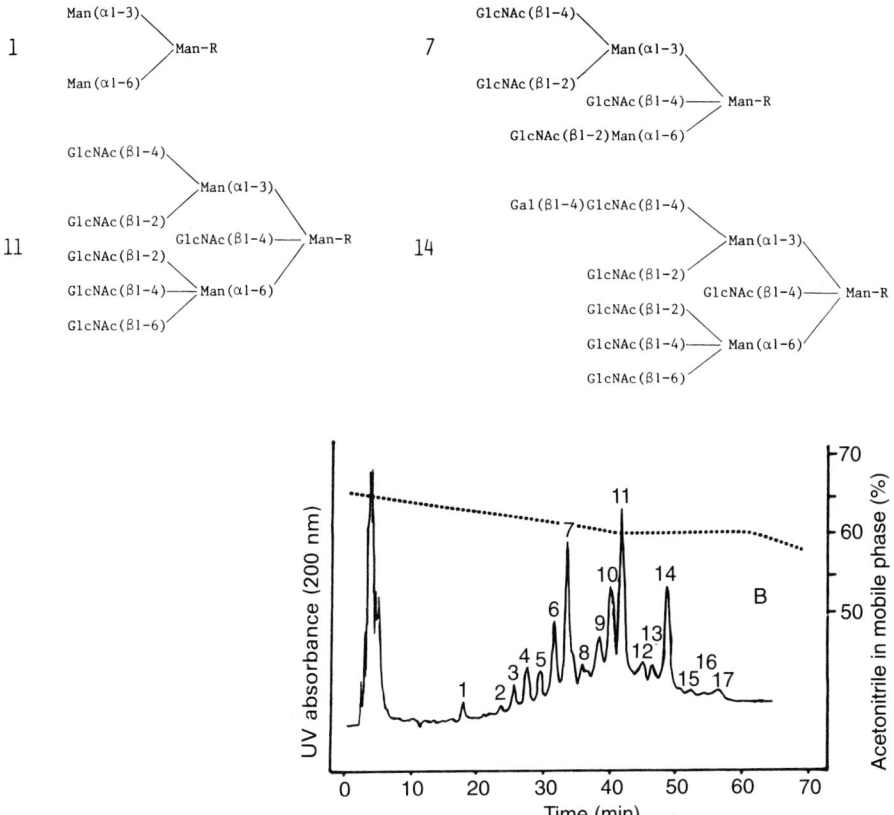

**Figure 10.** HPLC on 5 µm Amino AS-5A of glycans liberated by hydrazinolysis of hen ovomucoid. Primary structure of glycans, 1, 7, 11, and 14 are given above the elution diagram. R = (β1 → 4)GlcNAc(β1 → 4)GlcNAc-ol.

described above. Dissolve 1 mg of oligosaccharides in 10 µl of distilled water and inject into a 5 µm Amino AS-5A (Brownlee Labs, Santa Clara, CA, USA) column (25 cm × 4 mm i.d.) equilibrated with the initial solvent (acetonitrile/ phosphate buffer 15 mM, pH 5.2; 4:1 v/v). Develop the chromatogram isocratically for 25 min with 4:1 (v/v) acetonitrile/15 mM phosphate buffer (pH 5.2), then apply a linear gradient to decrease the acetonitrile concentration in the mobile phase. Use a flow rate of 1 ml/min. Oligosaccharides are detected at 200 nm with detector sensitivity 0.35 and integrator attenuation 4.

An example of separation of six oligosaccharides liberated from Cad glycophorin A by β-elimination is given in *Figure 11*.

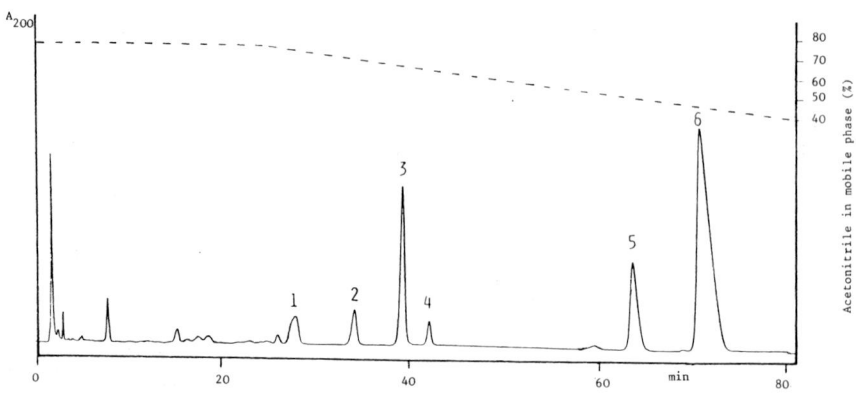

**Figure 11.** HPLC on 5 μm Amino-5A of glycan-alditols liberated by β-elimination from Cad erythrocyte membrane glycophorin. 1: NeuAc(α2 → 3)Gal(β1 → 3)GalNAc-ol; 3: NeuAc (α2 → 3)GalNAc(β1 → 4)Gal(β1 → 3)GalNAc-ol; 5: NeuAc(α2 → 3)Gal(β1 → 3)[NeuAc (α2 → 6) ]GalNAc-ol; 6: NeuAc(α2 → 3)[GalNAc(β1 → 4) ]Gal(β1 → 3) [NeuAc(α2 → 6)] GalNAc-ol.

### 3.2.3 Partition HPLC of oligosaccharides of the oligomannosidic-type on alkylamine-bonded silica (20)

Oligosaccharides of the oligomannosidic-type (see *Figure 4*) liberated from *N*-glycosylpeptides by hydrazinolysis (see Section 2.3), by alkaline cleavage (see Section 2.1 and 2.2), or by endoglycosidases (see Section 8.3.3), as well as oligomannosides present in urine of patients with mannosidosis are easily separated by partition HPLC on alkylamine-bonded silica.

The procedure performed in our laboratory is as follows, using the Spectra-Physics apparatus described above.

(a) Use 5 μm Supelcosyl LC-NH$_2$ column (0.4 × 25 cm) (Supelco Inc., Bellefonte, Pennsylvania).

(b) Dissolve 1 mg oligosaccharide in 50 μl water and inject on the column equilibrated with the initial solvent (acetonitrile/water, 70:30). After injection, maintain the isocratic conditions (acetonitrile/water, 70:30) for 15 min.

(c) Elute with a linear gradient of acetonitrile/water in the ratio 60:40 for 60 min for the separation of M$_2$GN to M$_5$GN oligosaccharides (*Figure 12A*), or in the ratio 50:50 for the separation of M$_5$GN to M$_9$GN oligo-saccharides (*Figure 12B*).

### 3.2.4 Reverse-phase HPLC of neutral oligosaccharides on C$_{18}$-bonded silica (21)

Perform HPLC on a 3 μ C$_{18}$-column (10 cm × 4.6 mm i.d.; Brownlee Labs, Santa Clara, CA, USA) using a Spectra-Physics apparatus as described

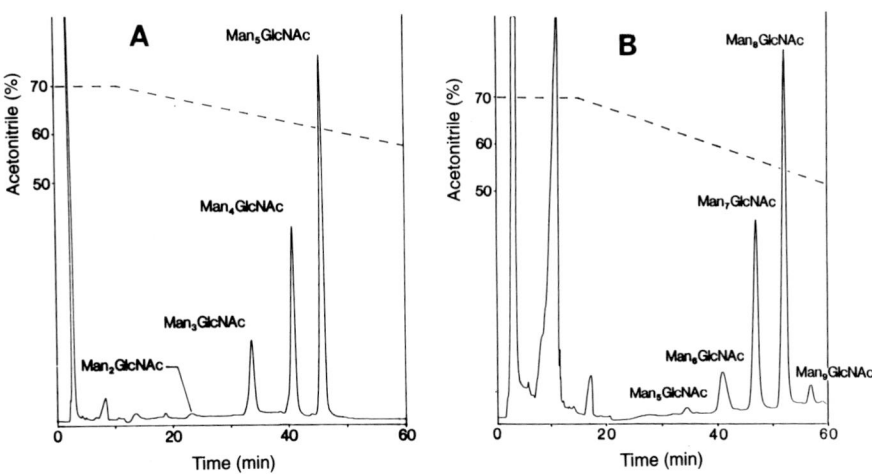

**Figure 12.** HPLC elution profile of oligosaccharides of the oligomannosidic-type. Linear gradient of acetonitrile/water in the ratio 60:40 (A) and in the ratio 50:50 (B).

above. Inject into the column 1 mg of oligosaccharides dissolved in 10 µl of distilled water. Chromatograph by isocratic elution with distilled water for 15 min, followed by a linear gradient to water/acetonitrile (90:10 v/v) for 25 min, after which apply isocratic conditions with the same solvent for 20 min. Use a flow rate of 1 ml/min. Oligosaccharides are detected at 200 nm with detector sensitivity 0.8 and integrator attenuation 8.

As an example, from fraction 9 shown in *Figure 10*, 15 fractions were obtained by reverse-phase chromatography. This indicates the high degree of microheterogeneity of glycans of hen ovomucoid and the resolution power of the method (*Figure 13*).

### 3.2.5 High pH anion-exchange chromatography of oligosaccharides (HPAE–Dionex)

The recent introduction of HPLC using pellicular ion-exchange resins under high pH conditions and detection of the sugars with a pulsed amperometric detector (PAD) has simplified the separation and analysis of both mono- and oligosaccharides (22). The method may be used for analytical (i.e. for the mapping of oligosaccharides released from glycoproteins) or for preparative purposes, and constitutes a final step in oligosaccharide purification. HPAE chromatography has been successfully applied for separating glycans originating both from *N*-glycosylproteins and mucins and from glycolipids (23).

### i. Equipment

(a) HPAE is realized using the Dionex LC system (Dionex, Sunny Vale, CA) fitted with a PAD detector with gold electrode. The following pulse

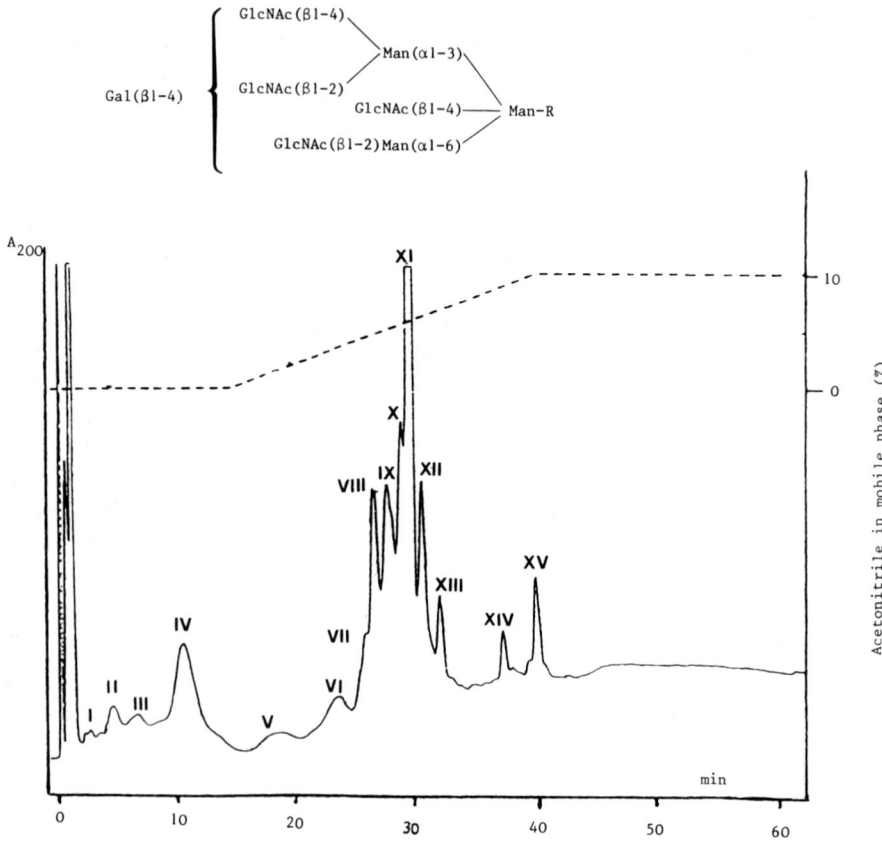

**Figure 13.** Reverse-phase HPLC on a 3 µm C$_{18}$-column of peak 9 from *Figure 10* obtained after preparative HPLC of the glycan-alditols liberated from hen ovomucoid by hydrazino-lysis. Only the structure of oligosaccharide XI is given in the figure. R: ($\beta1 \rightarrow 4$)GlcNAc ($\beta1 \rightarrow 4$)GlcNAc-ol.

potentials and durations are used for detection of oligosaccharides: $E_1 =$ 0.05 V/300 ms; $E_2 = 0.60$ V/120 ms; $E_3 = 0.60$ V/60 ms working at 300 nA full scale.

(b) CarboPac PA-1 columns (4.6 × 250 mm) equipped with a CarboPac PA-100 guard column (Dionex).

(c) Elution is carried out at a flow rate of 1 ml/min at ambient temperature. Eluents are made by suitable dilutions of a 50% NaOH stock solution (Baker) and by increasing ionic strength with sodium acetate.

*ii. Separation of neutral oligosaccharides*

(a) Separation of oligosaccharides from *N*-glycosylproteins: use the gradient described in *Table 1*.

(b) Separation of oligosaccharides-alditols released by reductive alkaline treatment of mucins (24). Due to the lower retention times of the alditols as compared to the native oligosaccharides, a very low base concentration (isocratic elution with 15 mM NaOH) is used for fractionating the oligosaccharide-alditols.

### iii. Separation of acidic oligosaccharides

According to Hardy and Towsend (23), sialyloligosaccharides can be separated either at low pH (pH 4.0 to 6.0) or at pH 13 leading to very variable retention times (*Tables 2 and 3*).

The principle of the methods is based on the formation of sugar oxyanions under alkalkine conditions. The contribution of the anomeric oxyanion to the retention time is dominant as demonstrated by the lower phase-interactions observed with oligosaccharide-alditols. Chromatographic selectivity for oligo-saccharide oxyanions is also related to the accessibility of stationary phase of the readily ionizable hydroxyl groups. This phenomenon is observed with fucosylated compounds which are less retained as compared to their non-fucosylated counterparts. Although neutral oligosaccharides require alkaline

**Table 1.** Elution gradient for fractionating neutral oligosaccharides by HPAE–Dionex chromatography

| Time (min) | % Eluent 1 (0.1 M NaOH) | % Eluent 2 (0.15 M sodium acetate in 0.1 M NaOH) |
|---|---|---|
| 0 | 100 | 0 |
| 10 | 100 | 0 |
| 60 | 40 | 60 |
| 65 | 40 | 60 |
| 125 | 100 | 0 |

**Table 2.** Low pH elution gradient for fractionating acidic oligosaccharides by HPAE–Dionex chromatography

| Time (min) | % Eluent 1 (5 mM sodium acetate pH 4.65) | % Eluent 2 (0.2 M sodium acetate buffered to pH 4.65 with 0.165 M acetic acid) |
|---|---|---|
| 0 | 100 | 0 |
| 5 | 100 | 0 |
| 30 | 95 | 5 |
| 60 | 70 | 30 |
| 70 | 70 | 30 |
| 80 | 100 | 0 |

**Table 3.** Alkaline pH elution gradient for fractionating acidic oligosaccharides by HPAE–Dionex chromatography

| Time (min) | % Eluent 1 (0.1 M NaOH) | % Eluent 2 (0.1 M sodium acetate in 0.1 M NaOH) |
|---|---|---|
| 0 | 95 | 5 |
| 25 | 95 | 5 |
| 85 | 91 | 9 |
| 95 | 85 | 15 |
| 102 | 95 | 5 |

conditions for separation by HPAE, acidic oligosaccharides can be separated both at low and high pH due to the ionization of acidic groups.

### 3.2.6 Separation of derivatized oligosaccharides by HPLC as pyridylamino derivatives

Numerous procedures have been described for the derivatization of oligosaccharides with UV absorbing or fluorescent chromophores. Coupling is generally achieved by the reductive amination method. Such derivatives may generally improve chromatographic HPLC separation of oligosaccharides and provide much greater detection sensitivity in comparison to measurment of intrinsic UV absorption. Generally, the derivatized oligosaccharides may be directly analysed by FAB–MS after separation and the fluorescent probe has an enhancement effect of the ion intensity (25).

---

**Protocol 10.** Pyridylamination of oligosaccharides

1. Add to 100 nmol of oligosaccharide lyophilized powder 10 µl of a coupling reagent prepared by dissolving 100 mg of 2-aminopyridine (2-AP) in 50 µl of acetic acid and 60 µl of methanol.[a]

2. Heat the solution under constant stirring at 90 °C for 15 min. Remove the excess reagent by evaporation under a stream of nitrogen.

3. Add 10 µl of a reducing reagent (6 mg of borane–dimethylamine complex dissolved in 100 µl acetic acid). Heat the solution at 90 °C for 30 min, and evaporate the reaction mixture under a stream of nitrogen with methanol and toluene to remove excess reagents.

4. Dissolve in water the 2-PA derivatives for HPLC analysis.

5. HPLC of PA oligosaccharides may be achieved either by reverse-phase on C18 or amino-bonded silica columns.

---

(a)  Reverse-phase chromatography on C18 columns

- For reverse-phase HPLC on C18 silica column, use the same columns as previously described in Section 3.2.4.

- Solvents: Solvent A: 50 mM acetic acid adjusted to pH 5.0 with triethylamine.
  Solvent B: solvent A containing 0.5% 1-butanol.

- After injection of the sample on the column equilibrated with solvent A, increase linearly the ratio of solvent B over a 50 min period. Flow rate is 1 ml/min. Detect by using a fluorescence detector, with excitation and emission wave-lengths of 320 and 400 nm, respectively.

(b)  Chromatography on amino-bonded silica

- Use column described in Section 3.2.2.

- Solvents: Solvent A: 25:75 mixture (v/v) of 200 mM acetic acid adjusted to pH 7.3 with triethylamine and aceto-nitrile.
  Solvent B: 50:50 mixture (v/v) of the same two components.

- Equilibrate the column with solvent A, and increase linearly the ratio of solvent B to 50% over a 60 min period. Detect PA oligosaccharides by their fluorescence with excitation and emission wavelengths of 310 and 380 nm, respectively.

[a] The reaction is never complete, the coupling yield being around 90%. For this reason, before HPLC analysis it is better to remove non-derivatized oligosaccharides and excess of 2-AP as follows. After dissolving the PA derivatives in water, the pH is adjusted to 8.5–9.0 by addition of an equal volume of saturated bicarbonate. Excess of 2-AP is removed by successive extraction (seven times) with 2 ml of benzene. Further purification may be achieved by gel filtration on a Sephadex G-15 column with 10 mM acetic acid is eluent.

# 4. Use of immobilized lectins

## 4.1 Introduction to the use of lectins

Lectins are 'sugar-binding proteins or glycoproteins of non-immune origin which agglutinate cells and/or precipitate glycoconjugates'. They bear at least two sugar-binding sites, the presence of which explains the ability of lectins to precipitate polysaccharides, glycoproteins, and glycolipids, and to agglutinate cells. As these interactions are often reversed by monosaccharides or glycosides, lectins are powerful tools for isolating sugar components and cells either in their soluble or immobilized form (for reviews, see refs. 26–28).

**Table 4.** Lectins commonly used for glycoprotein study and classified according to their monosaccharide specificity

| | |
|---|---|
| α-D-Mannose, α-D-glucose | |
| *Canavalia ensiformis* | Con A |
| *Lens culinaris* | LCA |
| β-D-Galactose, *N*-acetyl-β-D-galactosamine | |
| *Ricinus communis* | RCA$_I$ |
| | RCA$_{II}$ |
| *Glycin max* (soybean) | SBA |
| *Arachis hypogaea* (peanut) | PNA |
| α-D-Galactose, *N*-acetyl-α-D-galactosamine | |
| *Griffonia simplicifolia I* | GSA$_I$ |
| *Dolichos biflorus* | DBA |
| *N*-Acetyl-β-D-glucosamine | |
| *Triticum vulgare* (wheat germ) | WGA |
| α-L-Fucose | |
| *Lotus tetragonolobus* | LTA |
| *Ulex europeus I* | UEA$_1$ |
| *Aleuria aurantia* | AAA |
| α-*N*-Acetylneuraminic acid | LPA |
| *Limulus polyphemus* (limulin) | |

The carbohydrate specificity of a given lectin is usually established by the method of fixation-site saturation in which different sugars are tested for their ability to inhibit either haemagglutination or precipitation of polysaccharides and/or glycoconjugates by the lectin. Another procedure consists in the determination of the association constant by equilibrium analysis.

The specificity of a lectin is defined in terms of the best monosaccharide inhibitor. The results from such studies lead to the classification of lectins into groups some of which are given in *Table 4*.

Originaly, it was thought that lectins recognized monosaccharides in the non-reducing external position. However, it was further demonstrated that in most cases complex oligosaccharides are several thousand-fold more potent as 'haptens' toward lectins than monosaccharides themselves so that the concept of the 'lectino-dominant monosaccharide' must be replaced by that of the 'lectino-dominant oligosaccharide' (29–34). Moreover, we know that the lectins are able to bind 'laterally' to glycoprotein glycans. The structures which are actually recognized by lectins commonly used in biochemical and biological research are described in *Table 5*. In addition, we must keep in mind that the affinity of immobilized lectins toward free and/or conjugated sugars is absolutely unforseeable and cannot be deduced at all from the results obtained with the lectins in solution from haemagglutination experiments (29, 30). For example, oligosaccharides are able to inhibit the haemagglutination by LCA (see *Tables 4* and *5* for lectin abbreviations) but only glycoasparagines, glycopeptides, and glycoproteins are fixed on the immobilized LCA. So, oligosaccharides liberated by hydrazinolysis or alkaline cleavage are not retained by this latter material. In the same way, the presence of the 'bisecting' *N*-acetylglucosamine residue (see *Figure 5*) does not influence the inhibitory effect of oligosaccharides on the haemagglutination by Con A, but hinders the fixation of glycans on the immobilized Con A (32).

In conclusion, the only method for determination of the binding specificity of any immobilized lectin consists of the study of the behaviour of free and conjugated oligosaccharides of well defined primary structure such as those listed in *Table 6*.

## 4.2 Immobilization of lectins

Numerous methods have been proposed for preparing immobilized lectins, many of which are now available from different suppliers (Pharmacia, Miles, E.Y. Laboratories, I.B.F.). Lectins directly coupled to CNBr activated agarose are the most popular. However, immobilized lectins can be easily prepared using agarose, Sepharose 4B, or Ultrogel, at a concentration of 2–10 mg of lectin per millilitre of settled gel. Immobilization and subsequent use of the immobilized lectins requires some essential rules.

(a) During the coupling reactions, lectin binding sites must be protected by addition of the monosaccharide specific of the lectin (hapten sugar).

(b) In some cases, non-specific (hydrophobic or ionic) adsorption of glycoproteins on the immobilized lectins has been observed. In order to avoid this phenomenon, new types of immobilization matrices have been developed, such as polyacrylic-hydrazido-agarose on which lectins can be coupled with glurataldehyde, giving a stable, non-leaking, uncharged, and hydrophilic absorbent.

(c) Immobilized lectins must be stored at 4 °C in buffer containing 0.02% sodium azide as a bacteriostatic agent, in order to avoid any loss of activity for several years.

(d) Some lectins possess metal-binding sites and the presence of metal ions is important to induce a proper conformation of the lectin needed for carbohydrate binding. For example, Con A needs $Mn^{2+}$ and $Ca^{2+}$ for full activity so that the buffers used for the affinity chromatography on this lectin must contain $MnCl_2$ and $CaCl_2$ (1 mM each).

(e) In order to limit non-specific interactions between glycoproteins and immobilized lectins, such as ionic interactions, the buffers must possess a moderate ionic strength (0.1–1 M in NaCl).

**Table 5.** Specificity of commonly used lectins towards oligosaccharide sequences belonging to *N*-glycosylproteins

Concanavalin A (Con A)

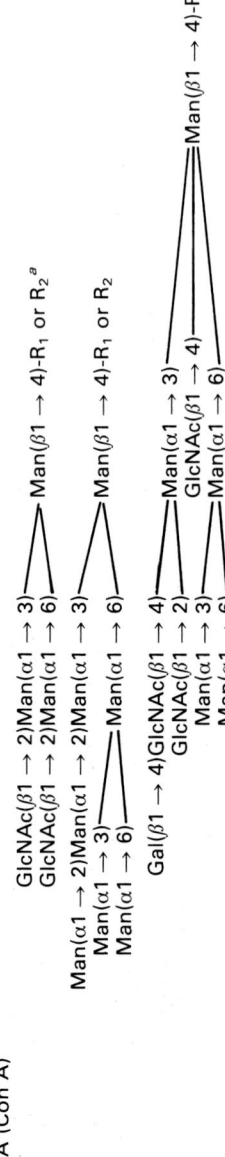

$$GlcNAc(\beta1 \rightarrow 2)Man(\alpha1 \rightarrow 3)$$
$$GlcNAc(\beta1 \rightarrow 2)Man(\alpha1 \rightarrow 6)$$
$$Man(\beta1 \rightarrow 4)\text{-}R_1 \text{ or } R_2{}^a$$

$$Man(\alpha1 \rightarrow 2)Man(\alpha1 \rightarrow 3)$$
$$Man(\alpha1 \rightarrow 3)$$
$$Man(\alpha1 \rightarrow 6)$$
$$Man(\beta1 \rightarrow 4)\text{-}R_1 \text{ or } R_2$$

$$Gal(\beta1 \rightarrow 4)GlcNAc(\beta1 \rightarrow 4)$$
$$GlcNAc(\beta1 \rightarrow 2)$$
$$Man(\alpha1 \rightarrow 3)$$
$$GlcNAc(\beta1 \rightarrow 4)$$
$$Man(\alpha1 \rightarrow 6)$$
$$Man(\beta1 \rightarrow 4)\text{-}R_1$$

*Lens culinaris* (LCA)

$$GlcNAc(\beta1 \rightarrow 2)Man(\alpha1 \rightarrow 3)$$
$$GlcNAc(\beta1 \rightarrow 2)Man(\alpha1 \rightarrow 6)$$
$$Man(\beta1 \rightarrow 4)\text{-}R_2$$

$$Gal(\beta1 \rightarrow 4)GlcNAc(\beta1 \rightarrow 2)Man(\alpha1 \rightarrow 3)$$
$$Gal(\beta1 \rightarrow 4)GlcNAc(\beta1 \rightarrow 2)Man(\alpha1 \rightarrow 6)$$
$$Gal(\beta1 \rightarrow 4)GlcNAc(\beta1 \rightarrow 6)$$
$$Man(\beta1 \rightarrow 4)\text{-}R_2$$

*Ricinus communis* agglutinin[b]
(RCA₁)

$$\boxed{\begin{array}{l} Gal(\beta1 \rightarrow 4)GlcNAc(\beta1 \rightarrow 4) \\ Gal(\beta1 \rightarrow 4)GlcNAc(\beta1 \rightarrow 2) \\ Gal(\beta1 \rightarrow 4)GlcNAc(\beta1 \rightarrow 2) \\ Gal(\beta1 \rightarrow 4)GlcNAc(\beta1 \rightarrow 6) \end{array}}$$
$$Man(\alpha1 \rightarrow 3)$$
$$Man(\alpha1 \rightarrow 6)$$
$$Man(\beta1 \rightarrow 4)\text{-}R_1 \text{ or } R_2$$

Leukoagglutinating lectin from *Phaseolus vulgaris* (L₄-PHA)

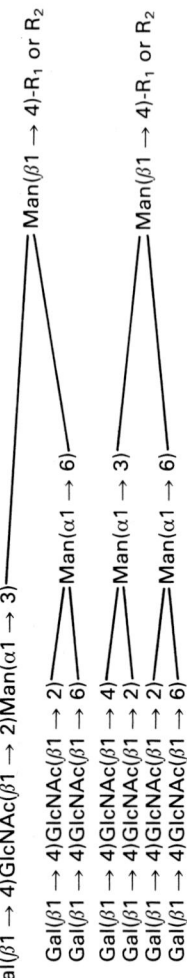

$$Gal(\beta1 \rightarrow 4)GlcNAc(\beta1 \rightarrow 2)Man(\alpha1 \rightarrow 3)$$
$$Gal(\beta1 \rightarrow 4)GlcNAc(\beta1 \rightarrow 2)$$
$$Gal(\beta1 \rightarrow 4)GlcNAc(\beta1 \rightarrow 6)$$
$$Man(\alpha1 \rightarrow 6)$$
$$Man(\beta1 \rightarrow 4)\text{-}R_1 \text{ or } R_2$$

$$Gal(\beta1 \rightarrow 4)GlcNAc(\beta1 \rightarrow 4)$$
$$Gal(\beta1 \rightarrow 4)GlcNAc(\beta1 \rightarrow 2)$$
$$Man(\alpha1 \rightarrow 3)$$
$$Gal(\beta1 \rightarrow 4)GlcNAc(\beta1 \rightarrow 2)$$
$$Gal(\beta1 \rightarrow 4)GlcNAc(\beta1 \rightarrow 6)$$
$$Man(\alpha1 \rightarrow 6)$$
$$Man(\beta1 \rightarrow 4)\text{-}R_1 \text{ or } R_2$$

Wheat germ agglutinin (WGA)[b]

NeuAc(α2 → 6)Gal(β1 → 4)GlcNAc(β1 → 2)Man(α1 → 3)⟍
$$\boxed{\text{GlcNAc}(\beta1 \to 4)}\;\; \boxed{\text{Man}(\beta1 \to 4)\text{-R}_1}$$
NeuAc(α2 → 6)Gal(β1 → 4)GlcNAc(β1 → 2)Man(α1 → 6)⟋

Gal(β1 → 4)GlcNAc(β1 → 4)
Gal(β1 → 4)GlcNAc(β1 → 2)Man(α1 → 3)
Man(α1 → 3)
$$\boxed{\text{GlcNAc}(\beta1 \to 4)}\;\; \boxed{\text{Man}(\beta1 \to 4)\text{-R}_1}$$
Man(α1 → 6)

Erythroagglutinating lectin from *Phaseolus vulgaris* (E₄-PHA)

Gal(β1 → 4)GlcNAc(β1 → 2)Man(α1 → 3)⟍
$$\boxed{\text{GlcNAc}(\beta1 \to 4)}\;\; \text{Man}(\beta1 \to 4)\text{-R}_1\ \text{or R}_2$$
Gal(β1 → 4)GlcNAc(β1 → 2)Man(α1 → 6)⟋

Gal(β1 → 4)GlcNAc(β1 → 4)
Gal(β1 → 4)GlcNAc(β1 → 2)Man(α1 → 3)
GlcNAc(β1 → 4)
Gal(β1 → 4)GlcNAc(β1 → 2)Man(α1 → 6)   Man(β1 → 4)-R₁ or R₂

*Datura stramonium* agglutinin (DSA)

[Gal(β1 → 4)GlcNAc(β1 → 3)]ₙGal(β1 → 4)GlcNAc(β1 → 2)Man(α1 → 3)⟍
[Gal(β1 → 4)GlcNAc(β1 → 3)]ₙGal(β1 → 4)GlcNAc(β1 → 2)Man(α1 → 6)   Man(β1 → 4)-R₁
[Gal(β1 → 4)GlcNAc(ƒ1 → 3)]ₙGal(β1 → 4)GlcNAc(β1 → 6)⟋

[Gal(β1 → 4)GlcNAc(β1 → 3)]ₙGal(β1 → 4)GlcNAc(β1 → 4)
[Gal(β1 → 4)GlcNAc(β1 → 3)]ₙGal(β1 → 4)GlcNAc(β1 → 2)Man(α1 → 3)
[Gal(β1 → 4)GlcNAc(β1 → 3)]ₙGal(β1 → 4)GlcNAc(β1 → 2)
[Gal(β1 → 4)GlcNAc(β1 → 3)]ₙGal(β1 → 4)GlcNAc(β1 → 6)Man(α1 → 6)   Man(β1 → 4)-R₁

[a] R₁: GlcNAc(β1 → 4)GlcNAc(β1 → N)Asn; R₂: GlcNAc(β1 → 4) [Fuc(α1 → 6)]GlcNAc(β1 → N)Asn.
[b] Sequences in boxes are the minimal oligosaccharide structure necessary for lectin recognition.

**Table 6**. Structure of *N*-glycosylpeptides used for studying immobilized lectin specificity

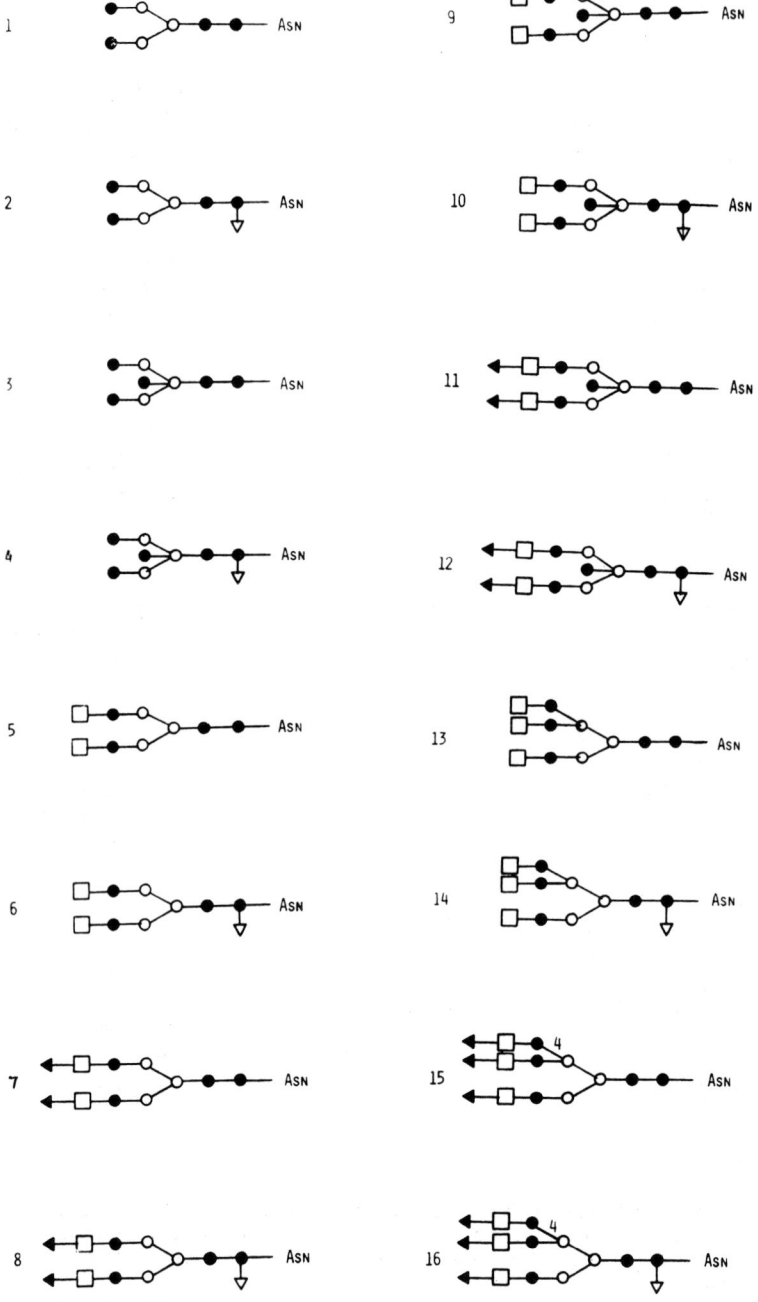

**210**

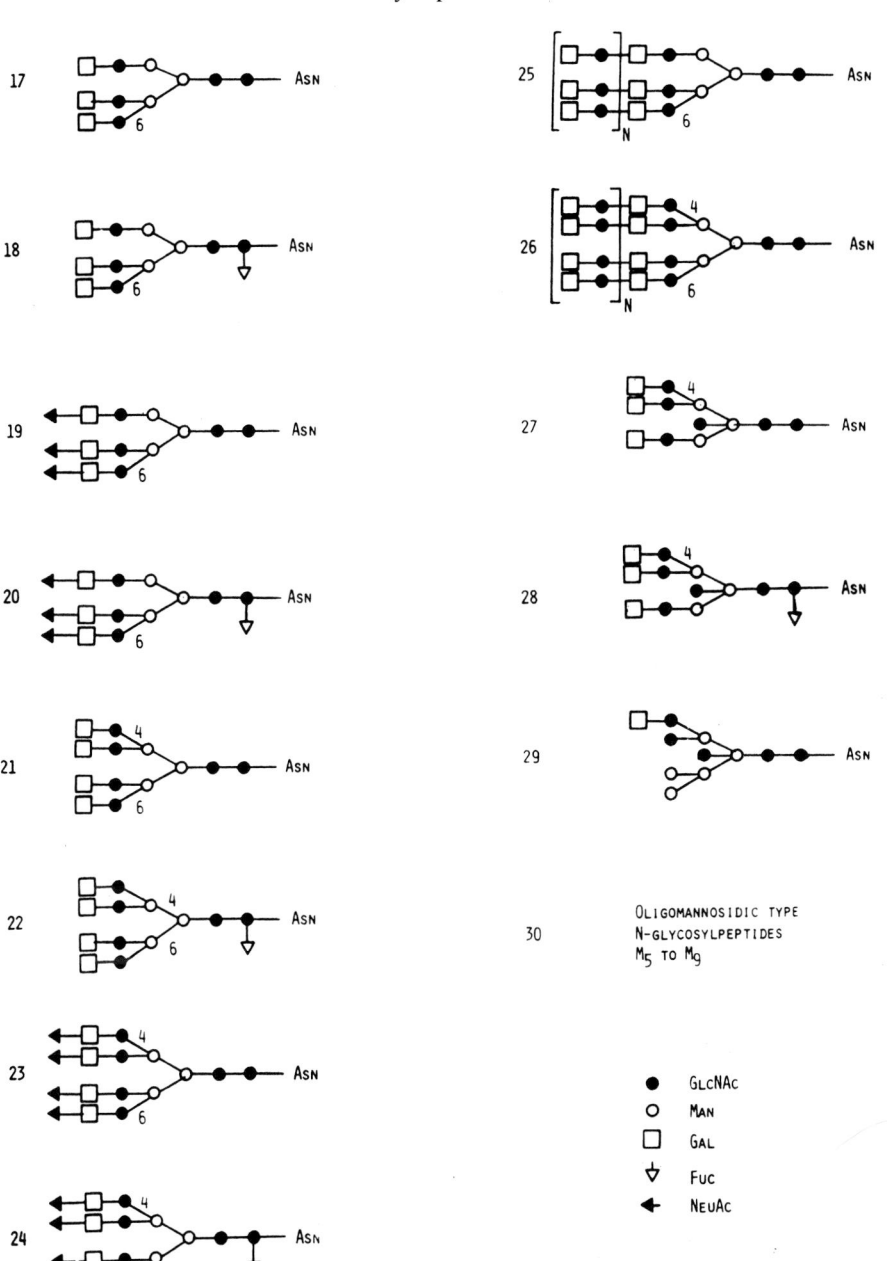

Some of the structures are detailed in *Figures 4–7* and in *Table 5*. Numbers given in the schemes refer to glycosidic linkages: 4: $\beta 1 \rightarrow 4$; 6: $\beta 1 \rightarrow 6$. In all schemes, the antennae linked to the $\alpha$-1,3 and $\alpha$-1,6-mannose residues are the upper and the lower ones, respectively.

**Protocol 11.** Activation of Sepharose 4B with CNBr according to March *et al.* (35) and coupling of lectins

1. Wash 50 ml of wet Sepharose 4B beads (Pharmacia) on a sintered glass filter with 1 litre of chilled distilled water and transfer into a beaker cooled in an ice-bath.

2. Mix 100 ml of ice-cold 2 M $K_2CO_3$ solution with the wet filtered Sepharose beads, under gentle magnetic stirring. Add 6.25 ml of a freshly prepared CNBr solution in acetonitrile (2 g/ml), and maintain at 0 °C for 2 min in a well ventilated hood.

3. Quickly pour the entire reaction mixture into a sintered glass filter funnel and wash extensively with 3 litres of ice-cold distilled water. Transfer the wet activated Sepharose into a 200 ml measuring cylinder and add the ice-cold lectin solution to be coupled (0.25 g of lectin dissolved in 0.5 M NaCl/0.2 M $NaHCO_3$/0.2 M haptenic sugar, pH 8.5). Stopper the measuring cylinder and gently stir for 24 h at 4 °C. Filter the reaction mixture through a sintered glass filter funnel and wash the coupled gel extensively with 1 litre each of ice-cold distilled water, 0.1 M $NaHCO_3$ and water, successively.

4. Block the unreacted iminocarbonate groups by suspending the preparation in 100 ml of a 1 M glycine solution, under gentle stirring, for 3 h at room temperature.

5. Estimate the approximate amount of uncoupled lectin by measuring the absorbance at 280 nm (protein content) of the reaction mixture and recover the immobilized lectin by filtration, after a 24 h coupling period.

6. Wash the glycine-blocked gel extensively with cold distilled water and then with the equilibration buffer containing 0.02% sodium azide. Store the immobilized lectins at 4 °C in this equilibration buffer until used.

**Protocol 12.** Immobilization of lectins on polyacrylhydrazido-Sepharose (36)

1. To washed polyacrylhydrazido-Sepharose (Miles), add 3 vol. of 10% (w/v) glutaraldehyde solution, with gentle stirring, for 4 h at 4 °C.

2. Dissolve the lectin to be coupled in 0.1 M $NaHCO_3$/0.9 M NaCl/0.1 M haptenic monosaccharide (4–5 mg of lectin/ml of coupling buffer) and add to the derivatized gel (4–5 mg of lectin/ml of settled gel beads). Stir the mixture slowly overnight.

**3.** Wash the gel with phosphate-buffered saline (PBS) and add NaBH$_4$ dissolved in a gel volume of PBS (0.5 mg naBH$_4$/ml of settled gel), for 3 h at 4 °C. After washing with PBS containing 0.02% sodium azide, store the immobilized lectin at 4 °C.

## 4.3 Fractionation of glycoproteins

### 4.3.1 General remarks

Affinity chromatography on different immobilized lectins has been widely used to purify various glycoproteins and many examples have been reviewed elsewhere (37, 38) so that we shall restrict ourselves to some particular points which could be useful during such fractionations.

i. *Use of crossed immuno-affinity electrophoresis of glycoproteins as a guide for lectin affinity chromatography*

Crossed immuno-affinity electrophoresis (CIAE), introduced by Bøg-Hansen (39, 40), combines the interaction between lectins and glycoproteins in the first dimension electrophoretic step with electrophoresis into an antibody-containing gel in a second dimension. This sensitive and powerful technique can give important information: first, about interaction between a given lectin and the glycoprotein to be purified, a good correlation being generally observed between the results obtained by affinity electrophoresis and by affinity chromatography on the immobilized lectin; and secondly about the microheterogeneity of carbohydrate moieties of the glycoproteins to be purified. However, parameters such as column binding capacity and elution conditions cannot be predicted from the affinity electrophoresis experiments.

ii. *Binding capacity of the immobilized lectins and problems of the elution*

Variations in the amount, as well as in the quality of the lectin immobilized on a gel, is an important factor influencing the binding of glycoproteins. Affinity differences are often observed between different commercially available immobilized Con A preparations. Similarly, affinity of WGA–Sepharose 4B for glycoproteins varies with the density of lectin coupled to the gel. This implies the calibration of the column before using a new batch of immobilized lectin with well known glycoproteins or glycopeptides such as $\alpha_1$-acid glycoprotein containing di-, tri,-, and tetra-antennary glycans (see *Figure 5.7*) and thyroglobulin containing glycans of the oligomannosidic type (see *Figure 4.1*). Three fractions are generally obtained, reflecting relative affinities of the immobilized lectins for the glycoproteins.

(a) The *non-reactive compounds* are eluted at the void volume of the column with the equilibration buffer. This fraction, when the exact capacity of the immobilized lectin is not known, must be submitted to a new cycle of adsorption and elution on the same column, to be sure that the immobilized lectin was not saturated during the first run.

(b) The *weakly reactive components* give a fraction which is obtained by elution with the equilibration buffer. The separation of weakly differently interacting glycoproteins can even be obtained by using a long and thin column which is more efficient than a wider column containing the same volume of immobilized lectin.

(c) A *strongly reactive fraction* which is specifically desorbed by addition of the appropriate sugar at a concentration between 0.1 and 0.5 M. According to the commercial origin of the immobilized lectin, a weakly reactive glycoprotein can be either bound and eluted with the lectin reactive fraction or unbound and eluted with the lectin non-reactive fraction.

The spatial conformations of the native glycoprotein or non-specific hydrophobic interactions may modulate the accessibility of some glycans to the lectin, which can explain artefactual lectin weakly reactive fractions. This inconvenience disappears when the glycoproteins are reduced and alkylated before fractionation on the immobilized lectin. In conclusion, the discovery of such lectin weakly reactive glycoprotein fractions must be considered carefully.

Usually, immobilized lectin columns are eluted, after extensive washing with the equilibration buffer, with the same buffer but containing the haptenic monosaccharide (0.01–0.5 M). However, when glycoproteins differing in their lectin reactive oligosaccharide residues are fractionated, it is interesting to elute the column with a concentration gradient of the haptenic sugar. Sometimes, the recovery of some bound glycoproteins is very low, even after elution with high concentration of haptenic sugar (0.5 M). This is the result either of a very high affinity of the lectin for some saccharidic determinants of the glycoproteins, or of multivalent interactions between the immobilized lectin and a very high proportion of the saccharide determinant on the glycoprotein. In the particular case of immobilized Con A, the recovery of high affinity glycoproteins can be improved by raising the temperature of the 0.5 M haptenic sugar solution to 37 °C or even 60 °C. It is noteworthy that most immobilized lectins are less efficient at 37 °C than at room temperature (20 °C) or 4 °C.

The specific displacement of glycoprotein by haptenic sugar is reversible and after extensive washing with the equilibration buffer, the immobilized lectin can be used again with the same efficiency. However, stronger lectin–glycoprotein interactions can be displaced only with non-specific desorption processes which very often cause the irreversible denaturation of the immobilized lectin. Such non-specific displacements can be performed by pH change or with borate buffers of 0.02–0.1 M (pH 8.0), without denaturation of the lectin. High yield recovery can also be obtained by heating the Con A–Sepharose–glycoprotein complex for 3 min at 100 °C, in buffer containing 5% (w/v) sodium dodecyl sulfate and 8 M urea, but with irreversible inactivation of the lectin.

### iii. Sequential affinity chromatography of glycoproteins on different immobilized lectins

As in the case of oligosaccharides and glycopeptides (see below) the sequential use of chromatography on immobilized lectins, possessing different and well defined specificities toward saccharidic determinants, is used to fractionate complex mixtures of glycoproteins into classes, depending on their affinity for the different lectins. Immobilized Con A and WGA are the most currently utilized.

---

**Protocol 13.** Affinity chromatography of glycoproteins on immobilized Con A

1. Equilibrate the lectin column (Con A–Sepharose 4B, 30 cm × 2.0 cm i.d.) in 5 mM sodium acetate buffer (pH 5.2) containing 0.1 M NaCl, 1 mM $CaCl_2$, and 1 mM $MnCl_2$ at a flow rate of 9 ml/h at room temperature. Use 0.02% sodium azide in all buffers.

2. Dissolve the glycoprotein mixture (30–50 mg) in 3 ml of the same buffer and centrifuge in order to remove any insoluble material.

3. Apply the clear glycoprotein solution to the column and elute with the equilibration buffer until the effluent is free from protein. Monitor the effluent absorbance at 280 nm and collect 2 ml fractions. In certain cases, Con A weakly reactive variants can be recovered as a retarded fraction by elution with the equilibration buffer.

4. Remove the weakly retained glycoproteins with 0.01 M methyl α-D-glucoside in the equilibration buffer and the strongly Con A reactive glycoproteins with 0.3 M methyl α-D-glucoside solution.

5. Regenerate the Con A column by extensive washing with the equilibration buffer.

6. Pool the separated glycoprotein fractions and extensively dialyse against distilled water before freeze-drying. When the exact capacity of the Con A column is not known, the unretained fraction must be submitted to a new fractionation cycle on the regenerated column.

---

---

**Protocol 14.** Affinity chromatography of glycoproteins on
immobilized WGA

1. Purify the WGA from wheat germ by affinity chromatography on a
   column of 2-acetamido-*N*-ε-aminocaproyl-2-deoxy-*β*-glucopyrano-
   sylamine–Sepharose according to Lotan *et al.* (41). Couple the purified
   lectin to Sepharose 4B according to the procedure described above in
   order to obtain 5 mg immobilized WGA/ml of gel. Equilibrate the WGA
   column (20 cm × 2.5 cm i.d.) in 0.01 M PBS (pH 7.2), containing 0.02%
   sodium azide, at a flow rate of 9 ml/h.

2. Dissolve the glycoproteins (20 mg) in 3 ml of PBS and apply the clear
   glycoprotein solution to the column. Elute with PBS and monitor the
   effluent absorbance at 280 nm. As for Con A affinity chromatography,
   WGA weakly reactive glycoproteins can be recovered, as retarded
   fractions, by elution with the equilibration buffer.

3. Elute the retained glycoproteins with 0.1 M *N*-acetylglucosamine in
   PBS. Pool the separated glycoprotein fractions and dialyse extensively
   against distilled water before freeze-drying.

4. Regenerate the WGA column by extensive washing with PBS.

---

### 4.3.2 Limitations of lectin affinity chromatography for purification of glycoproteins

Affinity chromatography on immobilized lectins rarely succeeds in obtaining
pure glycoproteins from complex mixtures. Generally, lectin affinity chromatography results in an enrichment of classes of different heterogeneous
glycoproteins, possessing similar carbohydrate determinants recognized by
the immobilized lectin and called 'lectin receptors'. As different lectins are
able to recognize different saccharidic sequences belonging to the same
glycan class and as these glycans are likely to be common to numerous
glycoproteins, the different lectins will react in fact with a broad spectrum of
glycoproteins.

In addition, non-specific interactions between glycoproteins and lectins
often restrict the use of immobilized lectins for the fractionation of
glycoproteins.

## 4.4 Fractionation of glycopeptides and oligosaccharides

If purification and fractionation of glycoproteins present some limitations
which restrict the use of the affinity chromatography on immobilized lectins
for isolating glycoproteins, the procedure represents, in contrast, a powerful
tool for the fractionation of glycopeptides and glycans. This is essentially due
to the fact that the above mentioned non-specific interactions often observed

between glycoproteins and lectins are very rare in the case of glycans and glycopeptides so that these compounds interact specifically with the lectins and can be fractionated on the basis of an 'actual' affinity chromatography. Consequently, this implies, even more than in glycoprotein fractionations, a very precise knowledge of the exact specificities of the lectins to be used. That is to say that, once the precise specificity of an immobilized lectin is well defined, it becomes possible to predict the primary structure of bound glycopeptides or oligosaccharides.

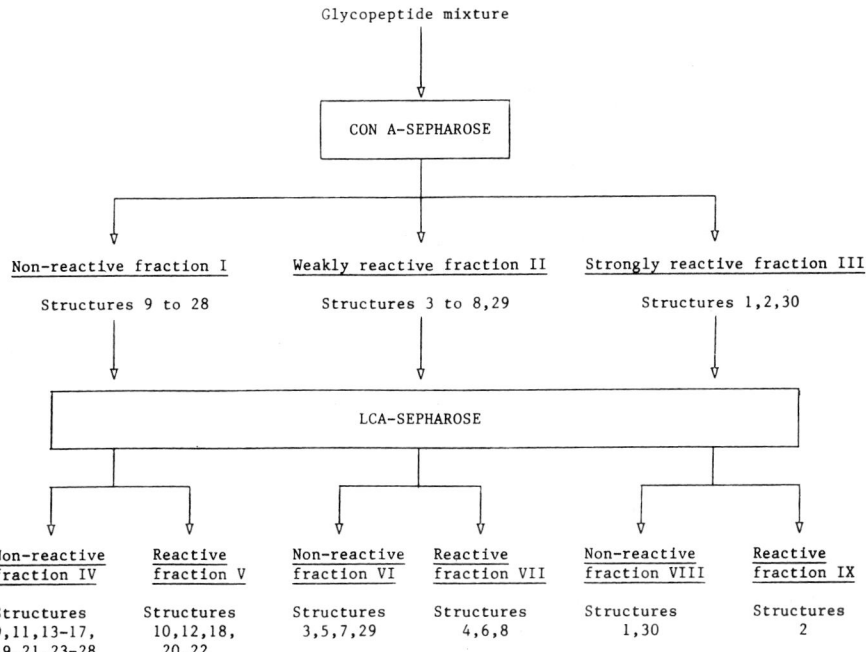

**Figure 14.** Scheme of fractionation of *N*-glycosylpeptides by combining affinity chromatography on immobilized ConA– and LCA–Sepharose. Arabic numerals refer to glycopeptide structures of *Table 6*.

---

**Protocol 15.** Affinity chromatography of glycopeptides and oligosaccharides on immobilized Concanavalin A

1. Equilibrate the Con A–Sepharose column (10 cm × 1.0 cm i.d.) in 5 mM sodium acetate buffer (pH 5.2), containing 0.1 M NaCl, 1 mM CaCl₂, 1 mM MnCl₂ and 1 mM MgCl₂, at a flow rate of 9 ml/h at room temperature. Use 0.02% sodium azide in all buffers.

**Protocol 15.** *Continued*

2. Dissolve the glycopeptides (or the oligosaccharides) in 1 ml of equilibration buffer and apply the clear solution to the column.

3. Elute (1.5 ml fractions) successively with five column volumes of equilibration buffer, then with three column volumes of 0.01 M methyl α-D-glucoside and finally with five column volumes of 0.3 M methyl α-D-glucoside. In this way, oligosaccharides and glycopeptides are fractionated into three classes (*Figure 14*).

   (a) *Non-reactive glycopeptides* (or *oligosaccharides*), eluted at the void volume of the column with the equilibration buffer.

   (b) *Weakly reactive derivatives* obtained by eluting with the equilibration buffer itself. However, depending on the amount of immobilized ConA/ml of Sepharose, the same type of compounds can be weakly retained and may be eluted as a sharp peak with low concentration of the haptenic sugar derivative (0.01 M methyl α-D-glucoside).

   (c) *Strongly reactive components*, eluted as a sharp or broad tailing peak with 0.3 M methyl α-D-glucoside.

4. Purify the fractions from salts and haptenic sugar by gel filtration on a Bio-Gel P-2 column (50 cm × 2 cm i.d.) equilibrated with distilled water. After elution with the last haptenic sugar solution, regenerate the Con A column by extensive washing with the equilibration buffer.

*Notes*

(a) The saccharidic structures (listed in *Table 6*) constituting each of the obtained fractions are described in *Figure 14*.

(b) Glycans released from glycosylpeptides, either by hydrazinolysis or by the action of various endo-*N*-acetyl-*β*-D-glucosaminidases, present the same behaviour on immobilized Con A as glycopeptides.

(c) Sometimes, high affinity oligosaccharides are eluted from the column only by raising the temperature of the sugar solution to 60 °C or by irreversible denaturation of the lectin with two column volumes of a 1% (w/v) solution of SDS. In such a case, the column cannot be reused.

(d) Before using a new batch of immobilized Con A, calibration of the immobilized lectin with well-defined glycopeptides must be performed.

(e) Extensive use leads to a decrease of the binding capacity of Con A–Sepharose due to lectin leakage.

(f) When the exact binding capacity of the gel is not known, the unretained fraction must be recycled using the regenerated column. Commercially available Con A–Sepharose (Pharmacia) is able to bind

50–75 µg of biantennary *N*-glycosylpeptides, obtained from human serotransferrin, per millilitre of gel.

(g) Before chromatography, glycopeptides can be $^{14}$C- or $^{3}$H-labelled by *N*-acetylation according to the general procedure described above (see *Protocol 2*), by using labelled acetic anhydride. In the same way, oligosaccharides can be labelled by reducing the terminal monosaccharide residue with tritiated sodium borohydride by applying the general procedure of reduction and purification as described in *Protocol 1*.

---

**Protocol 16.** Affinity chromatography on immobilized *Lens culinaris* agglutinin

LCA–Sepharose 4B can be purchased from different manufacturers or easily prepared from *Lens culinaris* seeds (42) and coupled to CNBr-activated Sepharose 4B, at a concentration of 5 mg of lectin/ml of gel. Equilibrate the lectin column (15 cm × 2.5 cm i.d.) in PBS (pH 7.4) containing 0.02% sodium azide (divalent cations such as $Ca^{2+}$ or $Mn^{2+}$ are not required), at a flow rate of 9 ml/h at room temperature.

*Method*

1. Dissolve the glycopeptides in 1 ml of PBS and apply the clear solution to the column.

2. Elute (1.5 ml fractions) the non-reactive glycopeptides (at the void volume) by passing five column volumes of PBS, and the lectin reactive glycopeptides with five column volumes of PBS containing 0.15 M methyl α-D-glucoside. As for immobilized Con A, fractionation can be easily followed by labelling the glycopeptides by *N*-acetylation with [$^{14}$C] or [$^{3}$H]acetic anhydride.

3. Purify the fractions from salts and haptenic sugar by gel filtration on a Bio-Gel P-2 column (50 cm × 2 cm i.d.) equilibrated with distilled water.

4. Regenerate the LCA column by extensive washing with PBS.

*Notes*

(a) The saccharidic structures constituting each of the fractions are described in *Figure 14*. From these results, it clearly appears that the α-1,6-linked fucose residue is determinant for the binding of glycopeptides and that affinity chromatography on immobilized LCA is a valuable tool for separating glycopeptides into two classes: α-1,6-fucosylated and non-fucosylated.

**Protocol 16.** *Continued*

(b) Only glycopeptides can be fractionated on immobilized LCA, not oligosaccharides. So, glycans released from *N*-glycosylpeptides by hydrazinolysis/re-*N*-acetylation or from *O*-glycopeptides by reductive *β*-elimination do not interact with immobilized LCA and are eluted in the void volume.

(c) As in the above case, immobilized *Vicia faba* and *Pisum sativum* lectins interact only with fucosylated glycopeptides. However, their affinity is weaker, so that the fractionation into the two classes is achieved by using only the equilibration buffer.

---

**Table 7.** Composition of the equilibration and elution buffers and of the lectin reactive fractions obtained

| Immobilized lectin | Mode of elution | | Composition of the lectin reactive fractions[a] |
|---|---|---|---|
| | Equilibration buffer | Elution buffer | |
| L₄-PHA[b] | PBS | PBS | 17, 18, 21, 22 |
| E₄-PHA[c] | PBS | PBS | 9, 10, 27, 28 |
| WGA | PBS | 0.1 M GlcNAc in PBS | 11, 29 |
| RCA₁ | PBS | 0.15 M Gal in PBS | 5, 6, 13, 14, 17, 18, 21, 22 |
| DSA | PBS | 8 mg/ml of a mixture of *N,N'*-diacetyl-chitobiose and *N,N',N''*-triacetylchito-triose in PBS | 25, 26 |
| AAA | PBS | 0.05 M fucose in PBS | 2, 4, 6, 8, 10, 12, 14, 16, 18, 20, 22, 24, 28 |

[a] Numbers refer to the structures in *Table 6*.
[b] *Phaseolus vulgaris* leukoagglutinating lectin.
[c] *Phaseolus vulgaris* erythroagglutinating lectin.

---

### 4.4.1 Affinity chromatography on other lectins

Other immobilized lectins are also used, but to a lesser extent, for fractionating glycopeptides or oligosaccharides: PHA, WGA, RCA, and DSA. In all cases, the general procedures are identical to that described for immobilized Con A and LCA. They differ only by the method of elution (*Table 7*).

The following general comments apply to this procedure:

(a) *N, N'*-Diacetylchitobiose-Asn inner-core is an important determinant for binding to immobilized WGA and LCA, so that free glycans and oligosaccharides, devoid of the Asn residue, do not interact with the immobilized lectins. Substitution by an α-L-fucose residue at the C-6 position of the N-acetylglucosamine residue involved in the N-glycosylamine bond reduces the affinity of immobilized WGA for glycopeptides.

(b) Immobilized WGA can also interact with glycopeptides possessing numerous clustered O-glycosidically linked sialyloligosaccharides.

(c) Asialo, bi-, tri-, and tetra-antennary glycans of the N-acetyllactosamine type can be fractionated on immobilized $RCA_I$. However, the separation depends upon the amount of bound $RCA_I$/ml of gel so that immobilized $RCA_I$ columns must be carefully calibrated with known glycopeptides before use.

(d) Some lectins possess a specificity directed toward the terminal non-reducing sugars and are interesting tools for the fractionation of particular glycopeptides. For example, the immobilized $A_4$ tetrameric isolectin from *Griffonia simplicifolia* agglutinin I ($GSA_I$-$A_4$) presents a strict specificity toward N-acetyl-α-D-galactosamine in the terminal non-reducing position (43), while the immobilized $B_4$ tetrameric isolectin from *G. simplicifolia* agglutinin I ($GSA_I$-$B_4$) is specific for glycopeptides with terminal non-reducing α-D-galactose residues (44). Another lectin ($GSA_{II}$), also isolated from the seeds of *G. simplicifolia*, is a very useful tool for fractionating glycopeptides or oligosaccharides with N-acetyl-α-D-glucosamine residues in a terminal non-reducing position (45). The same specificity has been recently found in the lectin of the mushroom *Psathyrella velutina* (PVA) (46).

Oligosaccharides and glycopeptides with terminal non-reducing β-D-GalNAc residues can be fractionated on the immobilized lectin isolated from *Wisteria floribunda* seeds (47).

Structures of the oligomannosidic-type possessing the sequence $Man(α1 \rightarrow 2)Man(α1 \rightarrow 6)Man(α1 \rightarrow 6)$ are specifically and strongly recognized by the lectin of *Bowringia milbreadii* (48). Sialylated oligosaccharides, glycopeptides, and glycoproteins can be characterized and fractionated by using the plant lectins of *Maackia amurensis* (49) specific for the terminal trisaccharide $NeuAc(α2 \rightarrow 3)Gal(β1 \rightarrow 4)$ GlcNac, of *Sambucus nigra* agglutinin (50) specific for the trisaccharide $NeuAc(α2 \rightarrow 6)Gal(β1 \rightarrow 4)GlcNAc$ or $NeuAc(α2 \rightarrow 6)GalNAc$, and of the beetle *Allomyrina* (51) specific for the sequence $NeuAc(α2 \rightarrow 6)$ $Gal(β1 \rightarrow 4)GlcNAc$.

(e) Two immobilized fucose-binding lectins (*Lotus tetragonolobus* (LTA) and *Ulex europeus* ($UEA_I$) agglutinins) have no affinity for fucosylated glycopeptides or oligosaccharides of the N-acetyllactosamine type. Only the $Fuc(α1 \rightarrow 6)GlcNAc(β1 \rightarrow N)Asn$ glycopeptide is retained on both immobilized lectins and is eluted with 0.1 M L-fucose in PBS. However, a new fucose-specific lectin, isolated from the mushroom *Aleuria aurantia*, is a very good tool for fractionating fucosylated glycopeptides and oligosaccharides (52, 53).

### 4.4.2 Sequential affinity chromatography on immobilized lectins

Most of the *N*-glycosylpeptides can be fractionated into relatively homogeneous classes by sequential affinity chromatography (29, 32, 54–56). Up to now, the associations Con A– and LCA–Sepharose, and of Con A–, LCA–, and L4-PHA–Sepharose are the most commonly utilized for fractionating the glycopeptides obtained from cell membrane glycoproteins. This methodology is rapid and sensitive, particularly when glycopeptides are radiolabelled by using $^{14}$C- and $^{3}$H-labelled precursors, or by *N*-acetylation carried out with [$^{14}$C]acetic anhydride. *Figure 14* presents the six classes of glycopeptides obtained by sequential affinity chromatography on immobilized Con A– and LCA–Sepharose and *Figure 15* shows the results obtained by subfractionating fractions IV and V isolated by this procedure on immobilized *Phaseolus vulgaris* leukoagglutinating lectin L4-PHA.

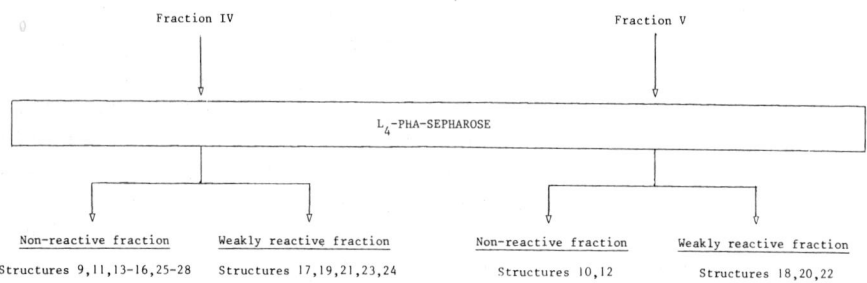

**Figure 15.** Scheme of subfractionation on immobilized L4-PHA of *N*-glycosylpeptide fractions IV and V from *Figure 14*. Arabic numerals refer to glycopeptide structures of *Table 6*.

# 5. Colorimetric assays of carbohydrates in glycoproteins and glycopeptides

## 5.1 Glycan composition

The composition of a glycoprotein or of a glycan in terms of neutral sugars, hexuronic acids, hexosamines, and sialic acids is usually determined by specific colorimetric procedures which have been described in Chapter 1 (see also refs 57–60).

The determination of neutral monosaccharides, of hexuronic acids and, in some cases, of sialic acids is carried out without prior hydrolysis of the glycosidic linkages since the conditions of the reactions developed in the presence of sulfuric acid ensure the complete breakdown of all glycosidic bonds. This is not the case for the determination of hexosamines which requires the prior liberation of the amino sugars, as well as for the determination of sialic acids by the thiobarbituric acid method. Therefore, one of the major problems concerns the risk of destruction of neutral mono-saccharides and hexuronic acids during the reaction in acidic medium, on the one hand, and of incomplete liberation of hexosamines and sialic acids by chemical or enzymatic hydrolysis, on the other hand. In order to limit and to control the destruction of neutral monosaccharides and hexuronic acids, the temperature of reaction mixtures must be identical in all the test-tubes during the assay and the rate of flow of the acid down the test-tube, the mixing of solutions, and the cooling must be standardized. In order to obtain maximal liberation of conjugated hexosamines and sialic acids without degradation of the liberated sugars, kinetics of hydrolysis using increasing concentrations of acidic or enzymatic agents have to be determined on each sample of glycan or glycoprotein.

The second major problem encountered during sugar determinations concerns the specificity of the reaction. In fact, the determination of neutral sugars and hexuronic acids is based on the production of furan derivatives in hot acid solution followed by the condensation of the chromogens with a specific reagent. The formed chromophores develop a characteristic coloration which is proportional to the sugar concentration. However, the specific absorbances show a wide variation between the monosaccharides. In order to eliminate this cause of error, the internal standards must be prepared with a mixture of monosaccharides in the same molar ratio as expected in the analysed compound. In each series of determinations, appropriate blanks must be used containing the solution used for dissolving the analysed samples.

## 5.2 Colorimetric determination of neutral monosaccharides

Solutions of phenol, cysteine, and orcinol in sulfuric acid are the reagents most commonly used for quantitative determination of hexoses, 6-deoxyhexoses, methyl pentoses, and pentoses (see Chapter 1, Section 3).

## 5.3 Hexuronic acids

Naphthoresorcinol, sulhydryl compounds, and carbazole in hydrochloric or sulfuric acid solutions are the most common reagents used for the detection and the quantitative determination of hexuronic acids. We shall restrict ourselves to the description of the methods of Bitter and Muir's using carbazole reagent (see Chapter 1, *Protocol 8*) and Blumenkrantz and Asboe-Hansen using the *meta*-hydroxydiphenyl. This latter method is recommended for determining the acid mucopolysaccharides in biological materials.

---

**Protocol 17.** *Meta*-hydroxydiphenyl-sulfuric assay for acid
mucopolysaccharides

*Reagents*

- 0.15% (w/v) solution of *m*-hydroxydiphenyl in 0.5% (w/v) NaOH—the reagent is stable at 4 °C for more than one month

- 0.0125 M solution of sodium tetraborate in concentrated sulfuric acid
- Standard aqueous solution containing 100 µg of uronic acids/ml

*Method*

1. Add 1.2 ml of sulfuric–tetraborate to 0.2 ml of glycan, glycoprotein (0.5–20 µg of total uronic acids) or of standard solution.

2. Cool the tubes in crushed ice, shake vigorously.

3. Maintain the tubes in a boiling water-bath for 5 min, cool in a water-ice bath, and add 20 µl of the *m*-hydroxydiphenyl reagent.

4. Shake the tubes and determine the absorbances at 520 nm[a].

   [a] The absorbances given by glucuronic, iduronic, and galacturonic acids are very similar, while the absorbance given by mannuronic acid is about half this value. The interference of hexoses is very low. Therefore the method presents the advantage of the specificity.

---

## 5.4 Hexosamines and *N*-acetylhexosamines

All the procedures for the determination of free hexosamines derive from the Elson–Morgan reaction which involves a condensation with acetylacetone in alkaline solution leading to the formation of pyrrole derivatives which develop with the *p*-dimethylaminobenzaldehyde Ehrlich's reagent a specific pink coloration. The method proposed by Belcher *et al.* (61) will be described here.

Free *N*-acetylhexosamines are determined according to a different procedure proposed by Morgan and Elson and based on pink chromophore production by direct heating with the Ehrlich's reagent in alkaline solution (see Chapter 1, *Protocol 7*). On the same principle, free hexosamines can be determined after re-*N*-acetylation.

---

**Protocol 18.** Colorimetric determination of conjugated
hexosamines (61)

*Hydrolysis*

Hydrolysis of the very stable hexosaminyl linkages is usually carried out with 4 M HCl at 100 °C for 4 h. However, as the stability of this type of linkage depends on the structure of the glycans, this procedure cannot be considered as a general one. The best way is to determine for each glycoprotein, the optimal hydrolysis conditions by using HCl at concentrations ranging from 2 to 6 M and by determining the kinetics of hydrolysis (2–10 h). HCl is then removed under vacuum in the presence of NaOH. The dry product is dissolved in water and centrifuged.

### Reagents

- Acetylacetone reagent freshly prepared by adding 2 ml of acetylacetone to 48 ml of 0.625 M $Na_2CO_3$ (dried overnight at 100 °C)
- Ehrlich's reagent prepared by dissolving 1.6 g of recrystallized *p*-dimethylaminobenzaldehyde in 30 ml of concentrated HCl
- Standard aqueous solution containing 100 μg of free hexosamines/ml

### Method

1. Add 0.5 ml of acetylacetone reagent and 1 ml of distilled water to 0.5 ml of the hydrolysate corresponding to 5–50 μg of total hexosamines, or of standard solution.

2. Shake the tubes vigorously, cap with glass marbles to minimize evaporation, and heat in a boiling water-bath for 10 min.

3. Cool to room temperature, add 2.5 ml of ethanol, mix carefully, and maintain the tubes in a water-bath at 75 ° ± 2 °C for 5 min.

4. Add 0.5 ml of Ehrlich's reagent and heat at 75 °C for 30 min.

5. Cool the tubes to room temperature and add 2.5 ml of 95% ethanol.

6. Determine the absorbance at 520 nm after 30 min. The colour is stable for 24 h.

### Notes

(a) The absorption curve of hexosamines shows a maximum at 520 nm. As the relative extinction coefficients are different (glucosamine = 100; galactosamine = 140), standard solutions must contain a mixture of glucosamine and galactosamine in the same ratio as in the analysed sample.

(b) Muramic acid and *N*-acetylhexosamines interfere. Sialic acids react by producing a blue coloration. Hexoses and uronic acids do not interfere.

(c) Sodium tetraborate and sodium chloride decrease the intensity of the coloration. If the hydrochloric acid used to liberate the hexosamines is neutralized by NaOH, the standard solution of hexosamines has to be prepared in a NaCl solution having the same concentration as that of neutralized hydrolysates.

## 5.5 Sialic acids

Sialic acids are determined by two different types of procedures. The first group of methods leads to the determination of free and conjugated sialic acids and is based on the reaction of chromogens obtained in acidic medium with reagents such as resorcinol, *p*-dimethylaminobenzaldehyde, and diphenylamine. The second group of methods is applicable only to free sialic acids.

---

**Protocol 19.** Determination of total (free and conjugated) sialic acid (62)

*Reagents*

- Diphenylamine reagent (Dische's reagent) prepared by dissolving 1 g of diphenyl-amine (recrystallized from ethanol) in a mixture of 90 ml of glacial acetic acid and 10 ml of concentrated sulfuric acid—store the reagent at 4 °C in the dark and transfer

  to a water-bath at 40 °C just before use in order to redissolve the crystals
- Solution of 7.5% (w/v) trichloroacetic acid in water
- Standard aqueous solution containing 100 µg of sialic acid/ml

*Method*

1. Add 0.2 ml of trichloroacetic acid at 7.5% and 6 ml of diphenylamine reagent to 0.1 ml of glycan, glycoprotein (1–20 µg of sialic acid), or standard solution.

2. Mix and heat the tubes in a boiling water-bath for 30 min.

3. Cool the tubes in water and determine the absorbance of the violet-blue coloration at 530 nm after a 30 min stay in the dark. Coloration is stable for 6 h.

*Notes*

(a) The extinction coefficient of *N*-glycolylneuraminic acid at the $\lambda_{max}$ 530 nm is 83% of that of *N*-acetylneuraminic acid. Therefore, it is important to use standards containing the two sialic acids in the same ratio as that expected in the analysed solution.

(b) At high concentrations, hexoses yield a blue colour with absorption maxima at 530 nm and 650 nm, as well as ketohexoses whose absorption curve presents maxima at 530 nm and 640 nm. In both cases, molar absorption of neutral sugars and *N*-acetylneuraminic acid are identical. 2-Deoxypentoses, particularly 2-deoxyribose, produce a blue colour with a characteristic spectrum with an absorption maximum at 610 nm. In addition, the molar absorption is twice as high as that of *N*-acetylneuraminic acid. Therefore, the

---

complete absorption spectrum must be determined in order to avoid interference due to the presence of DNA. Methyl pentoses, hexosamines, and *N*-acetylhexosamines do not interfere in the reaction. Glycerol produces a green colour.

### 5.5.1 Quantitative determination of free sialic acids

Due to the external position of sialic acids in glycans and to the lability of their glycosidic linkages, these sugars can be easily removed under very mild acidic conditions or by using neuraminidases. Oxidation of free *N*-acetyl or *N*-glycolylneuraminic acids with periodic acid leads to the formation of formyl-pyruvic acid which reacts with 2-thiobarbituric acid and produces a pink colour with an absorption maximum at 549 nm (see Chapter 1, *Protocol 9*). Details of the colorimetric assays of *N,O*-acylneuraminic acids are described in ref. 63.

### 5.5.2 Hydrolysis of sialyl linkages

*i. Chemical hydrolysis*

Liberation of sialic acids is usually carried out with 0.1 N sulfuric acid at 80 °C for 1 h, or with 0.01 N sulfuric acid at 100 °C for 1 h, or with 2 M acetic acid at 80 °C for 3 h. Under these conditions, all types of sialic acids are removed without any degradation, except *O*-deacetylation. Moreover, the other monosaccharides, with the exception of fucose, are not liberated. Hydrochloric acid is not recommended since lactonization and partial destruction of sialic acids have been observed.

*ii. Enzymatic hydrolysis*

Enzymatic hydrolysis of sialic acid linkages can be achieved by use of neuraminidases from *Clostridium perfringens*, *Vibrio cholerae* or Influenza virus. The specificity of these enzymes as well as the hydrolysis conditions are described in *Protocol 26*.

## 6. Identification and determination of glycan monosaccharides by gas–liquid chromatography

### 6.1 Glycoprotein hydrolysis

For a long time, the determination of the molar composition of glycans encountered an important problem: that of the quantitative liberation of monosaccharides which must be achieved before any chromatographic analysis. In fact, all the glycosidic bonds must be split without any destruction of the liberated sugars. In addition, these bonds vary in stability depending on the nature of the monosaccharides and on the type of glycosidic linkage; the sialyl bonds being the most labile and the uronosidyl and glucosaminyl ones being the most stable toward hydrochloric and sulfuric acids. Therefore, the use of these acids necessitates the determination, for each type of sugar (i.e. hexoses, methyl pentoses, hexosamines, uronic acids, and sialic acids), of different conditions of acid molarity, time, and temperature of hydrolysis. A compromise between the liberation of the monosaccharides and their destruction is obtained. The failure of aqueous hydrochloric or sulfuric acid hydrolysis has been overcome by the introduction, on the one hand, of hydrolysis by trifluoroacetic acid even if sialic acids are destroyed and, on the other hand, the methanolysis in which anhydrous methanol–HCl mixtures are used. All kinds of monosaccharides, including sialic acids, are stable to methanolysis. In any case, the gas-liquid chromatography is the best method for identification and determination of liberated monosaccharides (for reviews, see refs 60, 64, and 65).

## 6.2 The use of alditol acetates

GLC of alditol acetates is widely used for determining the composition of monosaccharide mixtures. In fact, the main difficulty encountered in the separation of the monosaccharides themselves is the formation of at least four glycosides per monosaccharide resulting from anomeric and ring isomerization, each of which giving a peak on the chromatogram. Thus, in a complex mixture containing several monosaccharides, the multiplicity of peaks leads to overlappings so that accurate determinations cannot be achieved (see also Chapter 1, Section 7). On the contrary, an alditol acetate originating from one monosaccharide gives rise to only one peak.

### 6.2.1 Hydrolysis of glycoproteins

Hydrolyse amounts of glycoproteins corresponding to 10 µg of total sugars with 100 µl of 4 M trifluoroacetic acid in the presence of 2 µg of mesoinositol used as internal standard, at 100 °C for 4 h, under helium, in glass tubes fitted with Teflon-lined screw caps (Sovirel, Paris, France; also Pierce). After cooling, evaporate the solution in a desiccator, under vacuum. Do not forget that, in these conditions, sialic acids are totally destroyed.

### 6.2.2 Reduction and analysis of monosaccharides

This is described in Chapter 1, *Protocol 15*.

## 6.3 The use of methylglycoside trifluoroacetates

In the first step, the *O*-glycosidic bonds are quantitatively split by methanolysis, leading to the formation of *O*-methylglycosides which are trifluoroacetylated in a second step. The trifluoroacetates of *O*-methylglycosides (TFA derivatives) are easily separated by GLC with little overlapping. Sialic acids and uronic acids are stable toward methanolysis. We currently use a method derived from the procedure described by Zanetta *et al.* (for review, see ref. (64).

---

**Protocol 20.** Preparation of methylglycoside trifluoroacetates

1. The methanolysis reagent is prepared in Chapter 1, *Protocol 14*.

2. Freeze-dry carefully (complete dehydration of samples is the main condition of success) amounts of glycoproteins corresponding to 10 μg of total sugars to which 2 μg of mesoinositol are added as internal standard.

3. Treat the residue with 200 μl of 0.5 M methanol–HCl for 24 h at 80 °C in glass tubes fitted with Teflon-lined screw caps.

4. Cool, remove HCl and methanol under a stream of nitrogen at 50 °C.

5. Trifluoroacetylate by dissolving the dried residue in a mixture of dichloromethane (50 μl) and trifluoroacetic anhydride (50 μl). Stopper the tube quickly and keep at room temperature overnight.

6. GLC of TFA-*O*-methylglycosides is carried out on 10 μl samples by using a glass column (300 cm × 0.3 cm i.d.) filled with 5% (w/w) Silicone OV 210 on Chromosorb W-HP (mesh 120–140) and a Varian Aerograph model 1400 chromatograph equipped with a flame ionization detector. The carrier gas is helium at a flow rate of 7.5 ml/min. The oven temperature is programmed from 110 °C to 220 °C at 2 °C/min with injector and detector temperatures of 225 °C and 250 °C, respectively. A typical gas–liquid chromatogram of the methylglycoside trifluoroacetates obtained from $\alpha_1$-acid glycoprotein (orosumucoid) is presented in *Figure 16*.

---

## 6.4 The use of trimethylsilylated methylglycosides

This is described in Chapter 1, *Protocol 14*. Before carbohydrate analysis, purify all glycoproteins (10 μg) by rapid gel filtration on Aquagel type P-2 10 μm (Polymer Laboratories, Shropshire, England) column (30 cm × 0.7 cm i.d.) fitted to a Spectra-Physics Model 8770 liquid chromatograph (with a 500 μl sample loop), eluting with water (0.4 ml/min). Detect glycoproteins with an UV detector (L.K.B. 2138 Uvicord S) at 206 nm. Collect the excluded fraction in a glass tube fitted with Teflon-lined screw cap and freeze-dry in the presence of 0.1 μg of internal standard mannitol. A typical gas–liquid chromatogram is given in *Figure 17*.

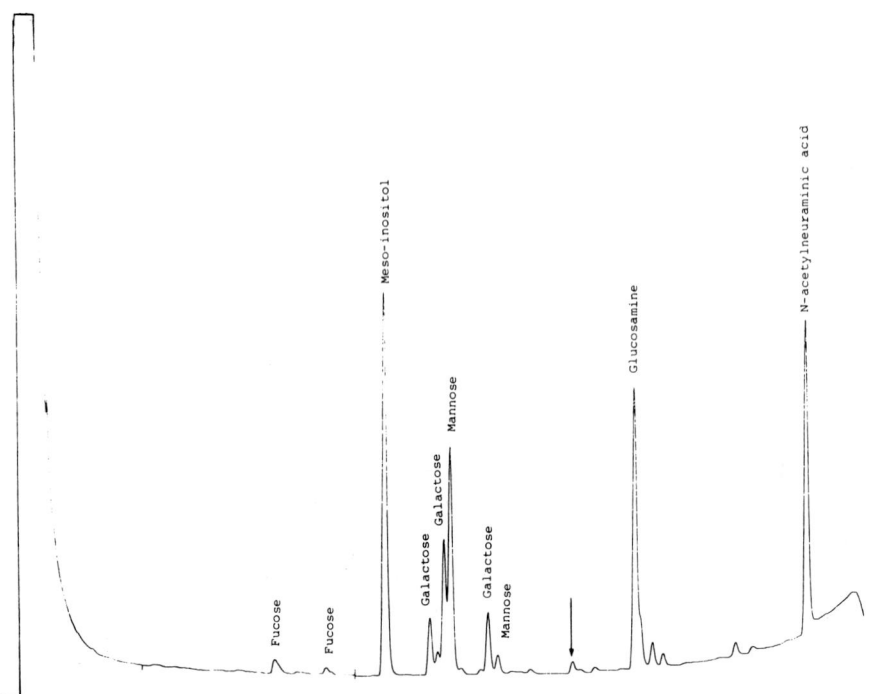

**Figure 16.** GLC (glass column of 5% (w/w) Silicone OV210) of methylglycoside trifluoroacetates obtained from $\alpha_1$-acid glycoprotein. Arrow indicates the trifluoroacetate derivative of glucosamine.

### 6.4.1 General comments on the GLC of glycoprotein derived monosaccharides

(a) Methanolysis offers, as compared to hydrolysis, the considerable advantage of cleaving all *O*-glycosidic linkages in a one-step procedure. However, we must keep in mind the following failures:

    *i.* Under the usual conditions (0.5 M methanolic HCl, 24 h, 80 °C), the linkage between GlcNAc and Asn is split to only a very limited extent and the glucosamine is liberated as a free monosaccharide (see *Figure 16*) and not as a methyglycoside.

    *ii.* Dehydration of monosaccharides can occur. For example, *N*-acetyl-neuraminic acid gives 3% of 2,7 anhydro-*N*-acetylneuraminic acid, and alditols, like 2-acetamido-2-deoxy-D-galactitol coming from glycan removed by *β*-elimination in reductive conditions (see Section 2.1), also give anhydro-derivatives.

(b) Methanolysis must be carried out under anhydrous conditions. Thus, carefully dry all materials and samples, and be careful not to introduce water in the methanol–HCl reagent!

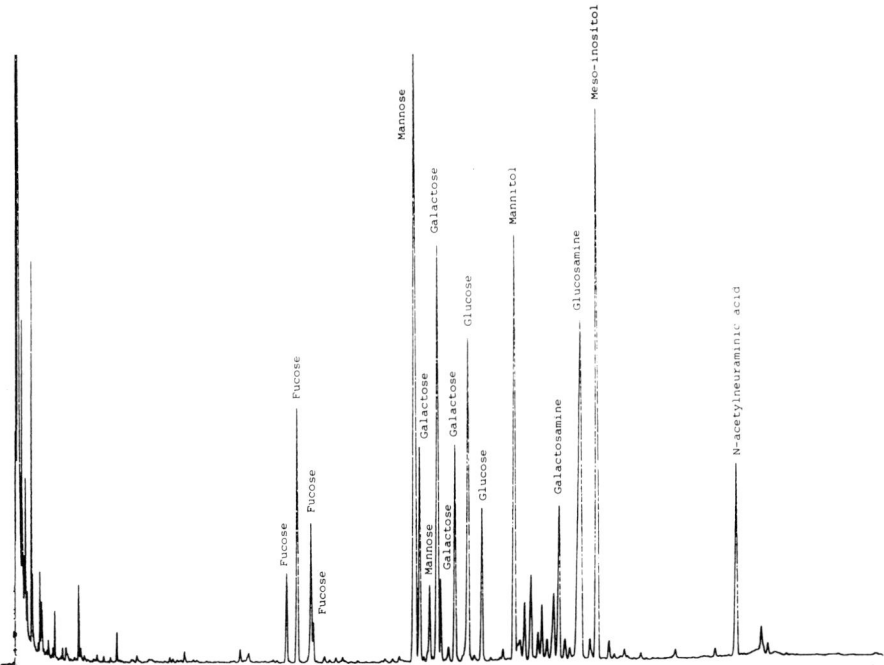

**Figure 17.** GLC (glass capillary column of Silicone OV101) of trimethylsilylated methylglycosides of fucose, mannose, galactose, glucose, galactosamine, glucosamine, and *N*-acetylneuraminic acid.

(c) The lower limit of sensitivity is not due to the detector response but to contaminants which accumulate during the operations (e.g glucose and xylose released from dialysis membranes, glucose from the Sephadex columns or from sucrose gradients used for fractionation of glycoproteins, and various materials from insufficiently cleaned glassware).

# 7. Methylation

Methylation analysis is one of the most powerful procedures for structural analysis of polysaccharides and glycoprotein glycans (for reviews, see refs 65–67). It involves the permethylation of all free hydroxyl groups of the sugars followed by the liberation of methylated monosaccharides by hydrolysis or methanolysis and the qualitative and quantitative analysis of the methylated derivatives. The positions of the free hydroxyl groups of the partially methylated monosaccharides gives the positions in which the sugar residues was glycosylated. Many methylation methods have been reported since the first procedure described in 1903 by Purdie and Irvine but most of the proposed methods led to an incomplete methylation of saccharides. In this connection, Hakomori's procedure (68) using sodium hydride and dimethyl sulfoxide represented real progress by producing a more powerful nucleophile than the bases previously used. The method leads to a rapid and complete *O*-methylation and *N*-methylation (of the acetamido group of hexosamine residues) without any loss of *N*-acetyl groups.

Because the sodium hydride contains impurities (mainly due to the oil in which it is dispersed), cleaner reagents were proposed to perform the methylation of small amounts of sugars. In this connection, we have produced a procedure based on the use of dimethyl sulfoxide and *n*-butyl lithium which gives less background in GLC and, thus, can be used to methylate microquantities of glycoprotein glycans ($<$ 10 µg total sugars) (69).

---

**Protocol 21.** Methylation using potassium *tert*-butoxide in dimethyl sulfoxide

A. *Preparation of potassium tert-butoxide*

1. Treat 5 g of potassium with 25 ml of dry *n*-pentane in an atmosphere of helium and add 25 ml of freshly distilled *tert*-butyl alcohol until the evolution of hydrogen starts. Repeat the washing twice (yield: 4.7 g of clean potassium).

2. Place the potassium (4.7 g), under a stream of helium, in a one-necked flask containing 100 ml of *tert*-butyl alcohol and equipped with a reflux condenser connected to a drying tube packed with calcium chloride.

3. Heat the mixture for 30 min at 40–50 °C using an electric heating mantle. Reflux the mixture for 2–3 h increasing the temperature gradually to 80–81 °C. When the reaction is complete cool the flask rapidly to room temperature.

4. Evaporate the *tert*-butyl alcohol at 40 °C in a carefully washed and dried rotary evaporator. During the evaporation, the receiver flask of rotary evaporator is cooled in an ice-acetone bath. Eliminate the remaining *tert*-butyl alcohol by lyophilization in a helium atmosphere. The yield of potassium *tert*-butoxide is 14 g.

B. *Preparation of potassium methylsulfinyl carbanion*

1. Add rapidly 14 g of potassium *tert*-butoxide in 42 ml of dry and freshly distilled dimethyl sulfoxide (DMSO) and place in an ultrasonic bath for 30 min. The base concentration of the reagent is 3 M.

2. Before using, heat at 60 °C for 30 min.

C. *Micromethylation procedure*

- Methylation is carried out in small glass tubes (6 × 80 mm) placed in Teflon-lined screw-cap tubes (13 × 100 mm)

1. Dissolve 10 µg of glycopeptides or reduced oligosaccharides in 50 µl of DMSO and 50 µl of potassium methylsulphinyl carbanion under an inert atmosphere.

2. Sonicate the mixture for 60 min. After cooling at −4 °C, add 100 µl cold methyl iodide.

3. Sonicate the solution at 20 °C for 45 min. An important point is to alkylate at room temperature to avoid the decomposition of methyl iodide to iodine.

4. Stop the methylation by the addition of 0.5 ml water and of a few crystals of sodium thiosulfate.

5. Extract the permethylglycan once with 0.3 ml of chloroform and twice with 0.5 ml of chloroform. Wash the chlorofom phase (1.3 ml) five times with 0.5 ml water.

6. Dry the chloroform extract by the addition of anhydrous sodium sulfate, filter, and concentrate in a rotary evaporator. Eliminate the last traces of DMSO by lyophilization.

---

**Protocol 22.** Methylation using lithium methylsulfinyl carbanion (69)

A. *Preparation of lithium methylsulfinyl carbanion*[a]

*Apparatus*

The apparatus used for preparing the lithium methylsufinyl carbanion consists of a 500 ml two-necked flask provided with a magnetic stirrer, fitted with a 250 ml dropping funnel and with a distillation system equipped with a drying tube packed with calcium chloride, a manometer, and a one-necked 1 litre trapflask. The apparatus is connected to a water-pump. Before the reaction procedure, the system is purged using a helium stream.

*Method*

1. Introduce 120 ml of freshly dried and distilled DMSO (Merck, Darmstadt, FRG) into the two-necked flask under a helium atmosphere. Place a solution of *n*-butyl lithium (Janssen Chimica, Beerse, Belgium; 1.6 M) in hexane (250 ml) in the dropping funnel.

2. Add this solution gradually at room temperature, under reduced pressure (30 mm) with constant stirring.

3. Remove the *n*-butane (produced during the reaction of *n*-butyl lithium with DMSO) and hexane under vacuum. The addition is carried out in the course of about 2 h after which the colour of the reaction mixture

**Protocol 22.** *Continued*

changes to dark green. After complete addition of the *n*-butyl lithium solution, maintain the mixture at low pressure for 2 h at room temperature.

4. Bring the system back to atmospheric pressure with the admission of helium. Store the reagent at 4 °C under helium atmosphere in Teflon-lined screw-cap tubes (13 × 100 mm).

B. *Micromethylation procedure*

1. Dissolve 10 µg of reduced oligosaccharides in 20 µl of DMSO in small glass tubes (6 × 80 mm) placed in Teflon-lined screw-cap tubes (13 × 10 mm).

2. Add 20 µl of lithium methylsulfinyl carbanion under an inert atmosphere (argon) and sonicate the mixture at 20 °C for 60 min.

3. Cool to −4 °C, add cold methyl iodide (40 µl), and sonicate the solution at 20 °C for 45 min.

4. Stop the methylation by the addition of water (0.4 ml) and sodium thiosulfate. Extract the permethylated oligosaccharide-alditols three times with 0.4 ml of chloroform.

5. Wash five times with 0.5 ml of water, filter, concentrate, and evaporate the chloroform phase at 25 °C.

[a] We use successfully a simpler method in which the lithium methylsufinyl carbanion is replaced by sodium hydroxide (71). The method may also be used on a larger scale.

(a) Dissolve 10 µg of oligosaccharides or oligosaccharide-alditols in 20 µl of DMSO as described above.

(b) Add 2 mg of sodium hydroxide powder freshly prepared by grinding in a mortar NaOH pellets under inert atmosphere (argon).

(c) Add cold (−4 °C) methyl iodide and sonicate the solution at 20 °C for 45 min.

(d) Wash as described above.

## 7.1 Analysis of monosaccharide methyl ethers

The combination of gas–liquid chromatography with mass spectrometry allows an easy identification of methylated monosaccharides obtained by hydrolysis or methanolysis of permethylglycans. In addition, chemical ionization mass spectrometry greatly enhances the sensitivity of detection of methylated sugars by removing the background noise from the gas chromatograms. The methylated sugars are analysed as their partially methylated and acetylated methylglycosides (*Protocol 23*).

---

**Protocol 23.** Methanolysis of permethylglycans

1. Treat the permethyloligosaccharide-alditols (1–10 μg) with 0.5 M methanol–HCl (150 μl) in a small glass tube (6 × 80 mm) placed in Teflon-lined screw-cap tube (13 × 100 mm) for 20 h at 80 °C.

2. Eliminate HCl–methanol reagent under a stream of nitrogen.

3. Acetylate the methylated methylglycosides in the same tube with 40 μl pyridine/acetic anhydride (1:1 v/v) at 37 °C for 24 h.

4. Evaporate the reagent under a stream of nitrogen.

5. Dissolve the methylated acetylated methylglycosides in methanol and inject into a gas chromatograph coupled to a mass spectrometer.

---

The methylglycoside methyl ethers are analysed by GLC–MS under the following experimental conditions in our laboratory: Girdel model 300 GLC apparatus (Suresnes, France); glass capillary column (60 m × 0.35 mm i.d.) coated with Silicone OV 101; helium pressure: 0.4 bar; column temperature: 110–240 °C at 3 °C/min. The retention times of the methylglycoside methyl ethers are calculated relative to the palmityl methyl ester. Mass spectra are recorded on a Riber-Mag 10–10 mass spectrometer (Rueil-Malmaison, France) using an electron energy of 70 eV and an ionizing current of 0.2 mA for electron impact (EI) mode. Retention times of methylated and acetylated derivatives of methyl-2-deoxy-2-(*N*-methyl acetamido)-D-glucopyranoside, methyl-D-galactoside, and D-mannoside are given in *Tables 8* and *9*, respectively (see also Chapter 1, Section 8).

---

**Table 8.** Retention time of *O*-acetyl-*O*-methyl derivatives of methyl-2-deoxy-2-(*N*-methylacetamido)D-glucosidic relative to palmityl methyl ester

| Position of substituent | | Retention times (min) | |
|---|---|---|---|
| **Methyl** | **Acetyl** | **Peak 1** | **Peak 2** |
| 3,4,6 | | 0.80 | 0.96 |
| 4,6 | 3 | 1.06 | 1.08 |
| 3,6 | 4 | 1.03 | 1.15 |
| 3,4 | 6 | 1.11 | 1.21 |
| 6 | 3,4 | 1.18 | 1.21 |
| 4 | 3,6 | 1.35 | 1.40 |
| 3 | 4,6 | 1.29 | 1.42 |
| Palmityl methyl ester | | | 34.40 |

**Table 9.** Retention time of *O*-acetyl-*O*-methyl derivatives of methyl D-galactosides and D-mannosides relative to palmityl methyl ester

| Position of substituent | | Retention times (min) | | |
|---|---|---|---|---|
| | | Galactose | | Mannose |
| **Methyl** | **Acetyl** | $\alpha$ | $\beta$ | $\alpha$ |
| 2,3,4,6 | | 0.39 | 0.36 | 0.35 |
| 2,3,4 | 6 | 0.562 | 0.561 | 0.53 |
| 2,3,6 | 4 | 0.48 | 0.47 | 0.54 |
| 2,4,6 | 3 | 0.59 | 0.54 | 0.51 |
| 3,4,6 | 2 | 0.535 | 0.65 | 0.44 |
| 2,3 | 4,6 | 0.68 | 0.79 | 0.78 |
| 2,4 | 3,6 | 0.80 | 0.76 | 0.74 |
| 2,6 | 3,4 | 0.73 | 0.70 | 0.68 |
| 3,4 | 2,6 | 0.77 | 0.856 | 0.64 |
| 3,6 | 2,4 | 0.64 | 0.71 | 0.63 |
| 4,6 | 2,3 | 0.705 | 0.73 | 0.65 |
| 2 | 3,4,6 | 0.89 | 0.86 | 0.89 |
| 3 | 2,4,6 | 0.85 | 0.91 | 0.85 |
| 4 | 2,3,6 | 0.91 | 0.96 | 0.93 |
| 6 | 2,3,4 | 0.80 | 0.82 | 0.80 |
| Palmityl methyl ester | | | 34.30 | |

EI mass spectra are analysed according to Zolotarev *et al.* (72).

*Table 10* gives the main fragments leading to the characterization of partially methylated and acetylated methylglycosides (73).

*Figure 18* gives the GLC pattern of partially methylated and acetylated methylglycosides obtained after methylation, with lithium methylsulfinyl carbanion reagent, and methanolysis of 10 μg of oligosaccharide alditol-11 from hen ovomucoid. It can be observed that GLC on a Silicone OV 101 capillary column with flame ionization detector allows identification of all methyl ethers without use of MS so showing that the lithium methylsulfinyl carbanion reagent gives less background in GLC. Analysis of samples containing 1 μg or less of total sugars necessitates the use of chemical ionization (ionizing gas: ammonia) with specific ions: $(M + 18)^+$ for neutral methyl ethers (m/z 268 for permethyl derivatives; m/z 296 for tri-*O*-methyl-mono-*O*-acetyl derivatives; m/z 324 for di-*O*-methyl-di-*O*-acetyl derivatives; m/z 352 for mono-*O*-methyl-tri-*O*-acetyl derivatives); $(M + 1)^+$ for hexosamine methyl ethers (m/z 292 for permethyl derivatives; m/z 320 for di-*O*-methyl-mono-*O*-acetyl derivatives; m/z 348 for mono-*O*-methyl-di-*O*-acetyl derivatives); $(M + 1)^+$ for methyl ethers of *N*-acetylneuraminic acid (m/z 408).

**Table 10.** Main fragments leading to the characterization of partially methylated and acetylated methylglycosides

A1 m/z 219: Tetra-O-methyl hexoside ⟶ 2,3,4,6-Me

J1 m/z 75 < K2 m/z 71
B1 = 0%; m/z 176, 204, 232 ⟶ 2,4,6-Me-3-Ac

A1 m/z 247: Tri-O-methyl Hexoside
Mono-O-acetyl

J1 m/z 75 > K2 m/z 71 ⟶
B1 m/z 176
D1 m/z 177 ⟶ 2,3,4-Me-6-Ac

K2 m/z 71 < 10% ⟶ 2,3,6-Me-4-Ac
B1 m/z 204
K2 m/z 71 > 40% ⟶ 3,4,6-Me-2-Ac

K2 m/z 71 ⩽ J1 m/z 75
or
K2 m/z 71 < 5%
when J1 is absent ⟶
J1 m/z 75 = 0% ⟶ 2,6-Me-3,4-Ac

J1 m/z 75 ⟶
B1 m/z 204 ⟶ 2,3-Me-4,6-Ac
B1 m/z 232 ⟶ 3,6-Me-2,4-Ac

A1 m/z 275: Di-O-methyl Hexoside
Di-O-acetyl

K2 m/z 71 = J1 m/z 75
or
K2 m/z 71 > 20%
when J1 is absent ⟶
J1 m/z 75 = 0% ⟶ 4,6-Me-2,3-Ac

J1 m/z 75 > 30% ⟶
B1 m/z 204 ⟶ 3,4-Me-2,6-Ac
B1 m/z 204 = 0% ⟶ 2,4-Me-3,6-Ac

**Table 10.** *Continued*

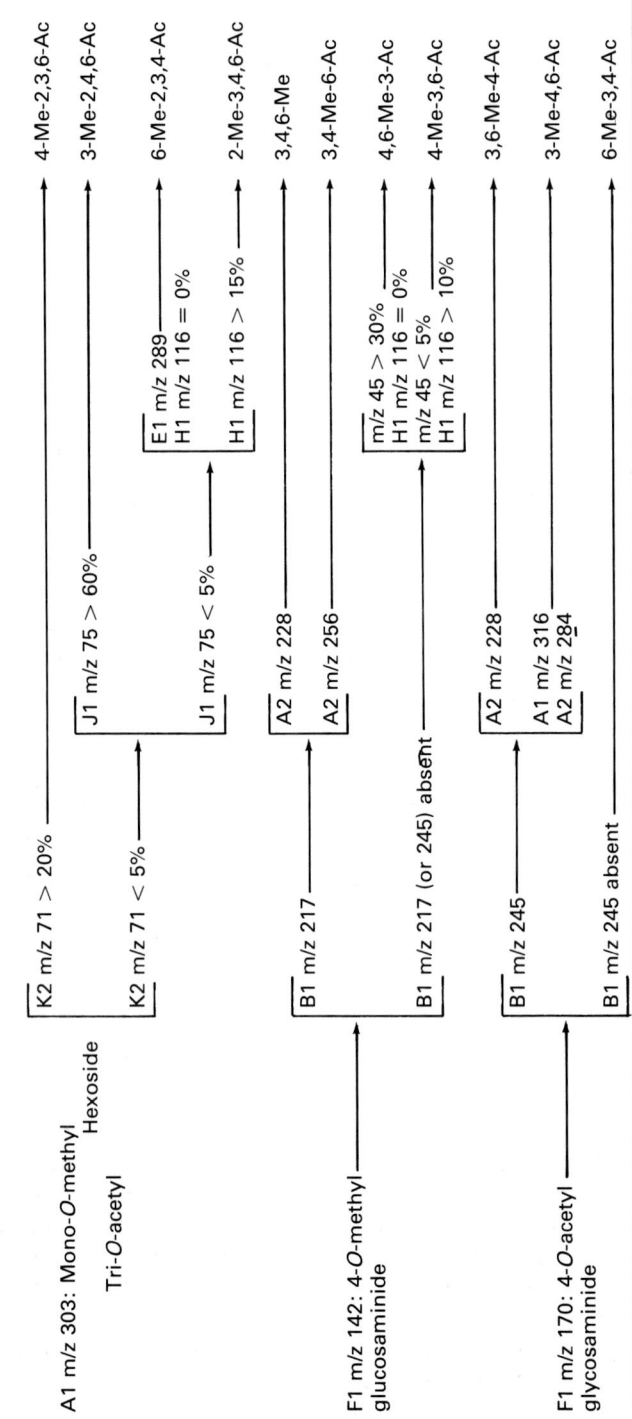

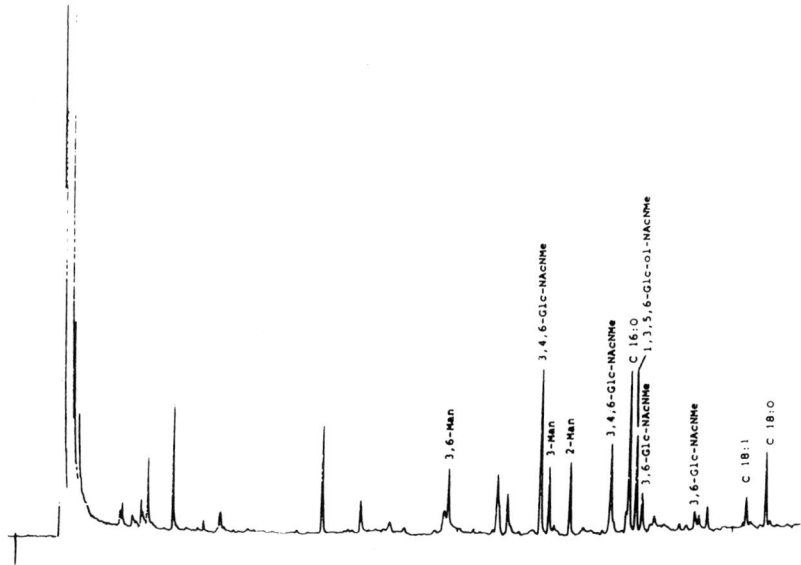

**Figure 18.** GLC (glass capillary column wall coated with OV101; flame ionization detection) of partially methylated and acetylated methylglycosides obtained by methanolysis of 10 μg of permethylated oligosaccharide-alditol 11 from hen ovomucoid (see *Figure 10*). 1.5 μg of the sample was injected.

## 7.2 Sequence and molecular weight determination of (permethylated) oligosaccharides by mass spectrometry

In the field of carbohydrate structural studies, mass spectrometry plays a predominant role especially by its ability to characterize heavy molecular weight molecules by FAB ionization and, more recently, by matrix-assisted laser desorption (MALD) associated to time of flight mass analyser.

### 7.2.1 FAB mass spectrometry

Fast atom bombardment mass spectrometry (FAB–MS) method is based on the bombardment of samples dissolved in a viscous matrix, either by a beam of atoms like argon or xenon or of the caesium ion in which case the method is called liquid secondary ionization mass spectrometry (LSIMS).

The FAB ionization method may be applied for analysis of oligosaccharides either in their native or derivatized form. In both cases the mass spectra exhibit pseudomolecular ions formed by different ways: loss of proton: $[M - H]^-$, proton attachment: $[M + H]^+$; anion attachment: $[M + Cl]^-$, cation attachment: $[M + Na]^+$, or $[M + NH_4]^+$.

Depending on the nature of sample and matrix and of the energy of the incident particle beam, fragmentations also occur in the FAB source. The obtained fragments originating from different pathways are described in

*Figure 19*. The fragment labellings refer to the nomenclature of Domon and Costello (74).

### i. *FAB-MS of native oligosaccharides*

Greater sensitivity is generally obtained by applying negative ion mass spectrometry to the native oligosaccharides since they give rise preferentially to $[M - H]^-$ deprotonated pseudomolecular ions.

(a) The most commonly used matrix is glycerol. However some additives may be employed to improve the signal-to-noise ratio. We use glycerol/TFA 5% in the case of polycarboxylated or sulfated oligosaccharides in order to avoid the formation of several pseudomolecular ions $[M - H]^-$, $[M + Na - 2H]^-$, $[M + 2 Na - 3H]^-$. We have often noticed that well desalted and detergent-free samples, even if they are rich in acid functional groups, give an increased response by using an alkaline matrix such as glycerol/water/triethanolamine (3:2:1). Indeed, these contaminants, such as salts of detergents, are more surface active in the matrix than the oligosaccharides themselves. Consequently, the samples have to be carefully desalted and free of detergents prior FAVB analysis.

(b) The amount of injected compound is of great importance. Good quality spectra are acquired with 1 to 10 µg of native oligosaccharides considering that the higher the molecular weight, the greater the sample amount required.

(c) Native carbohydrates exhibit poorer response than their derivatized forms. For this reason, the recording of mass spectra from the native state is advised only when chemically labile substitutents are suspected. Moreover, ambiguous fragmentations may occur during FAB ionization of such native sugars making the interpretation of spectra unfruitful. As an example, combination of pathways B and C of *Figure 19* during fragmentation of unreduced native compounds makes the resulting fragments indistinguishable.

### ii. *FAB–MS of derivatized oligosaccharides*

Like many groups involved in mass spectrometry of carbohydrates we use, in most cases, permethylated (see *Protocols 21* and *22*) or peracetylated oligosaccharides. As mentioned by Dell (9) the derivatization step presents many advantages: greater sensitivity (0.1 to 5 µg are sufficient) in FAB ionization, both the sequence and branching pattern are unambiguously determined, derivatization step is a 'cleaning procedure' and contaminated samples do not consequently constitute a critical point, accessible molecular weight is extended to higher mass (< 5000 Da). The most important step is the choice of the matrix, the quality of which determines the quality of spectra.

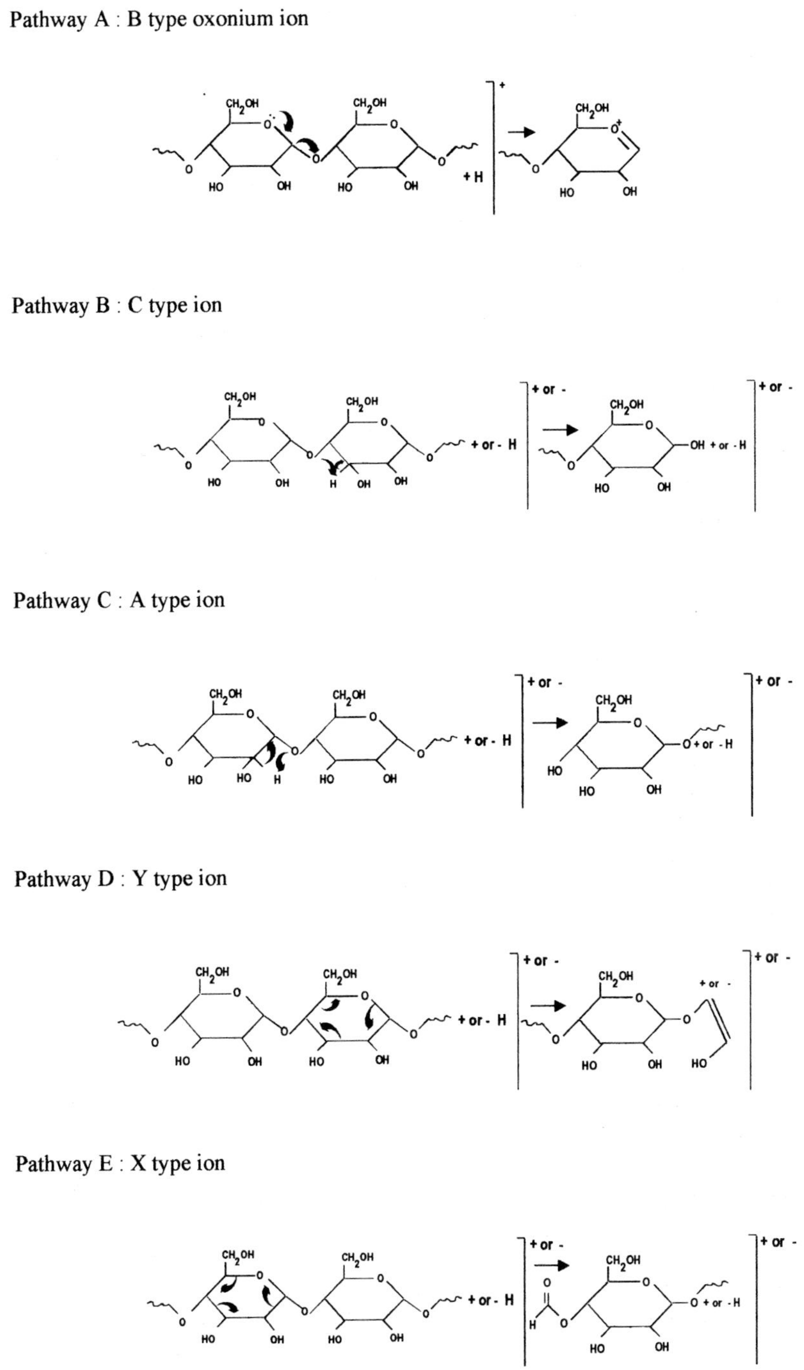

Pathway A : B type oxonium ion

Pathway B : C type ion

Pathway C : A type ion

Pathway D : Y type ion

Pathway E : X type ion

**Figure 19.** Main fragmentation pathways on FAB mass spectra of oligosaccharides.

241

(a) Thioglycerol always gives excellent data. However, when a long lasting signal is needed for instance during the acquisition of raw data (accumulated spectrum) or tandem mass spectrometry experiments, one may also use the thioglycerol/glycerol (1:1) mixture.

(b) *Meta*-nitrobenzyl-alcohol has been proposed (75) in order to discriminate between the 1,3- and 1,6-linkages of the common inner-core of *N*-glycans by inducing specific fragmentations.

(c) Similarly to the negative spectra, some additives greatly influence the spectrum pattern. For example, if relatively high amounts of material are available (1–5 µg or more) a thioglycerol/TFA 1% furnishes interesting spectra in which both intense B oxonium fragments (*Figure 19*) and [M + H]⁺ pseudomolecular ions are produced. This cleavage occurs preferentially up to the *N*-acetylhexosamine residues.

This type of fragmentation which is really predominant for derivatized oligosaccharides, represents a powerful basis to elucidate some aspects of the carbohydrate structure, especially the primary sequence and the branching points. An example of such a typical FAB mass spectrum, that of a fucosylated milk oligosaccharides, is given in *Figure 20*. The

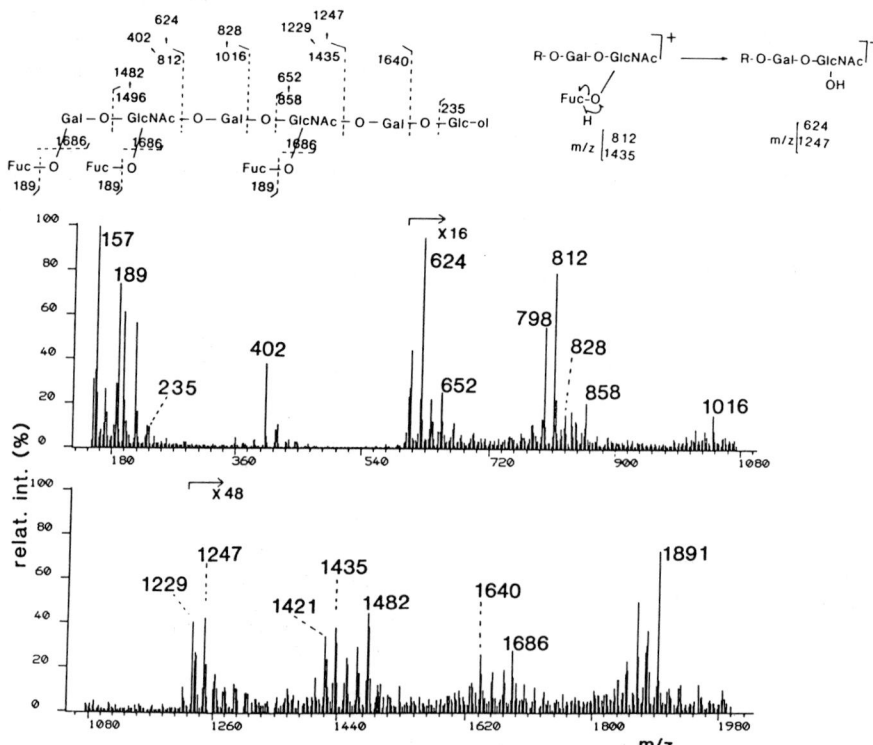

**Figure 20.** FAB mass spectrum of a human milk fucosylated oligosaccharide (76).

242

molecular ion $[M + H]^+$ at m/z 1891 is consistent with a nonasaccharide (Fuc 3, Gal 3, GlcNAc 2, Glc-ol 1) as determined by GLC of the trifluoroacetyl derivatives. Key ions are m/z 1435 together with the daughter ion at m/z 1229, which was produced by the preferred elimination of the fucose residue linked to C-3 of GlcNAc-III, and m/z 812 ($Fuc_2$ + Gal + GlcNAc). An intense ion at m/z 402 is indicative of the elimination of the 1,3-linked Fuc-Gal sequence from fragment m/z 812, that shows the fucose to be α-1,4-linked to GlcNAc-V. The elimination of C-3-linked substituent arises from proton transfer and is quite exclusively specific from B fragments up to the HexNAc residue. The non-specific loss of (Fuc + OH) is also observed from m/z 812 and 1435 whch arise from the fragmentation of fucose (*Figure 20*). The other significant fragments are depicted in *Table 11*. On the basis of these results, the structure of the oligosaccharide may be established as described in *Figure 20*.

(d) It is possible to enhance the pseudomolecular ion signal by drying a 0.1% solution of sodium acetate in methanol prior to loading the target with the matrix. This leads mainly to the formation of $[M + Na]^+$ ions.

(e) We have from time to time noticed that a thioglycerol/NaI 0.5 M matrix allows characterization with a greater sensibility of the $[M + Na]^+$ ions due to complete inhibition of fragmentation of the molecule in the FAB source.

---

**Table 11.** Carbohydrate composition and relative intensity of major ions of diagnostic importance observed in the FAB-MS spectra of reduced and permethylated nonasaccharide described in *Figure 20*

| m/z | Relative intensity (% of base peak) | Carbohydrate composition |
|---|---|---|
| 1891 | 1.6 | $Fuc_3Gal_3GlcNAc_2Glc$-ol + H |
| 1686 | 0.6 | $Fuc_2Gal_3GlcNAc_2Glc$-ol |
| 1640 | 0.6 | $Fuc_3Gal_3GlcNAc_2$ |
| 1496 | 0.2 | $Fuc_2Gal_2GlcNAc_2Glc$-ol |
| 1482 | 1.1 | $Fuc_2Gal_2GlcNAc_2Glc$-ol − 14 |
| 1435 | 0.9 | $Fuc_3Gal_2GlcNAc_2$ |
| 1421 | 0.8 | $Fuc_3Gal_2GlcNAc_2$ − 14 |
| 1247 | 1 | $Fuc_2Gal_2GlcNAc_2$ + $H_2O$ [a] |
| 1229 | 1 | $Fuc_2Gal_2GlcNAc_2$ [b] |
| 858 | 1.2 | Fuc. GlcNAc.Gal.Glc-ol |
| 812 | 5 | $Fuc_2Gal$. GlcNAc |
| 652 | 1.5 | GlcNAc. Gal. Glc-ol [b] |
| 624 | 6 | Fuc. Gal. GlcNAc + $H_2O$ [a] |
| 402 | 42 | Fuc. GlcNAc [b] |
| 235 | 12 | Glc-ol |
| 189 | 72 | Fuc |

[a] These fragments are formed as illustrated in *Figure 20*.
[b] Secondary fragments.

## iii. Preparation of the samples
Since the acquisition of good quality data is widely conditioned by a higher concentration of the sample at the matrix surface, we recommend the following:

(a) Load the target with 1–2 µl of the chosen matrix using a 5 µl micropipette.

(b) Add 0.5–1 µl of the sample solution using a 5 µl microsyringe. A 10 µg per µl solution of native oligosaccharide in water and 1–5 µg per µl of derivatized oligosaccharide in methanol are the adequate concentrations considering the 500–3000 mass mass range.

If native oligosaccharide samples are placed in Eppendorf tubes, the solution should be prepared just before the mass spectra recording. In fact, we have sometimes observed the disappearance of FAB response when the analysis was performed one or several hours after the dissolution of the samples.

**Table 12.** Calculation of molecular weight of oligosaccharides[a]

| Compounds[b] | Native oligosaccharides | Derivatized oligosaccharides | |
| --- | --- | --- | --- |
| | | Peracetylated | Permethylated |
| Hex | 180 | 390 | 250 |
| Hex-ol | 182 | 434 | 266 |
| Hex-NAc | 221 | 389 | 291 |
| Hex-NAc-ol | 223 | 433 | 307 |
| NeuAc | 309 | 519 | 407 |
| Deoxy-Hex | 150 | 332 | 220 |

[a] The molecular weight is calculated as follows: sum of the mass of each monosaccharide, subtract (n × 18) for native, (n × 102) for peracetylated, (n × 46) for permethylated oligosaccharides where n is the number of glycosidic linkages.
[b] Hex: hexose; Hex-ol: hexitol; Hex-NAc: N-acetylhexosamine; Hex-NAc-ol: N-acetylhexosaminitol; NeuAc: N-acetylneuraminic acid; Deoxy-Hex: 6-deoxyhexose.

**Table 13.** Characteristic fragments and mass increments for sequence determination of oligosaccharides

| Compounds[a] | Characteristic fragments associated to non-reducing end of monosaccharides | | Mass increments to be added to non-reducing end fragment | |
| --- | --- | --- | --- | --- |
| | Acetylated | Methylated | Acetylated | Methylated |
| Hex | 331 | 219 | 288 | 204 |
| Hex-NAc | 330 | 260 | 287 | 245 |
| Deoxy-Hex | 273 | 189 | 230 | 174 |
| NeuAc | 460 | 376(344) | 417 | 361 |

[a] See Table 12.

*Table 12* indicates how to calculate the molecular weight of the native and derivatized oligosaccharides on the basis of FAB–MS data. *Table 13* gives the characteristic fragments and mass increments leading to the oligosaccharide primary structure.

## 7.2.2 Matrix-assisted laser desorption-time of flight mass spectrometry

The technique of matrix-assisted laser desorption ionization (MALDI) has been pioneered by Hillenkamp and Karas. From the technical point of view, the reader is invited to refer to a general presentation in *Methods in enzymology* by these authors (77).

Most of the instruments, are based on the following scheme:

- a short pulse (3–5 ns) nitrogen UV laser operating at 337 nm wavelength
- an optic system to focus the laser beam
- a sample probe equipped with a target on which several sample spots may be deposited
- a camera to vizualize the different spots and the impact of the laser beam
- a flight tube equipped or not with a reflectron whose aim is the improvement of mass resolution, especially for masses below 10 000

The matrix is a critical parameter in the acquisition of a LDI mass spectrum. In fact, the matrix has several important functions:

(a) A large excess of matrix relative to the sample ensures the separation of each analyte molecule thereby avoiding the formation of cluster ions harmful to the observation of molecular ion.

(b) The great amount of energy that is deposited into the matrix results in the projection in to the gas-phase of a small matrix-analyte portion. This leads to analyte molecules free and intact in the gas phase.

(c) The matrix, by a photochemical and photoionization mechanism, may also be implicated in the sample ionization.

---

**Protocol 24.** Sample preparation for MALDI

1. Dissolve the oligosaccharide sample in water to a concentration of 100 µg/ml (use a glass tube instead of an Eppendorf-type tube).

2. Mix the sample solution with 2,5-dihydroxybenzoic acid (10 µg/ml of 10% aqueous ethanol solution) in a ratio of 1:5.

3. Apply to the metal target 0.5 µl–1 µl of the sample matrix mixture which is then dried in a cold air stream.

---

The fact that no fragmentation is observed in MALDI-UV spectra makes this technique especially well suited for mixture analysis. Best results are obtained in the positive mode of detection during the analysis of carbohydrates, partly because cationized $[M + Na]^+$ and $[M + K]^+$ ions are most often the major species (*Figure 21*).

Another interesting aspect is the very high sensitivity of the method since order of picomoles of materials or less are required to perform experiments in the mass range up to 10 000 Da. Moreover, and in contrast to FAB ionization, observation of molecular ions is quite insensitive to contamination by salts or detergents. As an example of tolerance, you may use detergents 0.1%/Tris buffer 50 mM/SDS 0.05%.

Considering the easy use of the method, even for people not working routinely with other mass spectrometric instruments, its high sensitivity and rapid mass measurements, UV-MALDI technique will undoubtly take a more and more important place in oligosaccharide mass analysis beside the FAB ionization method.

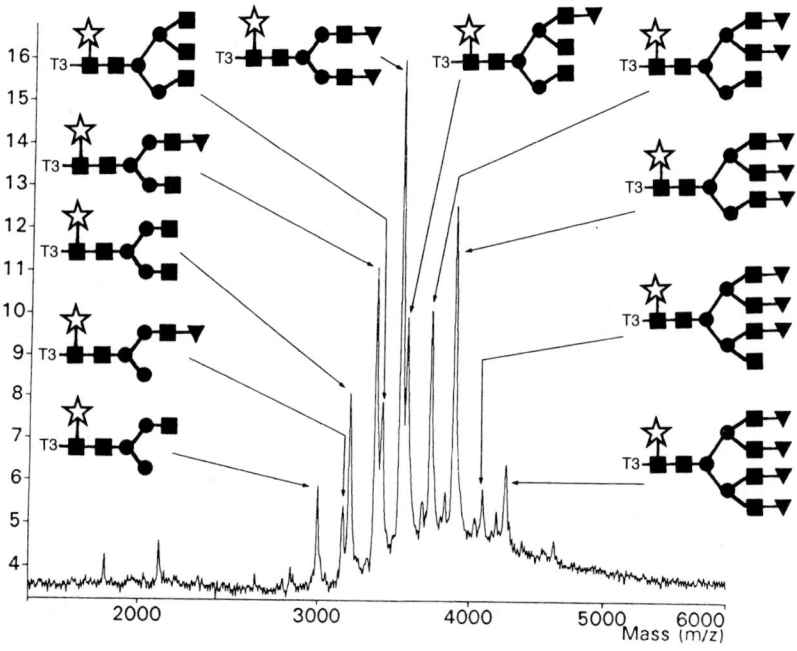

**Figure 21.** MALDI-TOF analysis of the $T_3$ tryptic glycopeptides from human tissue inhibitor of metalloproteinases (TIMP) after neuraminidase treatment. From Finnigan MAT, Poster compilation n°3 (78). ■: GlcNAc; ●: Man; ▼: Gal; ☆: Fuc.

# 8. Use of glycosidases

## 8.1 Experimental approach

Glycosidases are excellent tools for:

- the elucidation of the primary structure of glycans by sequential degradation
- the determination of the anomeric linkage of each conjugated monosaccharide
- the controlled modification of glycoprotein glycans, including membrane glycoproteins, in order to explore their biological role
- the preparation of specific acceptors for glycosyltransferase activity studies

Basically, two types of enzymes are used: *exoglycosidases* which hydrolyse glycosidic bonds of monosaccharides in terminal non-reducing position and may achieve a stepwise degradation of the glycan; and *endoglycosidases* which have been recently introduced and rapidly developed because of their very promising performances. They hydrolyse internal glycosidic bonds, so liberating oligosaccharides or the glycans themselves.

The use of glycosidases was limited for a long time by the following factors.

(a) Glycosidase activities depend on the origin of the enzyme (microorganisms, plants, or animals, cytoplasm, lysosome, or Golgi apparatus), on the type of glycoside linkage ($1 \rightarrow 2$, $1 \rightarrow 3$, $1 \rightarrow 4$, or $1 \rightarrow 6$), on the nature of the substrate (artificial substrates or natural substrates like oligosaccharides, glycopeptides, or glycoproteins), on the structure of substrates (branched, phosphorylated, or sulfated), and on the spatial conformation (i.e. mono- to penta-antennary glycans).

(b) Commercial sources of monospecific and pure glycosidases are relatively rare. In many cases, it is necessary to verify the absence of contaminant enzymes like proteases, phospholipases, or other glycosidases, and often to purify the commercial materials. In other cases, the lack of active enzymes has led the experimenters to prepare their own glycosidases.

(c) At the present time, only neuraminidases are commercially available in an immobilized form although immobilized glycosidases would be of interest in the production of hydrolysed substrates free from contaminating glycosidases and thus, avoiding artefacts due to their glycoprotein nature.

(d) Glycosidase activities are generally detected and measured by using synthetic *p*-nitrophenyl or methylumbelliferyl-sugar derivatives as substrates. This could be a cause of error since some glycosidases which are active on synthetic substrates are not active on natural structures, and *vice versa*.

---

**Protocol 25.** Determination of the activity of glycosidases

A. *Use of nitrophenyl derivatives as substrates*

*Reagents*

- 10 mM nitrophenylglycoside in aqueous solution
- 1 M sodium carbonate aqueous solution
- McIlvaine buffer (0.2 M sodium phosphate/ 0.1 M citric acid; pH adjusted to the optimal pH)

*Method*

1. Add 0.1 ml of nitrophenylglycoside to 0.1 ml of McIlvaine buffer and 0.2 ml enzyme solution.

2. Incubate for 5–15 min at the optimal temperature (previously determined) and stop the reaction by adding 0.6 ml of the sodium carbonate solution.

3. Determine the optical density of the liberated nitrophenol at 400 nm.

B. *Use of methylumbelliferyl derivatives as substrates*

*Reagents*

- 1 mM methylumbelliferyl glycoside in aqueous solution
- McIlvaine buffer (see part A)
- 0.2 M glycine buffer adjusted to pH 10.7 with 1 M NaOH

*Method*

1. Apply the same procedure as in part A.

2. Determine the amount of liberated methylumbelliferon by spectro-fluorimetry (excitation $\lambda$: 390 nm; emission: 450 nm).

---

**Table 14.** Properties of neuraminidases

| Origin of enzyme | Optimum pH | $Km^a$ (mM) | Effector | pI | Mw (kDa) |
|---|---|---|---|---|---|
| Vibrio cholerae | 5.6 | 1 | $Ca^{2+}$ | 4.8 | 66 |
| Clostridium perfringens | 4.3–5.2 | 0.2 | No | 5.1 | 63 |
| Arthrobacter ureafasciens | 5–5.5[b] | 0.6 | No | — | 39–51 |

[a] Substrate: $\alpha$-2,3-*N*-acetylneuraminyllactose (Neu5Ac$\alpha$-2,3 lactose).
[b] 4.3–4.5 with colominic acid as substrate.

---

**Protocol 26.** Use of neuraminidase (EC: 3.2.1.18; *N*-acetylneuraminyl hydrolase)

In spite of the existence of numerous sources of enzyme only neuraminidases from *Vibrio cholerae*, *Clostridium perfringens*, and *Arthrobacter ureafasciens* are commonly available. Their properties are described in *Table 14*.

*Method*

1. Dissolve the substrate in 50–100 µl of 50 mM sodium acetate buffer at the optimum pH of enzyme (in the case of the neuraminidase from *Vibrio cholerae* the buffer contains 10 mM $CaCl_2$ and 0.1 M NaCl in addition).

2. Add 0.1 U of neuraminidase per µmol of *N*-acetylneuraminic acid to be removed.

3. Incubate for from 10 min to 48 h at 37 °C, the time of hydrolysis depending on the nature of the substrate (oligosaccharide, glycopeptide, or glycoprotein), the type of linkage (α-2,3, α-2,6, or α-2,8), and the spatial conformation of glycans or glycoproteins (mucin-type or globular protein-type).

4. Stop the hydrolysis by heating in a boiling water-bath for 3 min.

5. Determine the amount of liberated sialic acid by applying colorimetric methods (see Section 5.5.1 and Chapter 1, *Protocol 9*).

*Notes*

(a) Enzyme specificity (for reviews, see ref. 79 and 80). Neuraminidases from bacterial origin are generally able to split α-2,3 and α-2,6 linkages easily. Enzymes from viruses (FPV, NDV, or A2) act only on α-2,3 linkages. In the case of *Arthrobacter ureafasciens*, the enzyme splits α-2,6 linkages much faster than α-2,3 linkages. The α-2,8 linkages of colominic acid form a poor substrate for bacterial neuraminidases. The rate of hydrolysis depends on the substitution of neuraminic acid. *N*-Glycolylneuraminyl linkages are resistant to hydrolysis except in the case of the enzyme from *Clostridium perfringens* which is able to split both α-2,3-*N*-acetyl and *N*-glycolyneuraminyl linkages. *O*-Substitution generally causes a 50% decrease of the enzyme activity, except in the case of enzyme from *Streptococcus sanguis* which splits 7-, 8-, or 9-*O*-acetyl-*N*-acetylneuraminic acid at the same rate as sialyllactose, but removes the 4-*O*-acetyl derivative five fold slower, similar to the *Vibro cholerae* enzyme.

**Protocol 26.** *Continued*

(b) Sulfate groups decrease the rate of hydrolysis. However, sulfation of the monosaccharide adjacent to that with which *N*-acetylneuraminic acid is linked, increases the rate of hydrolysis.

(c) *N*-Acetyneuraminyl bonds are cleaved differently according to the monosaccharide on which the *N*-acetylneuraminic acid is linked. For example, sialyl *N*-acetylgalactosamine linkages are relatively resistant to neuraminidase hydrolysis.

(d) The specificity of commercial neuraminidases is given in *Table 15*.

---

**Protocol 27.** Use of $\beta$-D-galactosidase (EC: 3.2.1.23; $\beta$-D-galacto-side galactohydrolase)

The most popular $\beta$-D-galactosidase preparations are from *Canavalia ensiformis* (Jack bean) (81) and almond emulsin (82).

*Method using Canavalia ensiformis enzyme*

1. Dissolve the substrate in 50–100 µl of 0.02 M sodium phosphate/0.01 M acetic acid buffer pH 3.5 (the pH can be increased to pH 4.0 when the substrate is not stable at the more acidic pH).

2. Add 0.4 U of $\beta$-D-galactosidase per µmol of galactose to be liberated. The final volume should not exceed 400 µl.

3. Incubate the mixture at 37 °C from 10 min to 48 h.

4. Stop the hydrolysis by heating in a boiling water-bath for 3 min.

5. Determine the free galactose by conventional methods such as HPLC, GLC, or the galactose dehydrogenase assay (see Chapter 1, Section 3.1).

*Notes*

(a) The enzyme activity is stable for five days at 37 °C.

(b) Glycopeptides and glycoproteins are hydrolysed 20-fold and 150-fold slower than oligosaccharides, respectively.

(c) The enzyme can act at high ionic strengths but not in chaotropic medium.

(d) Two groups of enzymes can be characterized. The first group present a high affinity toward the lactose (e.g. the enzymes from *E. coli, K. lactis,* or *K. fragilis*), are activated by $Na^+$, $K^+$, $Mg^{2+}$, or $Co^{2+}$ ions, and do not act on glycoconjugate derivatives. The second group are active on the $\beta$-galactosidic linkages present in glycoconjugates.

**Table 15.** Specificity of neuraminidase (results expressed as % of total hydrolysis)

| Origin of enzyme | Neu5Ac | | | Neu5Gc α-2,3 | Bovine submaxillary mucin | Human serum transferrin | $\alpha_1$-Acid glyco-protein |
|---|---|---|---|---|---|---|---|
| | α-2,3 | α-2,6 | α-2,8 | | | | |
| *Vibrio cholerae* | 100 | 53 | 31 | 25 | 55 | 100 | 100 |
| *Clostridium perfringens* | 100 | 44 | 44 | 20 | 20 | 100 | 100 |
| *Arthrobacter ureafasciens* | 63 | 100 | 44 | 8 | 52 | 100 | ND |

ND: not determined.

**Protocol 27.** *Continued*

(e) Generally, the β-D-galactosidases do not act when the monosaccharide to which the galactose is conjugated, is substituted as in lacto-*N*-fucopentaose II:

Gal(β1 → 3)[Fuc(α1 → 4) ] GlcNAc(β1 → 3)Gal(β1 → 4)Glc

(f) Some β-D-galactosidases present a very strict specificity. For example, the enzyme from *S. pneumoniae* hydrolyses Gal(β1 → 4)GlcNAc and lactose but not Gal(β1 → 3)GlcNAc or Gal(β1 → 6)GlcNAc.

(g) When the concentration of enzyme is low, the enzymes from almond emulsin and Jack bean hydrolyse the β-1,4 linkages 13-fold and 50-fold faster than β-1,3 linkages, respectively. On the contrary, when the enzyme concentration is high, there is no difference between the rate of hydrolysis by both enzymes. In the same way, bovine epididymis enzyme cleaves β-1,3 linkages more easily than β-1,4 or β-1,6 ones while the reverse situation is observed for calf intestine β-galactosidase. In *Table 16* the activities of β-galactosidase from Jack bean and *E. coli* toward different substrates are compared.

(h) Use of α-D-galactosidases (EC: 3.2.1.22; α-D-galactoside galacto-hydrolase) is limited to glycoproteins having blood group activity and to other rare glycoproteins possessing α-galactosyl residues. Experimental procedure is identical as for β-D-galactosidase. The main source of enzyme is the green coffee bean.

---

**Protocol 28.** Use of *N*-acetyl-β-D-hexosaminidase (EC: 3.2.1.52; 2-acetamido-2-deoxy-β-D-hexoside acetamido-deoxyhexohydrolase)

Generally, *N*-acetyl-β-D-hexosaminidases act on *N*-acetyl-β-D-glucosaminyl and *N*-acetyl-β-D-galactosaminyl linkages, the ratio of activities depending upon the origin of enzyme. The enzyme from Jack bean is the most utilized (83).

*Method*

1. Dissolve 1 μmol of substrate in 100–200 μl of McIlvaine buffer pH 5.0 (see *Protocol 25*).

2. Add 1.5–5.0 U of enzyme in 50–100 μl of buffer[a].

3. Incubate for 6 to 48 h at 37 °C.

**4.** Stop the reaction by heating in a boiling water-bath for 3 min.

[a] The enzyme is inhibited by acetate buffer. Enzyme activity is destroyed by denaturing medium. N-Acetyl-$\beta$-D-hexosaminidase activity is stable for three days at 37 °C. The rate of hydrolysis depends on the origin of enzyme, the type of linkage, and on the molecular conformation of the molecule (*Table 17*). The enzyme is more active on oligosaccharides than on glycopeptides and glycoproteins.

**Table 16.** Specificity of $\beta$-D-galactosidases from Jack bean and *E. coli* (results expressed as % of total hydrolysis)

| Substrates | Jack bean | *E. coli* |
|---|---|---|
| Gal($\beta$1 → 6)GlcNAc | 100 | 100 |
| Gal($\beta$1 → 4)GlcNAc | 75 | 55 |
| Gal($\beta$1 → 3)GlcNAc | 1 | 92 |
| Gal($\beta$1 → 4)Glc | 42 | 100 |
| Gal($\beta$1 → 6)Man | 19 | 0 |
| Asialofetuin | 100 | — |
| $\alpha_1$-Acid glycoprotein | 79 | 0 |

**Table 17.** Specificity of N-acetyl-$\beta$-D-hexosaminidases (results expressed as % of total hydrolysis)

| Substrates | Jack bean | *Aspergillus niger* |
|---|---|---|
| GlcNAc($\beta$1 → 2)Man | 100 | 62 |
| GlcNAc($\beta$1 → 4)Man | 35 | 56 |
| GlcNAc($\beta$1 → 6)Man | 23 | 100 |
| GlcNAc($\beta$1 → 4)Gal | 23 | 28 |
| GlcNAc($\beta$1 → 3)Gal($\beta$1 → 4)Glc | 68 | 40 |
| GlcNAc($\beta$1 → 2)Man($\alpha$1 → 3)Man | 47 | 53 |
| Man($\alpha$1 → 3)[GlcNAc($\beta$1 → 4) ]Man | 81 | 0 |
| GlcNAc($\beta$1 → 4)Man($\alpha$1 → 3)-[GlcNAc($\beta$1 → 4) ]Man | 71 | 0 |
| Asialo-agalacto serotransferrin glycopeptide | 80 | — |

**Protocol 29.** Use of $\alpha$-D-mannosidase (EC: 3.2.1.24; $\alpha$-D-mannoside mannohydrolase)

The most utilized is the enzyme from Jack bean (84).

*Method*

**1.** Dissolve the substrate in 50 mM sodium acetate buffer pH 4.2.

**2.** Add the enzyme in the ratio of 1.5 U/$\mu$mol of mannose to be liberated[a].

**Protocol 29.** *Continued*

**3.** Incubate the mixture from 30 min to 48 h at 37 °C.

**4.** Stop the hydrolysis by heating in a boiling water-bath for 3 min.

---

[a] All α-D-mannosidases, except *Aspergillus niger* enzyme, require from 0.1 to 5 mM of $Zn^{2+}$ salts in order to stabilize the enzyme toward pH and temperature. Enzyme activity is maximal in the presence of at least 0.01% (w/v) total proteins in the incubating medium. The Jack bean enzyme hydrolyses Man(α1 → 2)Man or Man(α1 → 6)Man linkages easily, but the Man(α1 → 3)Man linkages is cleaved 15-fold slower. The *A. niger* enzyme is very active on Man(α1 → 6)Man, Man(α1 → 4) Man, or Man(α1 → 4)GlcNAc. It is inactive on Man(α1 → 3)Man but weakly active on Man(α1 → 2)Man. Golgian α-mannosidases participating in the maturation of *N*-glycosylprotein glycans and having a very sharp specificity (α-1,2-mannosidases; α-1,3 and α-1-6-mannosidases, GlcNAc-dependent) have been described.

---

**Protocol 30.** Use of β-D-mannosidase (EC: 3.2.1.25; β-D-mannoside mannohydrolase)

We currently use a β-D-mannosidase we isolated from *A. niger* and which is, at the moment, the most active one described (85).

*Method*

**1.** Dissolve 1 μmol of substrate in 100 μl of 20 mM sodium-phosphate buffer (pH 3.5), add 5 mU of *A. niger* β-D-mannosidase (ensure that the final volume does not exceed 200 μl).

**2.** Incubate the mixture at 37 °C for 1–24 h.

**3.** Stop the reaction by heating in a boiling water-bath for 3 min.

*Notes*

(a) The enzyme is specific for β-1,4-mannosyl linkages (*Table 18*).

(b) Glycans and glycopeptides are hydrolysed four- to fivefold faster than synthetic substrates.

(c) The rate of hydrolysis decreases when a fucose residue is α-1,6-linked to the *N*-acetylglucosamine residue of *N*-glycosylamine linkage.

**Table 18.** Specificity of $\beta$-D-mannosidases (results expressed as % of total hydrolysis)

| Substrates | Hen oviduct | *Tremella fusiformis* | *A. niger* |
|---|---|---|---|
| p-Nitrophenyl-$\beta$-D-mannoside | 100 | 100 | 100 |
| Man($\beta$1 → 4)GlcNAc | 33 | 10 | 100 |
| Man($\beta$1 → 3)GlcNAc | — | — | 0.04 |
| Man($\beta$1 → 6)GlcNAc | — | — | 0.1 |
| Man($\beta$1 → 4)GlcNAc($\beta$1 → 4)-GlcNAc($\beta$1 → N)Asn | 25 | — | 100 |
| Man($\beta$1 → 4)GlcNAc($\beta$1 → 4)-[Fuc($\alpha$1 → 6) ]GlcNAc($\beta$1 → N)Asn | — | — | 80 |

---

**Protocol 31.** Use of $\alpha$-L-fucosidase (EC: 3.2.1.51; $\alpha$-L-fucoside fucohydrolase)

The enzyme is widely distributed in various living organisms and can be easily isolated in a pure form by affinity chromatography on fucosylamine–Sepharose columns.

*Method*

1. Dissolve the substrate in 50 μl of 100 mM phosphate-citrate buffer at the enzyme's optimum pH.

2. Add 50–100 μl of the enzyme preparation in the ratio of 1 U of $\alpha$-L-fucosidase/μmol of fucose to be liberated.

3. Incubate the mixture at 37 °C from 3 to 48 h.

4. Stop the reaction by heating in a boiling water-bath for 3 min.

*Notes*

(a) Fucosidases having a broad activity spectrum are weakly active toward fucose residues $\alpha$-1,6-conjugated to the asparagine-linked *N*-acetylglucosamine. They are unable to remove this kind of fucose residue when acting on native glycoproteins (for fucosidase specificity, see ref. 57 and 60).

(b) Free fucose is a powerful competitive inhibitor ($K_i$ = 100 μM).

(c) The rate of hydrolysis depends on the origin of the enzyme. For example, increasing activities of *Turbo cornutus* enzyme are as follows: Fuc($\alpha$1 → 4)GlcNAc > Fuc($\alpha$1 → 2)Gal > Fuc($\alpha$1 → 3) GlcNAc, and Fuc($\alpha$1 → 2)Gal > Fuc($\alpha$1 → 4)GlcNAc > Fuc($\alpha$1 → 3) GlcNAc for *Charonia lampas* enzyme.

**Protocol 31.** *Continued*

(d) In some cases, α-L-fucosidase is highly specific for α-1,2-fucosyl linkage (e.g. *Clostridium perfringens, Aspergillus niger*, or *Streptococcus sanguis*).

(e) Fucosidases from micro-organisms are able to remove fucose from native glycoproteins but are inactive toward synthetic substrates. On the contrary, mammalian enzymes act on synthetic substrates, oligosaccharides, and glycopeptides but do not act on native glycoproteins.

(f) The specificity of fucosidases of different origins is described in *Table 19*.

**Table 19.** Specificity of α-D-fucosidases (results expressed as % of total hydrolysis)

| Substrates | *Clostridium perfringens* | Almond | Mammalian tissues |
|---|---|---|---|
| p-Nitrophenyl α-L-fucoside | 0 | — | 100 |
| Fuc(α1 → 2)Gal | 100 | — | — |
| Fuc(α1 → 2)Fuc | 0 | — | — |
| Fuc(α1 → 2)Gal(β1 → 4)Glc | 79 | 1.5 | 28 |
| Fuc(α1 → 3)Gal(β1 → 4)Glc | — | 13 | — |
| Fuc(α1 → 2)Gal(β1 → 4)[Fuc-(α1 → 3) ]Glc | 39 | — | — |
| Lacto-N-fucopentaose I | 80 | 1 | 2 |
| Lacto-N-difucohexaose I | 7 | 17 | — |
| Lacto-N-fucopentaose II | 4 | 100 | 20 |
| Lacto-N-fucopentaose III | 9 | 93 | 6 |
| IgG glycopeptide | 0 | 0 | 54 |
| Porcine submaxillary mucin | 90 | — | — |
| Canine submaxillary mucin | 90 | — | — |
| Human ovarian cyst glycoprotein | 0 | — | — |
| H-antigen glycolipid | — | — | 100 |

## 8.2 Endoglycosidases

Endoglycosidases which release oligosaccharides from conjugated glycans are divided into two classes. In *class I*, the enzymes split the monosaccharide–amino acid or monosaccharide–peptide linkage, removing the complete glycan (aspartyl-N-acetylglucosaminidase; peptide-N-glycosidase). In *class II*, the enzymes cleave sugar–sugar glycosidic bonds, liberating part of the glycan (endo-N-acetyl-β-D-glucosaminidase or endo-β-D-galactosidase).

---

**Protocol 32.** Use of 4-*N*-(2-*β*-D-glucosaminyl)-L-asparaginase
(EC: 3.5.1.26; 4-*N*-(2-acetamido 2-deoxy-*β*-D-
glucopyranosyl-L-asparagine aminohydrolase)

The enzyme is of lysosomal origin and hydrolyses glycoasparagines
liberating the glycan, aspartic acid, and ammonia (86).

*Method*

1. Dissolve 3–20 μmol of glycoasparagine in 50 μl of 0.1 M phosphate
   buffer at the optimum pH.

2. Add 0.5–5 mU of enzyme and incubate the mixture (final volume
   100 μl) at 37 °C for from 30 min to 24 h.

3. Stop the hydrolysis by heating in a boiling water-bath for 3 min.[a]

   [a] The enzyme hydrolyses only the *N*-acetylglucosaminyl-asparagine linkage of
   glycoasparagines and is inactive on glycopeptides. It is active on glycoasparagines
   of the oligomannosidic and of the *N*-acetyllactosaminic type. The structure and size
   of the carbohydrate moieties does not seem to affect the enzyme activity.

---

### 8.2.1 Peptide-*N*-glycosidase

Peptide-*N*-glycosidase hydrolyses the GlcNAc-Asn bond not only of glycoasparagines, in a similar manner to the above enzyme, but also of
glycopeptides and some glycoproteins. The enzyme is present in almond
emulsin (87) and in the culture medium of *Flavobacterium meningosepticum*
(88).

*i. Action on glycopeptides*

Despite the fact that the enzyme is not currently used for the deglycosylation
of glycopeptide, we describe here a procedure allowing the characterization
of enzyme activity.

---

**Protocol 33.** Use of peptide-*N*-glycosidase on glycopeptides

1. Dissolve 50 nmoles of glycopeptide in 50 mM Tris–HCl buffer pH 8.5
   with 100 mU of enzyme. The final volume does not exceed 50 μl.

2. Incubate the mixture for 2 to 24 h.

3. Denature the proteins by heating at 100 °C for 3 min.

4. After centrifugation, the supernatants can be used for fractionation or
   analysis (HPAE–PAD, HPCE, GLC, HPLC, NMR, MS, PAGE, SDS–PAGE,
   PC, or TLC) without any further treatment. If the amount of salts is too
   high, the extracts are desalted on a mini-column containing AG50-X2
   and AG1-X2 resins or gel filtered on TSK HW40 S.

---

ii. *Action on denaturated glycoproteins*

The denaturation is carried out by three different ways.

(a) Heat denaturation: to 20–50 µg of denatured glycoprotein dissolved in 50 mM Tris–HCl buffer pH 8.5, add 50 mU of enzyme in a total reaction volume of 50 µl. Incubate the mixture overnight at 37 °C. Stop the reaction by heating at 100 °C for 3 min. The reaction products are analysed as above described.

(b) SDS denaturation without heating the glycoprotein: 50 µg of glycoprotein are suspended in 20 µl of 50 mM of Tris–HCl buffer pH 8.5 containing 2 mM phenanthroline and 0.1% SDS (w/v), 1% 2-mercaptoethanol (w/v), and 1% Nonidet P40 (w/v). 100 mU of *N*-glycanase F are added and incubated at 37 °C for 6 to 24 hours.

(c) SDS-denaturation and heating the glycoprotein: dissolve 50 µg of glycoprotein in 10 µl of 1% SDS (w/v) and 0.5% of 2-mercaptoethanol (w/v) by heating for 3 min at 100 °C, and mix with 25 µl of 50 mM Tris–HCl buffer pH 8.5 containing 1% Nonidet P-40 (w/v). Add 100 mU of *N*-glycanase F and incubate at 37 °C for one hour (or more, depending on the conformation of the protein).

iii. *Action of native glycopoproteins*

The efficiency of the enzyme dramatically decreases when native substrates are used. Consequently, 10 to 100 times more *N*-glycanase are required for the deglycosylation and the incubation time must be increased. Use reagents, substrates, and laboratory ware free from proteolytic activities. As a protection, include protease inhibitor in the incubation medium (e.g. PMSF, Pepstatin A, aprotinin, benzamidine). Chelating agents such as EDTA or phenanthroline can also be used.

*Notes on this procedure*

(a) Glycopeptides are generally more susceptible to endo-action than denaturated or native glycoprotein.

(b) The optimum pH of *N*-glycanase F is settled as pH 8.6 but at pH 6.5 and 9.5 the enzyme is still 60% active. The enzyme can be currently used in phosphate buffer pH 7.5 without any loss of reactivity.

(c) The optimal temperature of *N*-glycanase F is defined as 37 °C but it has been demonstrated that the reactivity of the enzyme is independent of temperature above 25 °C. The apparent lack of stimulation of deglyco-sylation at temperatures higher than 25 °C may be due to an increase in the heat inactivation rate of the enzyme. The deglycosylation of glycoconjugates with *N*-glycanase is better at 25 °C rather than at 37 °C.

(d) The optimum pH of the *N*-glycopeptidase A from almond emulsin is in the range of 4 to 6. Over pH 7, the activity falls rapidly. The optimum temperature is 37 °C.

(e) The specificity of the enzymes for oligosaccharidic structures is broad. Oligomannosidic-type, hybrid-type, and *N*-acetyllactosaminic-type structures (bi-, tri-, and tetra-antennary chains) are easily released.

(f) The reactivity does not decrease when fucose is α-1,6-attached to the Asn-linked GlcNAc residue, when glycans contain a bisecting GlcNAc or xylose residue, or when Gal-GlcNAc repeating units or polysialyl structures are present.

(g) *N*-Glycanase F is able to hydrolyse complex oligosaccharides containing sulfate residues but cannot release glycans with fucose residues α-1,3-conjugated to the asparagine-linked *N*-acetylglucosamine residue as does the *N*-glycopeptidase A.

(h) The GlcNAc (β1-*N*)Asn linkage obtained after endo H digestion is not hydrolysed by the *N*-glycanase F contrary to *N*-glycopeptidase A. The minimum structure required for *N*-glycanase F is the di-*N*-acetyl-chitobiose core.

(i) *N*-Glycanases do not release glycans located at the amino or carboxyl termini.

(j) The specificity of the enzymes for peptide structures is more restrictive. In fact *N*-glycanase F easily acts on glycoproteins or glycopeptides, *N*-glycopeptidase A splits preferentially glycopeptides having from 3 to 40 amino acid residues but can release oligosaccharides from glycoprotein if the concentration of *N*-glycanase is 1000 times that of the usual case. The enzyme is not able to split Asn-oligosaccharides or Asn-GlcNAc. It is recommended to use trypsin or pepsin instead of Pronase for proteolysis because pronasic-glycopeptides often possess a too short peptide moiety.

### 8.2.2 Endo-β-D-galactosidases

Endo-β-D-galactosidases cleave internal β-galactoside linkages as follows:

$$\text{Gal}(\beta1 \rightarrow 4)\text{GlcNAc}(\beta1 \rightarrow 3)\text{Gal}(\beta1 \rightarrow 4)(\text{monosaccharide})_n + H_2O$$
$$\rightarrow \text{Gal}(\beta1 \rightarrow 4)\text{GlcNAc}(\beta1 \rightarrow 3)\text{Gal} + (\text{monosaccharide})_n$$

The enzyme has been found only in the micro-organisms *Streptococcus pneumoniae*, *Escherichia freundii*, and *Bacteroides fragilis* (89).

---

**Protocol 34.** Use of endo-β-D-galactosidases

1. Dissolve 5–10 nmol of substrate in 100 μl of 50 mM sodium acetate buffer pH 5.8.

2. Add 5–20 mU of enzyme and incubate at 37 °C for from 1 to 24 h. When using the enzyme from *Bacteroides fragilis*, stabilize activity with 0.02% (w/v) serum albumin.

3. Stop the reaction by heating in a boiling water-bath for 3 min.

---

**Table 20.** Specificity of endo-N-acetyl-β-D-glucosaminidases

| Substrates | Endo-N-acetyl-β-D-glucosaminidase | | | |
|---|---|---|---|---|
| | H | D | C_II | B |
| **N-Acetyllactosaminic-type structures[a]** | | | | |
| NeuAc(α2 → 6)Gal(β1 → 4)GlcNAc(β1 → 2)Man(α1 → 3)⟍<br>NeuAc(α2 → 6)Gal(β1 → 4)GlcNAc(β1 → 2)Man(α1 → 6)⟋ Man(β1 → 4)R₁ | − | − | − | − |
| NeuAc(α2 → 6)Gal(β1 → 4)GlcNAc(β1 → 2)Man(α1 → 3)⟍<br>Gal(β1 → 4)GlcNAc(β1 → 2)Man(α1 → 6)⟋ Man(β1 → 4)R₁ | − | − | − | + |
| Gal(β1 → 4)GlcNAc(β1 → 2)Man(α1 → 3)⟍<br>Gal(β1 → 4)GlcNAc(β1 → 2)Man(α1 → 6)⟋ Man(β1 → 4)R₁ | − | − | − | + |
| Gal(β1 → 4)GlcNAc(β1 → 2)Man(α1 → 3)⟍<br>Gal(β1 → 4)GlcNAc(β1 → 2)Man(α1 → 6)⟋ Man(β1 → 4)R₂ | − | − | − | + |
| Gal(β1 → 4)GlcNAc(β1 → 2)Man(α1 → 3)⟍<br>GlcNAc(β1 → 2)Man(α1 → 6)⟋ Man(β1 → 4)R₂ | − | + | − | + |
| GlcNAc(β1 → 4)Man(α1 → 3)⟍<br>Man(α1 → 6)⟋ Man(β1 → 4)R₂ | − | − | − | + |
| Asialotriantennary glycopeptide[b] | − | − | − | + |
| Asialotetra-antennary glycopeptide[c] | − | − | − | − |

## Oligomannosidic-type structures

```
Man(α1 → 3)
Man(α1 → 6)
           Man(α1 → 3)
                       Man(β1 → 4)R₁
Man(α1 → 3)
Man(α1 → 6)
           Man(α1 → 6)
```

| Structure | | | | |
|---|:--:|:--:|:--:|:--:|

$$
\begin{array}{l}
\mathrm{Man}(\alpha 1 \to 3) \\
\mathrm{Man}(\alpha 1 \to 6)
\end{array}
\Big\rangle \mathrm{Man}(\beta 1 \to 4)\mathrm{R_1}
\qquad +\ \ +\ \ +\ \ +
$$

$$
\mathrm{Man}(\alpha 1 \to 3)\mathrm{Man}(\alpha 1 \to 6)\Big\rangle \mathrm{Man}(\beta 1 \to 4)\mathrm{R_1}
\qquad +\ \ +\ \ -\ \ +
$$

$$
\begin{array}{l}
\mathrm{Man}(\alpha 1 \to 3) \\
\mathrm{Man}(\alpha 1 \to 6)
\end{array}
\Big\rangle \mathrm{Man}(\alpha 1 \to 6)\ \ \mathrm{Man}(\beta 1 \to 4)\mathrm{R_1}
\qquad +\ \ +\ \ +\ \ +
$$

$$
\mathrm{Man}(\alpha 1 \to 2)\mathrm{Man}(\alpha 1 \to 3)\ \ \mathrm{Man}(\beta 1 \to 4)\mathrm{R_1}
\qquad +\ \ +\ \ +\ \ +
$$

$$
\begin{array}{l}
\mathrm{Man}(\alpha 1 \to 3) \\
\mathrm{Man}(\alpha 1 \to 6)
\end{array}
\Big\rangle \mathrm{Man}(\beta 1 \to 4)\mathrm{R_1}
\qquad +\ \ +\ \ -\ \ +
$$

$$
\mathrm{Man}(\alpha 1 \to 2)\mathrm{Man}(\alpha 1 \to 2)\mathrm{Man}(\alpha 1 \to 3)\ \ \mathrm{Man}(\beta 1 \to 4)\mathrm{R_1}
\qquad +\ \ -\ \ +\ \ +
$$

$$
[\mathrm{Man}(\alpha 1 \to 2)]_{0-1}\mathrm{M}(\alpha 1 \to 3)
$$

$$
\mathrm{Man}(\alpha 1 \to 6)
$$

$$
\mathrm{Man}(\alpha 1 \to 2)\mathrm{Man}(\alpha 1 \to 6)
$$

## Hybrid-type structure[d]

$$
+\ \ -\ \ -\ \ -
$$

[a] R₁: GlcNAc($\beta$1 → 4)GlcNAc($\beta$ → N)Asn or peptide; R₂: GlcNAc($\beta$1 → 4)[Fuc($\alpha$1 → 6) ]GlcNAc($\beta$ → N)Asn or peptide.

[b,c] Desialylated structures 5 and 7, respectively, from Figure 5.

[d] As an example, see Figure 7.

The enzymes are unable to split the galactosyl linkages when galactose residues are sulfated. Fucosylated glycans are less hydrolysed. The substitution by fucose at the C-2 position of the terminal galactose residues makes the substrate resistant to the enzyme. Similarly, the enzyme is inactive when galactose residues are substituted at the position C-6 by *N*-acetylglucosamine to form a branched point. Substitution of the terminal galactose residue at the C-3 position by *N*-acetylneuraminic acid or by α-galactosyl residues increases the enzyme activity.

### 8.2.3 Endo-*N*-acetyl-*β*-D-glucosaminidases

Endo-*N*-acetyl-*β*-D-glucosaminidases constitute an homogeneous family of enzymes from various sources that act on the *β*-1,4-*N*-acetylglucosaminyl bond of the *N,N'*-diacetylchitobiose residue present in the core of all *N*-glycosylproteins. Their specificity (see *Table 20*) allows us to classify these enzymes in three groups.

The first group comprises the following enzymes: endo H isolated from *Streptomyces plicatus* (90) and endo $F_1$ from *Flavobacterium meningosepticum* (91) acting on oligomannosidic-type and hybrid-type structures.

The second group comprises endo B isolated from *Sporotrichum dimorphosporum* (92), and endo $F_2$ and $F_3$ from *Flavobacterium meningosepticum* (93) acting on oligomannosidic-type and *N*-acetyllactosaminic-type structures.

The third group comprises the endo-enzyme D isolated from *Streptococcus pneumoniae* (94). This enzyme requires an unsubstituted α-1,3-mannose residue in the terminal non-reducing position or only substituted in the C-4 position in the case of *N*-acetyllactosaminic-type structures. In addition, the enzyme acts poorly on oligomannosidic-type structures and for these reasons it is not used for deglycosylating the glycoproteins.

---

**Protocol 35.** Action of endo-*N*-acetyl-*β*-D-glycosaminidases on glycopeptides

1. Dissolve 1 to 10 nmoles of glycopeptide in 20 μl of 0.01 M sodium citrate buffer pH 5.5 containing 0.1 mg/ml bovine serum albumin (the amount of substrate may be lowered if it is fluorescent or radio-labelled).

2. Add 2 mU of endo H. The final volume does not exceed 0.05 ml.

3. Incubate from 30 min to 12 h at 37 °C under toluene atmosphere in a shaker-bath, and then denature the proteins by heating at 100 °C for 3 min.

4. After centrifugation, the supernatants can be used for fractionation or analysis (HPAE–PAD, HPCE, GLC, HPLC, NMR, MS, PAGE, SDS–PAGE,

---

PC, or TLC) without any further treatment. If the amount of salts is too high, the extracts are desalted after dilution on mini-columns containing AG50-X2 and AG1-X2 resins or gel filtered on TSK HW40 S.

---

**Protocol 36.** Methods for denaturing glycoproteins for endo-*N*-acetyl-$\beta$-D-glucosaminidase treatment

A. *Heat denaturation*

1. Heat the glycoprotein (1 mg/ml) for 10 min at 100 °C in enzyme optimum pH buffer.

2. Add 2 to 5 nmoles of denatured glycoprotein to 5 mU of endo H in 0.01 M phosphate-citrate buffer pH 5.5. Incubation time is higher than in the case of glycopeptide: from 2 h up to 24 h at 37 °C under toluene atmosphere in a shaker-bath.

3. Denature the proteins by heating at 100 °C for 3 min.

4. After centrifugation the supernatants can be used for fractionation or analysis (HPAE–PAD, HPCE, GLC, HPLC, NMR, MS, PAGE, SDS–PAGE, PC, TLC) without any further treatment. If the amount of salts is too high, the extracts are desalted, after dilution on mini-column containing AG50-X2 and AG1-X2 resins or gel filtered on TSK HW40 S.

B. *Denaturation by chaotropic agent*

Treat the glycoprotein (1 mg/ml) with 0.25 to 1.5 M of chaotropic salt (generally sodium sulfocyanate) in the presence of 0.1 M 2-mercapto-ethanol for variable times.

*Method*

1. To 2 to 5 nmoles of native glycoprotein add 5 mU of endo H in 10 mM phosphate-citrate buffer pH 5.5 containing 0.5 M NaSCN.

2. Incubate the mixture for 2 h up to 48 h, and stop the reaction by heating.

3. Analyse the products as described above. If the NaSCN is troublesome it can easily be discarded by gel filtration prior to analysis.

C. *SDS denaturation*

1. Heat the glycoprotein (5 mg/ml) at 100 °C in buffer containing 1% SDS and 0.1 M 2-mercaptoethanol for 10 min. In these cases, excess detergent must be discarded or lowered prior to enzymatic deglyco-sylation.

**Protocol 36.** *Continued*

2. Add 10 μl of the denatured glycoprotein to 5 μl of 1% bovine serum albumin in 0.01 phosphate-citrate buffer pH 5.5. Heat the mixture at 100 °C for 5 min in order to achieve the interaction between SDS and albumin.

3. After cooling, add 5 mU of endo H and incubate at 37 °C for variable times (2 to 48 h). The incubation medium is then treated as previously described.

---

**Protocol 37.** Action of endo-*N*-acetyl-β-D-glucosaminidases on native glycoproteins

1. Mix the native glycoprotein (50 to 100 μg) solubilized in 20 μl of 0.01 M phosphate-citrate buffer pH 5.5 with largely increased quantities of endo-enzyme, the concentration of which is empirically determined (50 to 100 mU can be used in order to obtain deglycosylation).

2. In some cases, enzyme replenishment is required (10 to 25 mU are introduced at 24 h intervals) and the incubation time can be lengthened up to 72 h.

3. Although the enzymatic preparation is free of proteolytic enzymes, it is wise to prevent trace amounts of proteases by introducing 0.01 M EDTA, 100 U aprotinin/ml, 2 mM phenanthroline monohydrate, or 1 mM PMSF in the buffer just before adding the endo-enzyme. In practice it is very difficult to obtain a complete deglycosylation of native glycoprotein.

---

*Notes on the above procedures*

(a) Glycopeptides are generally more susceptible to endo-action than denatured or native glycoproteins.

(b) The endo-enzymes are able to hydrolyse the chitobiose unit in mannose-containing *N*-linked oligosaccharides with at least three mannose residues. The α-1,6-arm must be substituted in order to obtain high rate of splitting.

(c) The presence of a fucose residue in the core region is a limiting factor for hydrolysis. Linear substitution (hybrid-type structure) at the α-1,3-arm enables the oligosaccharides with fucose α-1,6-linked to the proximal *N*-acetylglucosamine to become accessible to endo H and endo $F_1$. In that case the reactivity of endo H is higher than that of endo $F_1$.

(d) The velocity of endo $F_2$ and endo $F_3$ on oligomannosidic-type structures is much lower (one tenth) than the velocity of endo $F_1$.

(e) The sulfated glycans are not hydrolysed by endo H and endo B but are susceptible to the action of endo $F_1$.

(f) Endo H and endo $F_1$ are able to hydrolyse oligomannosidic and complex-type structures.

  Endo B reacts with all oligomannosidic-type structures but only cleaves some of the hybrid-type structures: substitution on $\alpha$-1,3-arm with GlcNAc and presence of an intersecting GlcNAc or Xyl residue.

  Endo H and endo $F_1$ do not split complex-type structures as do endo B, and endo $F_2$, and $F_3$.

  Endo B and endo $F_2$ only split biantennary structures. Endo $F_3$ cleaves both bi- and triantennary structures. The velocity against biantennary structures is higher than against tetra-antennary structures. Neuraminic acid decreases the rate of hydrolysis. When the oligosaccharide moiety is fully sialylated, it becomes resistant to the action of endo B, endo $F_2$ and $F_3$.

(g) A tenfold excess of Triton X-100 or Nonidet P-40 protects endo F against the SDS. So, the deglycosylation of glycoproteins is possible without prior heat denaturation in SDS medium without any loss of endo F activity. In the case of endo B, the enzyme is inactivated by 0.01% SDS and the excess of SDS must be eliminated by addition of 1% bovine serum albumin. Free SDS may also be lowered by keeping the incubation mixture overnight at 4 °C and the precipitate obtained discarded by centrifugation.

(h) Commercial preparations of endo F are generally stabilized by 50% glycerol. It is important to note that a 1% glycerol concentration (v/v) is sufficient to produce high transglycosylation reactions. Consequently, the reducing end of the oligosaccharide is blocked and the molecular weight of the liberated oligosaccharides is enhanced. Endo H is also able to affect the incorporation of glycerol but to a lesser extent. Consequently, the glycerol must be discarded before deglycosylation.

# 9. $^1$H-NMR spectra as fingerprints for glycan primary structure determination

High field $^1$H-NMR was successfully introduced for the determination of oligosaccharides and glycan primary structure by Vliegenthart's and Montreuil's groups in 1977 (95). The method developed by Vliegenthart *et al.* (7, 8) is based on the set of structural-reporter groups, and is particularly well adapted for the structural elucidation of *N*-acetyllactosaminic and oligomannosidic-type glycans. The compilation of the $^1$H-NMR data that are presented in a general review (8) or in original papers we have selected and which concern both *N*-glycosyl (95–99) and *O*-glycosylproteins (100–107) furnishes the basis of interpretation of hundreds of different ¹H-NMR spectra. But, these data have less interest when new types of structures are investigated and the unambiguous assignment of ¹H-resonances requires new model compounds or the isolation of sufficient material for 2D NMR spectroscopy. 25 to 50 nmol of material are required for recording a 1D ¹H-NMR spectrum of adequate quality which is easy to interprete on the basis of the pattern of structural-reporter group signals.

## 9.1 The structural-reporter groups

The resonances of the structural-reporter groups which are clearly distinguishable for the interpretation of the ¹H-NMR spectra comprise:

- the anomeric protons
- the H-2 (and H-3, in some peculiar cases) atom of mannose
- the amino sugar *N*-acetyl-$CH_3$ protons
- the sialic H-3 atoms
- the fucose H-5 and $CH_3$ protons
- the galactose H-3 and H-4 atoms

The anomeric protons are recognizable owing to their resonances occurring in the low field region of the spectrum (4.3 to 5.4 p.p.m.). In the gluco- and galacto-series, the signals observed at the lower field (4.7–5.3 p.p.m.), with a $^3J_{1,2}$ value of 1–4 Hz are characteristic of the α-anomeric protons, whereas β-anomeric protons appear between 4.3–4.7 p.p.m. and possess a $^3J_{1,2}$ value of 8–10 Hz. In the manno-series, the equatorial position of the proton H-2 leads to equal dihedral angle H-1, H-2 (~ 60 °) for α- and β-anomers. Nevertheless, the α-anomer (δ H-1 4.8–5.4 p.p.m.) exhibits a $^3J_{1,2}$ value (~ 1.2 Hz) slightly higher than that observed for the β-anomer (δ = 4.7 p.p.m.; $^3J_{1,2}$ < 0.9 Hz.

## 9.2 *N*-Glycosylprotein glycans

### 9.2.1 Characterization of bi-, tri- and tetra-antennary glycans of the *N*-acetyllactosaminic-type

The asialo biantennary glycan isolated from various glycoproteins and particularly from human serotransferrin (95) has been chosen as the basic structure used for the comparative interpretation of NMR spectra. The anomeric proton resonance from each sugar unit was assigned using the model compounds listed in *Figure 22*.

The spectra of compounds **1** and **3** show identical doubtlets ($^3J_{1,2}$ = 9.08 Hz) at δ 5.071 p.p.m. which belong to the anomeric protons of GlcNAc-1. Consequently, the second doublet observed at 4.62 p.p.m. on the spectrum of **3** and **5** corresponds to GlcNAc-2 H-1 atom. By the same way, the comparison of the spectra from compounds **2** and **3** furnishes the assignment of the Man-4 and 4′ anomeric proton, respectively. Similarly, the comparison of **4** and **5** gives the H-1 resonances from Gal (δ = 4.70 p.p.m.) and GlcNAc (δ = 4.58 p.p.m.). To distinguish between Gal-6 and Gal-6′ H-1 resonances, the effect of anomerization affecting the structural-reporter group signal of the α-1,6-antenna was taken into account. As shown in *Figure 23*, two lower field doublets at δ 4.471 and 4.473 belong to H-1 of Gal-6′ in the α and β-anomers of **7**, whereas the highest field doublet, at δ 4.468, is attributed to Gal-6 in compound **8**.

The spectra of the tri-, tri′- and tetra-antennary glycopeptides and oligosaccharides (compounds **9**, **10**, **11**) exhibit new anomeric signals belonging to Gal-8. 8′ and GlcNAc-7. 7′. More characteristic are the resonance patterns of the mannose H-2 (and H-3) atoms, as well as the new *N*-acetyl-CH$_3$ protons of GlcNAc-7 and 7′ which can be used as 'fingerprints' for recognition of the type of branching (*Figures 24, 25*, and *Table 21*).

The presence of the bisecting GlcNAc-9 (compound **12**) has a profound influence on the chemical shift of several reporter groups, as shown in *Figure 26*. Moreover, these shift increments are additive and can be used for the structural elucidation of complex carbohydrate sequences.

### 9.2.2 Sialylated glycans of the *N*-acetyllactosamine type

The presence of *N*-acetylneuraminic acid α-2,3- or α-2,6-linked to galactose is recognized owing two new signals relative to the H-3ax and H-3eq atom (*Figure 27*). Moreover, typical effects of attachment of NeuAc on the chemical shifts of structural-reporter groups of neighbouring residues are summarized in *Tables 22* and *23*.

The α-2,3- and α-2,6-linked sialic acids can be distinguished on the basis of the set of chemical shifts of the H-3ax and H-3eq. The attachment of NeuAc at *O*-3 of Gal has strong effect on the H-1 atom resonance of Gal, and also, to a certain extent, on the GlcNAc and Man H-1 resonance. In the case of a α-2,6-linkage, strong influence is observed on GlcNAc and Man H-1 chemical shifts. The *N*-acetyl-CH$_3$ resonances of the GlcNAc residues are also modified by these two types of sialylation. Furthermore, the H-2 signals of *O*-3 sialylated galactose undergoes a large downfield shift (Δδ 0.45 p.p.m.) and becomes a structural-reporter group for the NeuAc(α 2 → 3) Gal(β1 → ) sequence. These observations allow a precise determination of the structure of most of the sialylated polyantennary glycans, according to the rules defined by Vliegenthart *et al.* (8)

Incomplete glycan structures possessing non-reducing terminal GlcNAc or

**1**  GlcNAc(β1-N)Asn

**2**  Man(β1-4)GlcNAc
/
Man(α1-3)

**3**  Man(α1-6)
\
Man(β1-4)GlcNAc(β1-4)GlcNAc(β1-N)Asn

GlcNAc(β1-2)Man(α1-6)
\
**4**  Man(β1-4)GlcNAc
/
GlcNAc(β1-2)Man(α1-3)

Gal(β1-4)GlcNAc(β1-2)Man(α1-6)
\
**5**  Man(β1-4)GlcNAc(β1-4)GlcNAc(β1-N)Asn
/
Gal(β1-4)GlcNAc(β1-2)Man(α1-3)

Gal(β1-4)GlcNAc(β1-2)Man(α1-6)
\
**6**  Man(β1-4)GlcNAc
/
Gal(β1-4)GlcNAc(β1-2)Man(α1-3)

Gal(β1-4)GlcNAc(β1-2)Man(α1-6)
\
**7**  Man(β1-4)GlcNAc

**8**  Man(β1-4)GlcNAc
/
Gal(β1-4)GlcNAc(β1-2)Man(α1-3)

Gal(β1-4)GlcNAc(β1-2)Man(α1-6)
\
**9**  Man(β1-4)GlcNAc(β1-4)GlcNAc(β1-N)Asn
/
Gal(β1-4)GlcNAc(β1-2)Man(α1-3)
/
Gal(β1-4)GlcNAc(β1-4)

Gal(β1-4)GlcNAc(β1-6)
\
Gal(β1-4)GlcNAc(β1-2)Man(α1-6)
\
**10**  Man(β1-4)GlcNAc(β1-4)GlcNAc(β1-N)Asn
/
Gal(β1-4)GlcNAc(β1-2)Man(α1-3)

Gal(β1-4)GlcNAc(β1-6)
\
Gal(β1-4)GlcNAc(β1-2)Man(α1-6)
\
**11**  Man(β1-4)GlcNAc(β1-4)GlcNAc(β1-N)Asn
/
Gal(β1-4)GlcNAc(β1-2)Man(α1-3)
/
Gal(β1-4)GlcNAc(β1-4)

**268**

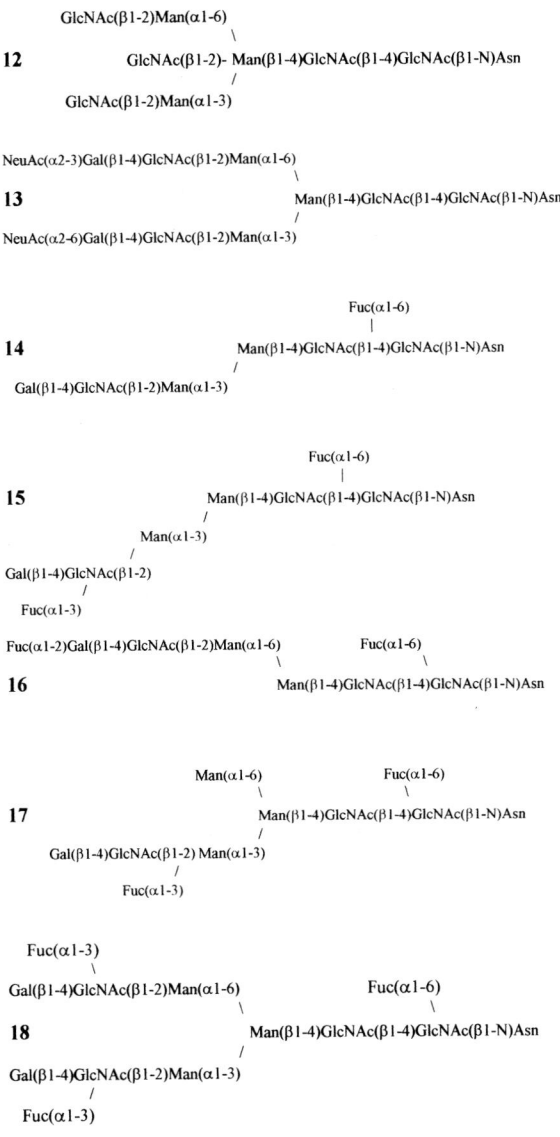

GlcNAc(β1-2)Man(α1-6)
      \
**12**            GlcNAc(β1-2)- Man(β1-4)GlcNAc(β1-4)GlcNAc(β1-N)Asn
      /
GlcNAc(β1-2)Man(α1-3)

NeuAc(α2-3)Gal(β1-4)GlcNAc(β1-2)Man(α1-6)
      \
**13**            Man(β1-4)GlcNAc(β1-4)GlcNAc(β1-N)Asn
      /
NeuAc(α2-6)Gal(β1-4)GlcNAc(β1-2)Man(α1-3)

                              Fuc(α1-6)
                                 |
**14**            Man(β1-4)GlcNAc(β1-4)GlcNAc(β1-N)Asn
                    /
Gal(β1-4)GlcNAc(β1-2)Man(α1-3)

                              Fuc(α1-6)
                                 |
**15**            Man(β1-4)GlcNAc(β1-4)GlcNAc(β1-N)Asn
                    /
            Man(α1-3)
         /
Gal(β1-4)GlcNAc(β1-2)
   /
 Fuc(α1-3)

Fuc(α1-2)Gal(β1-4)GlcNAc(β1-2)Man(α1-6)        Fuc(α1-6)
                              \                   \
**16**            Man(β1-4)GlcNAc(β1-4)GlcNAc(β1-N)Asn

            Man(α1-6)          Fuc(α1-6)
                \                 \
**17**            Man(β1-4)GlcNAc(β1-4)GlcNAc(β1-N)Asn
                    /
    Gal(β1-4)GlcNAc(β1-2) Man(α1-3)
                    /
            Fuc(α1-3)

   Fuc(α1-3)
      \
Gal(β1-4)GlcNAc(β1-2)Man(α1-6)          Fuc(α1-6)
                    \                     \
**18**            Man(β1-4)GlcNAc(β1-4)GlcNAc(β1-N)Asn
                    /
Gal(β1-4)GlcNAc(β1-2)Man(α1-3)
      /
 Fuc(α1-3)

**Figure 22.** Structure of the oligosaccharides used for the identification of the structural-reporter groups of bi-, tri-, and tetra-antennary glycans of *N*-acetyllactosaminic-type.

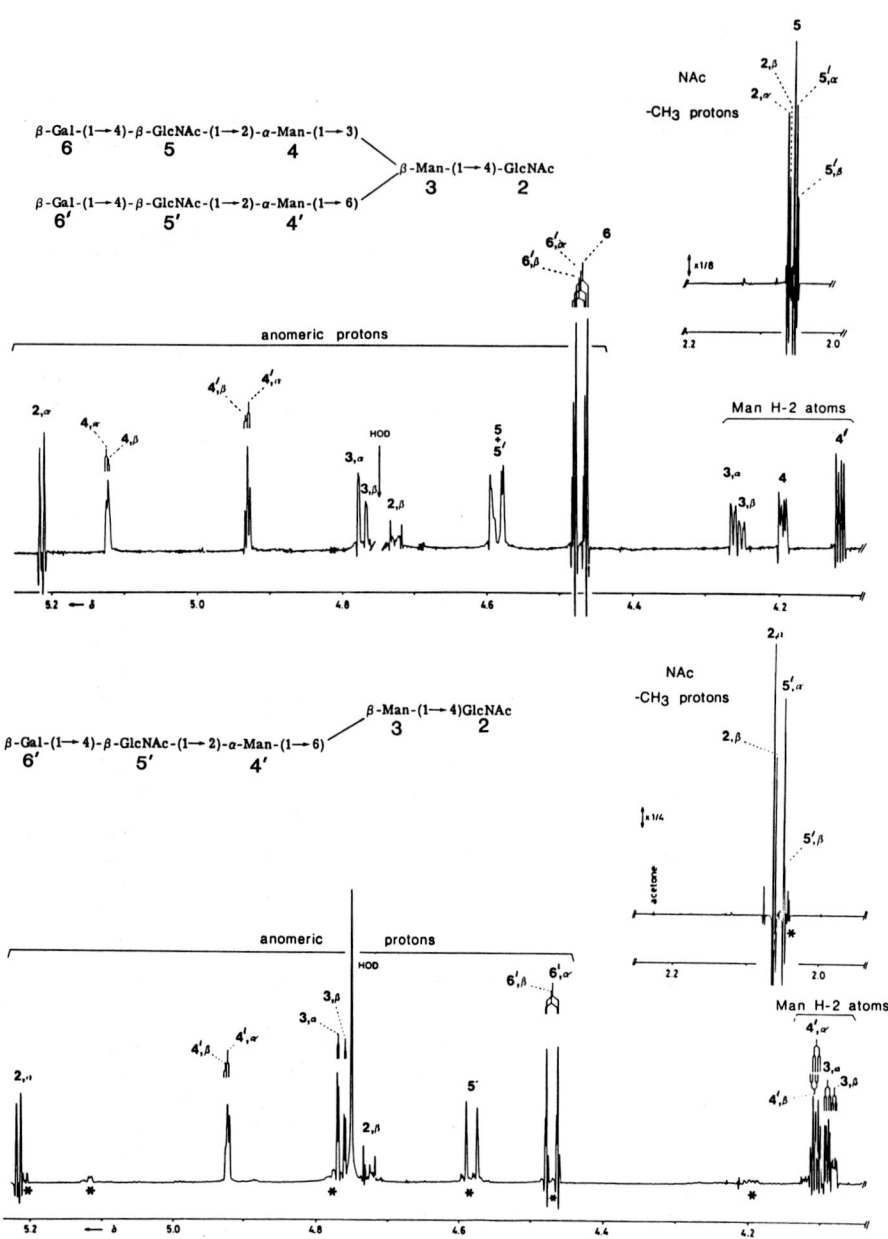

**Figure 23.** ¹H-NMR spectra of compounds **6** and **7** of *Figure 22*, showing the anomerization effect affecting the residues 4′, 5′, and 6′ of the 1,6-antenna.

**Table 21.** [1]H chemical shifts of structural-reporter groups of constituent mono-saccharides for asialo bi-, tri-, and tetra-antennary glycopeptides of the N-acetyllactosamine-type (8)

**Compound and schematic structure**[a]

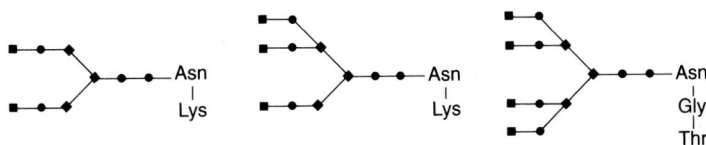

| Reporter group | Residue[b] | | | |
|---|---|---|---|---|
| H-1 of | 1 | 5.094 | 5.092 | 5.053 |
| | 2 | 4.616 | 4.614 | 4.614 |
| | 3 | 4.765 | 4.755 | 4.757 |
| | 4 | 5.121 | 5.120 | 5.129 |
| | 4' | 4.928 | 4.924 | 4.868 |
| | 5 | 4.582 | 4.570 | 4.573 |
| | 5' | 4.582 | 4.580 | 4.596 |
| | 6 | 4.467 | 4.468 | 4.470 |
| | 6' | 4.473 | 4.473 | 4.472 |
| | 7 | — | 4.545 | 4.547 |
| | 7' | — | — | 4.553 |
| | 8 | — | 4.462 | 4.465 |
| | 8' | — | — | 4.481 |
| H-2 of | 3 | 4.246 | 4.209 | 4.210 |
| | 4 | 4.190 | 4.218 | 4.224 |
| | 4' | 4.109 | 4.108 | 4.092 |
| H-3 of | 4 | <4.0 | 4.045 | 4.052 |
| | 4' | <4.0 | <4.0 | <4.0 |
| NAc of | 1 | 2.004 | 2.003 | 2.008 |
| | 2 | 2.079 | 2.078 | 2.078 |
| | 5 | 2.050 | 2.048 | 2.054 |
| | 5' | 2.046 | 2.045 | 2.042 |
| | 7 | — | 2.075 | 2.079 |
| | 7' | — | — | 2.041 |

[a] ■, Gal; ● GlCNAc, ◆, Man.
[b] Numbers refer to the structures of *Figure 24*.

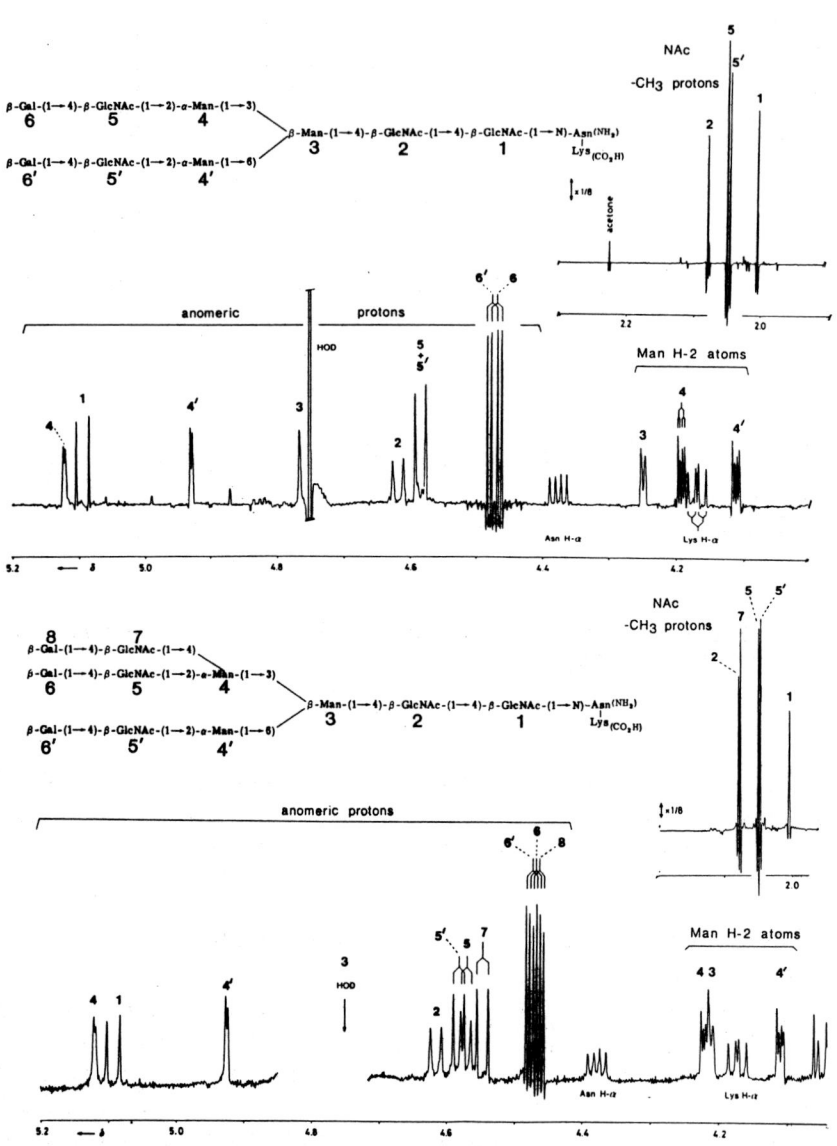

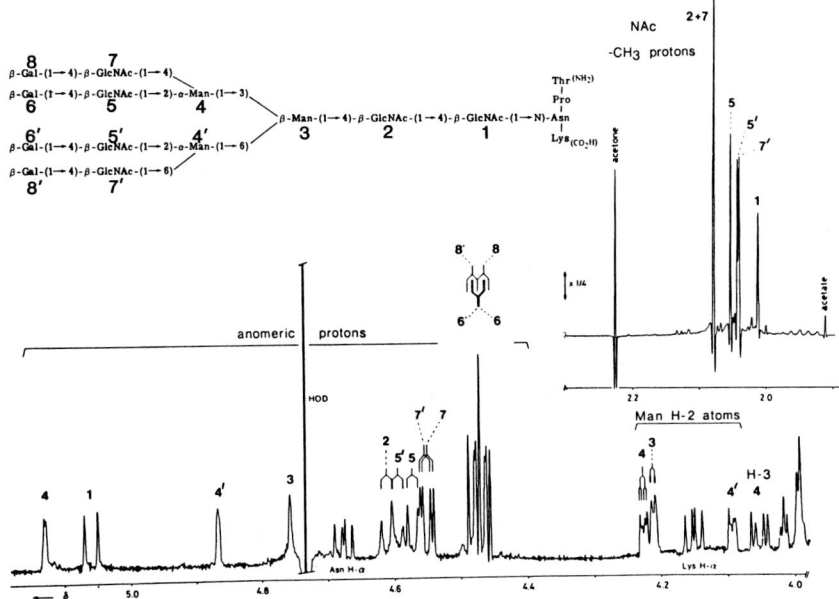

**Figure 24.** Structural-reporter groups characteristic of bi-, tri-, and tetra-antennary glycans of the *N*-acetyllactosaminic-type.

Man can be also determined according to the characteristic chemical shifts of GlcNAc H-1 or Man H-2 atom resonances. The H-2 atom resonance of non-reducing terminal Man residues is particularly shifted upfield.

### 9.2.3 Fucosylated glycans of the *N*-acetyllactosaminic-type

*i. Fuc α-1,6-linked to GlcNAc-1*

The structural-reporter group of Fuc α-1,6-linked to GlcNAc C-1 are H-1 ($\delta = 4.877$), H-5 ($\delta = 4.121$), and CH$_3$ ($\delta = 1.206$) which can be easily distinguished in the spectra of compounds **14** to **18**. The NAc resonance of GlcNAc-2 is shifted upfield from $\delta$ 2.072 (compounds **13**, **14**, **15**) to $\delta$ 2.095 (compounds **16**, **17**, **18**) when Man 4' is present (even in the non-reducing terminal position as in compound **17**).

*ii. Fuc α-1,3-linked to a peripheral GlcNAc*

This linkage is characterized by the following set of structural-reporter group signals $\delta$ H-1: 5.111 to 5.125, $\delta$ H-5: 4.822 to 4.830, $\delta$ CH$_3$: 1.172 to 1.176.

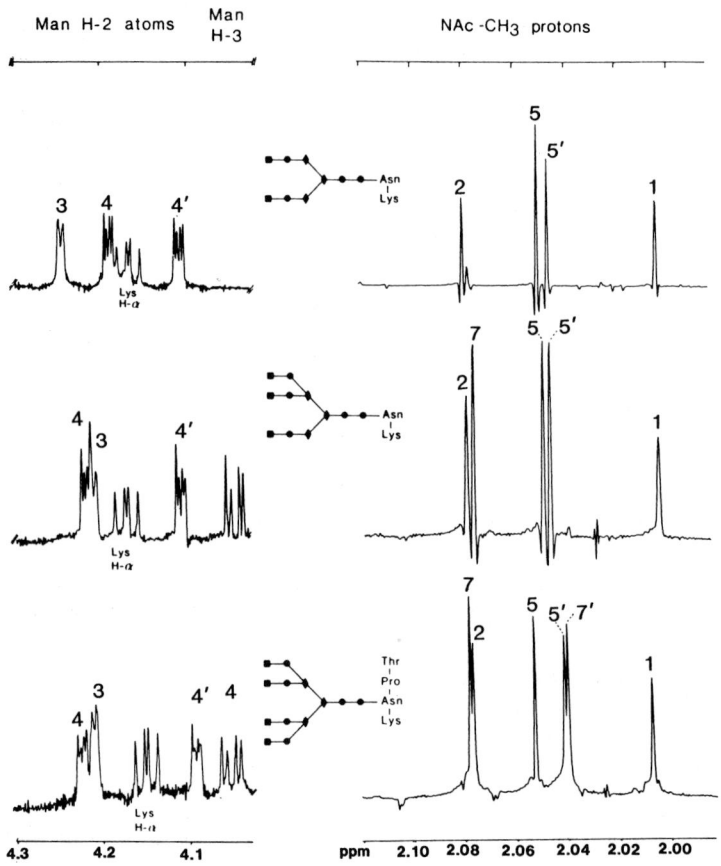

**Figure 25.** Basis for the characterization of the branching-type of a glycan of the *N*-acetyllactosaminic-type.

The introduction of this kind of Fuc linkage induces some shift effects on structural-reporter groups of neighbouring residues (*Figure 28* and *Table 24*). The H-1 doublet of Gal is shifted upfield ($\Delta\delta - 0.024$ to $- 0.026$ p.p.m.), as well as the *N*-acetyl signal of GlcNAc ($\Delta\delta - 0.006$ to $- 0.026$ p.p.m.), while the H-1 doublet of GlcNAc is shifted downfield ($\Delta\delta + 0.007$ to $0.014$ p.p.m.).

### iii. Fuc α-1,2-linked to Gal
Fuc α-1,2-linked to Gal can be recognized owing to the typical set of the H-1 ($\delta$: 5.299 to 5.303), H-5 ($\delta$: 4.206 to 4.230), and CH$_3$ ($\delta$ 1.206) resonances.

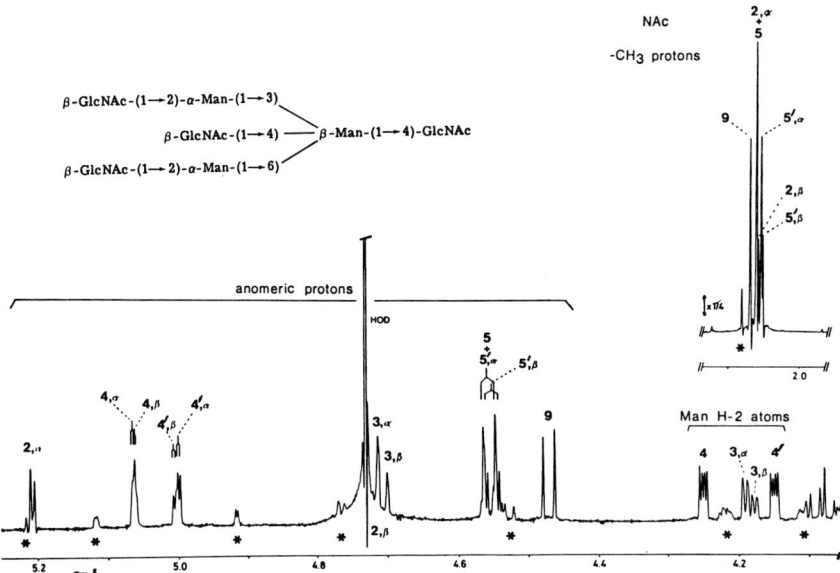

**Figure 26.** ¹H-NMR spectrum of glycan **12** of *Figure 22* with bisecting GlcNAc (asterisks are relative to an isomer devoid of the bisecting GlcNAc).

---

**Table 22.** Recognition of α-2,6-linked, terminal NeuAc in the sequence NeuAc(α2 → 6) Gal(β1 → 4) GlcNAc (β1 → x) Man(α1 → y) Man(β1-) (8)

¹H chemical shifts of structural-reporter groups of α-2,6-linked NeuAc (at pD ~ 7)

| Reporter group | δ ± **SD** |
|---|---|
| H-3a (x = 2; y = 3) | 1.718 ± 0.002 |
| H-3a (x = 2; y = 6) | 1.717 ± 0.001 |
| H-3a (x = 4; y = 3) | 1.706 ± 0.002 |
| H-3e (x = 2; y = 3) | 2.669 ± 0.001 |
| H-3e (x = 2; y = 6) | 2.672 ± 0.001 |
| H-3e (x = 4; y = 3) | 2.670 ± 0.002 |
| NAc (x = 2; y = 3 or 6) | 2.030 ± 0.001 |
| NAc (x = 4; y = 3) | 2.029 ± 0.002 |

Influence of α-2,6-linked NeuAc on the chemical shifts of structural-reporter groups of neighbouring residues

| Reporter group | Δδ (**p.p.m.**) ± **SD (p.p.m.)** |
|---|---|
| H-1 of Gal (x = 2 or 4; y = 3 or 6) | −0.024 ± 0.001 |
| H-1 of GlcNAc (x = 2 or 4; y = 3 or 6) | +0.024 ± 0.001 |
| NAc of GlcNAc (x = 2; y = 3 or 6) | +0.019 ± 0.001 |
| NAC of GlcNAc (x = 4; y = 3) | +0.025 ± 0.002 |
| H-1 of αMan (x = 2; y = 3) | +0.014 ± 0.002 |
| H-1 of αMan (x = 2; y = 6) | +0.020 ± 0.003 |
| H-1 of αMan (x = 4; y = 3) | ~0.0 |
| H-1 of βMan (x = 2; y = 6) | +0.008 ± 0.003 |
| H-2 of αMan (x = 2; y = 3) | +0.004 ± 0.002 |
| H-2 of αMan (x = 2; y = 6) | +0.005 ± 0.002 |
| H-2 of αMan (x = 4; y = 3) | ~0.0 |
| H-2 of βMan (x = 2; y = 6) | +0.007 ± 0.002 |

**Table 23.** Recognition of α-2,3-linked, terminal NeuAc in the sequence NeuAc(α2 → 3) Gal(β1 → 4) GlcNAc (β1 → x) Man(α1 → y) Man(β1-) (8)

¹H chemical shifts of structural-reporter groups of α-2,3-linked NeuAc (at pD ∼ 7)

| Reporter group | δ ± SD |
|---|---|
| H-3a (x = 2; y = 3) | 1.798 ± 0.002 |
| H-3a (x = 2; y = 6) | 1.800 ± 0.001 |
| H-3a (x = 4; y = 3) | 1.801 ± 0.001 |
| H-3e (x = 2; y = 3) | 2.758 ± 0.001 |
| H-3e (x = 2; y = 6) | 2.757 ± 0.001 |
| H-3e (x = 4; y = 3) | 2.757 ± 0.001 |
| NAc (x = 2 or 4; y = 3 or 6) | 2.030 ± 0.001 |

Influence of α-2,3-linked NeuAc on the chemical shifts of structural-reporter groups of neighbouring residues

| Reporter group | Δδ (p.p.m.) ± SD (p.p.m.) |
|---|---|
| H-1 of Gal (x = 2; y = 3 or 6) | +0.076 ± 0.001 |
| H-1 of Gal (x = 4; y = 3) | +0.083 ± 0.002 |
| H-3 of Gal (x = 2 or 4; y = 3 or 6) | +0.453 ± 0.003 |
| H-1 of GlcNAc (x = 2; y = 3 or 6) | −0.004 ± 0.002 |
| H-1 of GlcNAc (x = 4; y = 3) | +0.005 ± 0.002 |
| NAc of GlcNAc (x = 2 or 4; y = 3 or 6) | −0.003 ± 0.001 |

Strong shift effects on neighbouring residues are also observed which are summarized in *Table 24*.

Nevertheless, some differences in the shift effects can be observed between mono- and polyantennary structures. They may reflect the steric interaction between the different branch of the glycans. Despite these slight deviations from one model to another, these shift effects remain of high interest for determining the position of each Fuc residue.

## 9.2.4 Oligomannosidic-type glycans

In the series of $Man_nGlcNAc_2Asn$ (n = 2 to 9), 37 isomers could be identified on the basis of Man H-1 atom resonances. The number of Man residues is directly deduced from the integrated ¹H-NMR spectrum. Then, each Man residue can be located according to simple rules summarized in *Table 25*. Indeed, each Man H-1 resonance possesses two possible chemical shifts depending on the terminal or internal position of Man residues. *Figure 29* presents an example in which the successive upfield shifts of Man B H-1 and Man C H-1 can be easily observed, when Man $D_3$ and $D_1$ are lacking.

**Table 24.** Chemical shifts of Gal and GlcNAc $^1$H and of GlcNAc acetamido group of fucosylated antennae

**Monoantennary glycans**

**6**
Gal($\beta$1 → 4)GlcNAc($\beta$1 → 2)Man($\alpha$1 → 3)
H-1: 4.446   4.573                     5.122
NAc:         2.050

Fuc($\alpha$1 → 3)/
**5**
Gal($\beta$1 → 4)GlcNAc($\beta$1 → 2)Man($\alpha$1 → 3)
4.441     4.580                     5.110
          2.043

Fuc($\alpha$1 → 3)/
**5**
Fuc($\alpha$1 → 2)Gal($\beta$1 → 4)GlcNAc($\beta$1 → 2)Man($\alpha$1 → 3)
             4.534      4.534                     5.120
                        2.054

**6'**
Gal($\beta$1 → 4)GlcNAc($\beta$1 → 2)Man($\alpha$1 → 6)
H-1: 4.472   4.577                     4.917
NAc:         2.046

Fuc($\alpha$1 → 3)
Gal($\beta$1 → 4)GlcNAc($\beta$1 → 2)Man($\alpha$1 → 6)
4.448     4.583                     4.903
          2.040

Fuc($\alpha$1 → 2)Gal($\beta$1 → 4)GlcNAc($\beta$1 → 2)Man($\alpha$1 → 6)
             4.538      4.551                     4.907
                        2.052

**8**
Gal($\beta$1 → 4)GlcNAc($\beta$1 → 4)Man($\alpha$1 → 3)
H-1: 4.466   4.584                     5.110
NAc:         2.076

Fuc($\alpha$1 → 3)/
**7**
Gal($\beta$1 → 4)GlcNAc($\beta$1 → 4)Man($\alpha$1 → 3)
4.440     4.598                     5.109
          2.066

Fuc($\alpha$1 → 3)/
**7**
Fuc($\alpha$1 → 2)Gal($\beta$1 → 4)GlcNAc($\beta$1 → 4)Man($\alpha$1 → 3)
             4.540      4.560                     5.109
                        2.079

**8'**
**4'**
Gal($\beta$1 → 4)GlcNAc($\beta$1 → 4)Man($\alpha$1 → 6)
H-1: 4.476   4.574                     4.880
NAc:         2.053

Fuc($\alpha$1 → 3)
Gal($\beta$1 → 4)GlcNAc($\beta$1 → 4)Man($\alpha$1 → 6)
4.460     4.58                      4.880
          2.027

Fuc($\alpha$1 → 2)Gal($\beta$1 → 4)GlcNAc($\beta$1 → 4)Man($\alpha$1 → 6)
             4.541      4.558                     4.880
                        2.057

**Table 24** *Continued*

## Biantennary glycans

6'    5'

```
Gal(β1 → 4)GlcNAc(β1 → 2)Man(α1 → 6)
H-1: 4.473  4.582               5.121      ⎬
NAc:        2.046
Gal(β1 → 4)GlcNAc(β1 → 2)Man(α1 → 3)
6                               5
4.467       4.582               5.121
            2.050
```

```
              Fuc(α1 → 3)
                 /
Gal(β1 → 4)GlcNAc(β1 → 2)Man(α1 → 6)
H-1: 4.448  4.582               4.901      ⎬
NAc:        2.039
Gal(β1 → 4)GlcNAc(β1 → 4)Man(α1 → 3)
4.440       4.602               5.104
         /  2.067
Fuc(α1 → 3)
```

```
                            Fuc(α1 → 3)
                               /
Gal(β1 → 4)GlcNAc(β1 → 2)Man(α1 → 6)
4.442      4.589                4.909       ⎬
           2.040
Gal(β1 → 4)GlcNAc(β1 → 4)Man(α1 → 2)
4.448      4.589                5.107       ⎬
           2.040
           Fuc(α1 → 3)
```

```
Fuc(α1 → 2)Gal(β1 → 4)GlcNAc(β1 → 2)Man(α1 → 6)
          4.53  4.53                    4.908   ⎬
                     2.054
Fuc(α1 → 2)Gal(β1 → 4)GlcNAc(β1 → 2)Man(α1 → 3)
          4.54  4.55                    5.105   ⎬
                     2.054
```

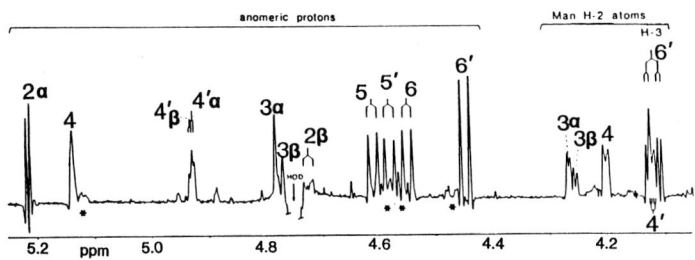

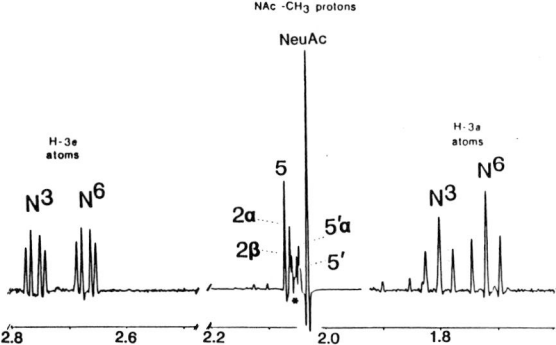

**Figure 27.** ¹H-NMR spectrum of compound **13** of *Figure 22* showing the discrimination between α-2,3- and α-2,6-NeuAc linkages.

**Table 25.** Chemical shifts of anomeric protons of oligomannosidic-type glycans

| Man residues[a] | Man$_9$GlcNAc | Man in terminal non-reducing position |
|---|---|---|
| 3 | 4.77 | |
| 4 | 5.333–5.345 | 5.092–5.111 |
| 4' | 4.868–4.871 | 4.899–4.915 |
| A | 5.398–5.401 | 5.076–5.093 |
| B | 5.141–5.145 | 4.908–4.909 |
| C | 5.296–5.308 | 5.050–5.059 |
| D$_1$,D$_2$,D$_3$ | 5.040–5.059 | 5.040–5.059 |

[a] For numbering of Man residues, see *Figure 29*.

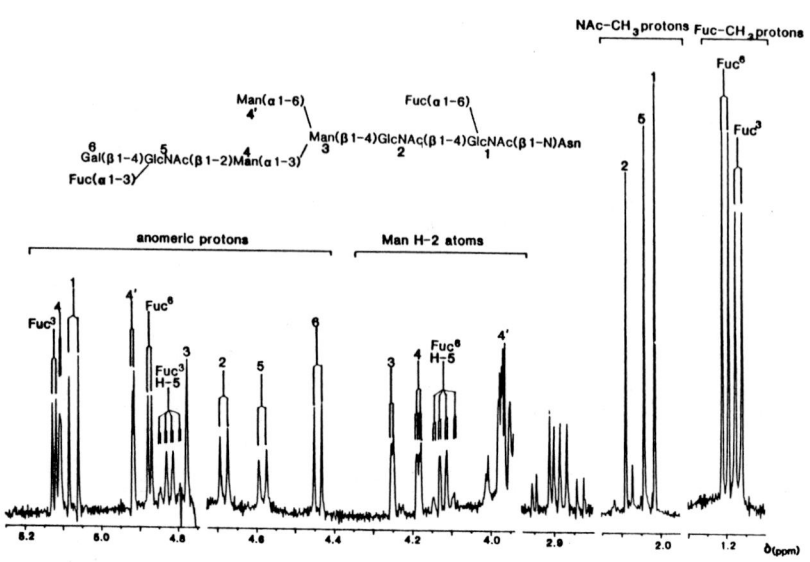

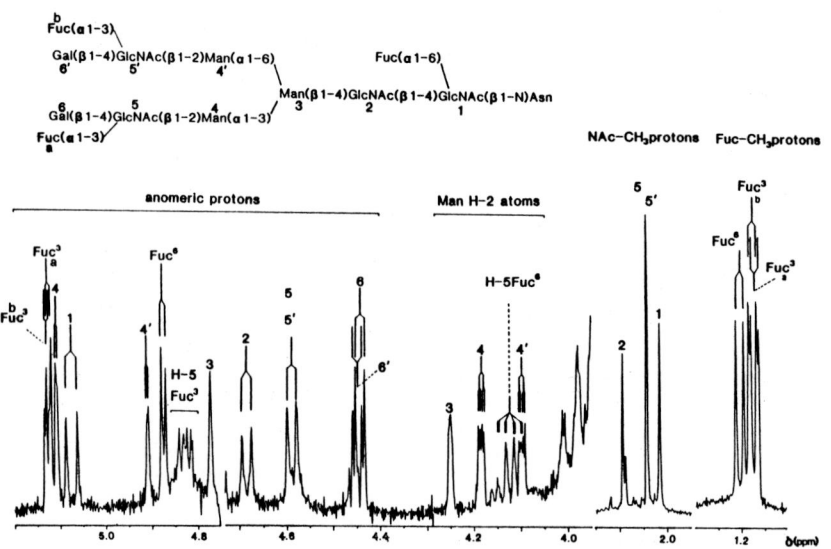

**280**

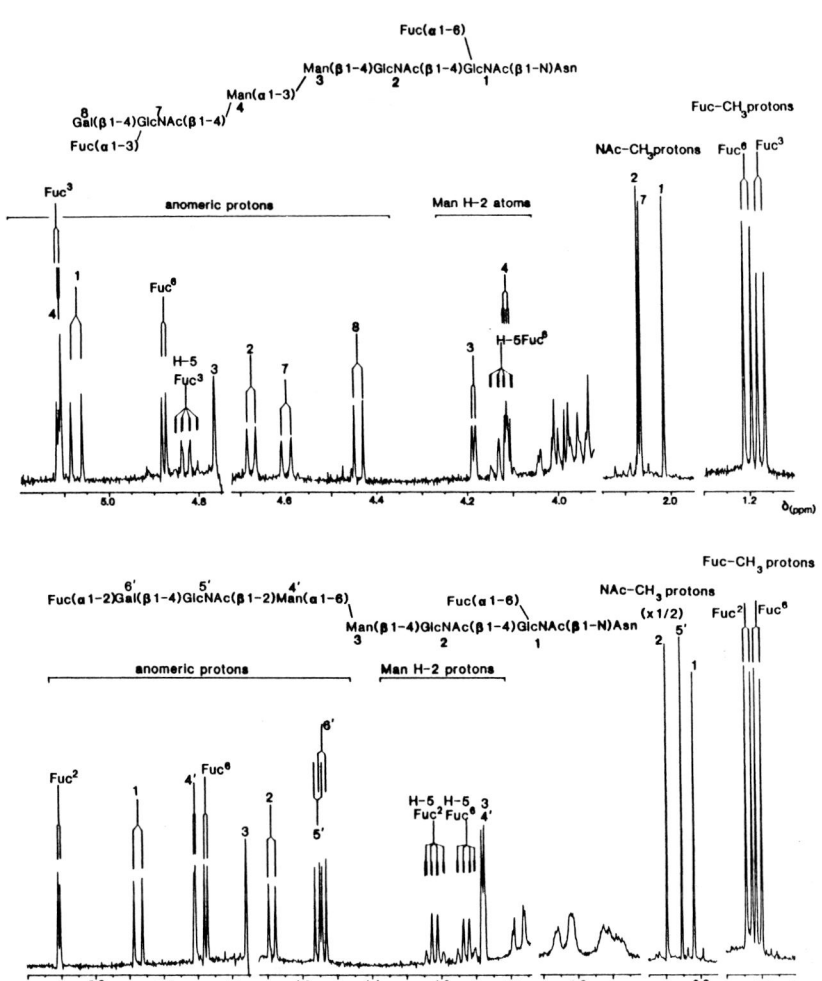

**Figure 28.** ¹H-NMR spectra of fucosyl glycoasparagines.

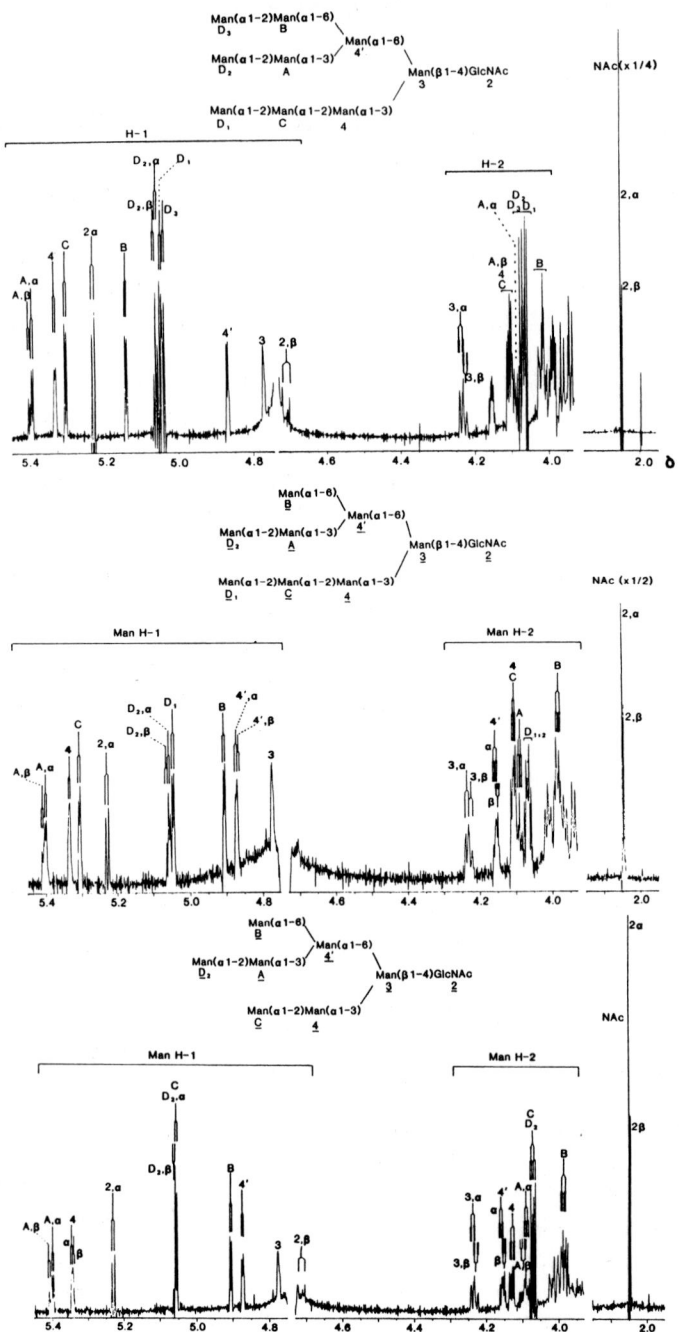

**Figure 29.** Comparison of the ¹H-NMR spectra of Man₉GlcNAc, Man₈GlcNAc, and Man₇GlcNAc, showing the upfield shift effect on Man **B** and **E** H-1 atom resonances induced by the removal of Man **D3** and **D1**.

282

## 9.3 *O*-Glycosylprotein glycans

### 9.3.1 Oligosaccharide-alditols released by reductive *β*-elimination: determination of the inner-core

The interpretation of the $^1$H-NMR spectra of oligosaccharide-alditols released by reductive *β*-elimination (see this chapter, Section 2.2) starts with the examination of the resonance positions of GalNAc-ol H-2 and H-5 atoms, which are sufficiently discriminative to locate the substituent monosaccharide at C-3 and C-6 of GalNAc-ol (*Table 26*). For compounds non-substituted in position C-3, the H-2 resonance is observed at δ = 4.24 p.p.m. (compounds **23**, **27** in *Figure 30*); when observed at δ = 4.37 ~ 4.40 p.p.m., it indicates the *O*-3 substitution with a *β*-Gal residue (compounds **22, 24, 28**).

The *O*-3 substitution by a *β*-GlcNAc residue leads to an intermediate value observed at δ = 4.26 ~ 4.28 p.p.m. The absence of *O*-3 substitution is also characterized by a typical value of the H-5 resonance at δ = 4.02 p.p.m. (compounds **23, 27**). The absence of *O*-6 substitution is characterized by a upfield shifted value of H-5 at δ = 4.193 p.p.m. for compound **22** with Gal at *O*-3 core type I, or 4.135 p.p.m. for compound **25**.

The attachment of NeuAc at *O*-6 of GalNAc-ol is attested by the relevant signals of NeuAc (H-3ax, eq) and the upfield shifted GalNAc-ol H-6′ resonance (δ = 3.48 ~ 3.54 p.p.m.) which is now observed on the right of the bulk of resonances (compounds **23, 24, 26**).

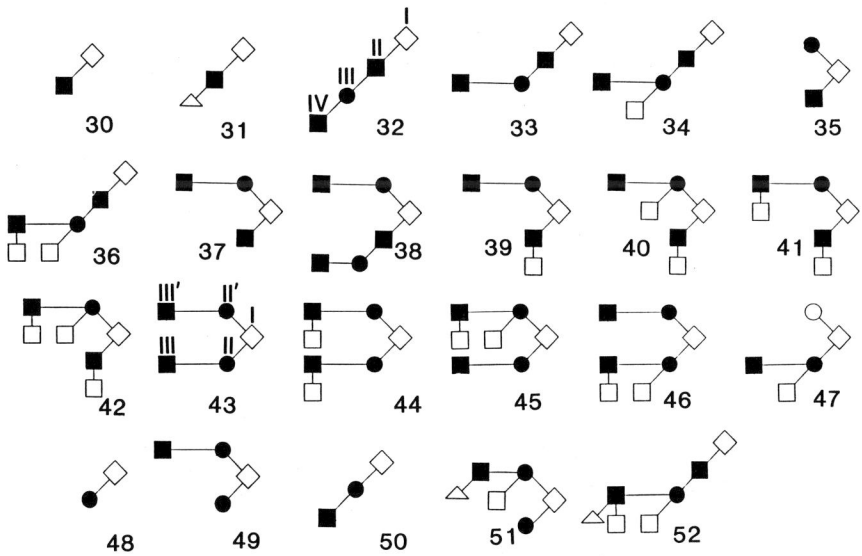

**Figure 30.** Structure of oligosaccharide-alditols described in *Table 27*. ◇: GalNAc-ol; ■: Gal, ●: GlcNAc; □: Fuc; △: NeuAc(α2 → 3); ○: NeuAc(α2 → 6).

**Table 26.** Chemical shifts of GalNAc-ol residues characteristic of the nature of oligosaccharide-alditols cores

|  | Gal(β1 → 3)\GalNAc-ol **22** | NeuAc(α2 → 6)\GalNAc-ol **23** | Gal(β1 → 3)\GalNAc-ol NeuAc(α2 → 6)/ **24** | NeuAc(α2 → 6)\GalNAc-ol GlcNAc(β1 → 3)/ **25** |
|---|---|---|---|---|
| H-2 | 4.393 | 4.246 | 4.378 | 4.283 |
| H-3 | 4.063 | 3.846 | 4.055 | 3.993 |
| H-4 | 3.506 | 3.411 | 3.534 | 3.540 |
| H-5 | 4.193 | 4.020 | 4.244 | 4.135 |
| H-6' | — | 3.533 | 3.486 | — |

|  | NeuAc(α2 → 6)\GalNAc-ol GlcNAc(β1 → 3)/ **26** | GlcNAc(β1 → 6)\GalNAc-ol **27** | GlcNAc(β1 → 6)\GalNAc-ol Gal(β1 → 3)/ **28** | Gal(β1 → 4)GlcNAc(β1 → 6)\GalNAc-ol GlcNAc(β1 → 3)/ **29** |
|---|---|---|---|---|
| H-2 | 4.260 | 4.242 | 4.391 | 4.279 |
| H-3 | 3.984 | 3.841 | 4.069 | 3.985 |
| H-4 | — | 3.379 | 3.468 | 3.516 |
| H-5 | 4.185 | 4.021 | 4.277 | 4.234 |
| H-6' | 3.490 | 3.933 | — | — |

## 9.3.2 Elongation of the cores with Gal–GlcNAc units

The NMR parameters of the Gal($\beta1 \rightarrow 3$)GlcNAc or Gal($\beta1 \rightarrow 4$)GlcNAc units attached at *O*-3 of Gal($\beta1 \rightarrow 3$)GalNAc have been defined by Mutsaers *et al.* (103). As shown in *Table 27*, the [1]H-atom resonances of Gal and GlcNAc possess too many similar values to allow a rapid characterization of an unknown glycan sequence. This investigation should begin by the

**Table 27.** Chemical shifts of Gal and GlcNAc anomeric protons from the most representative oligosaccharide-alditol structures[a]

| | | |
|---|---|---|
| Gal[3]II | —terminal | H-1 : 4.477 (30) |
| | —terminal, GalNAc-ol *O*-substituted with a GlcNAc unit | H-1 : 4.468 (35–37) |
| | —α-2,3-sialylated | H-1 : 4.54 (31) |
| | —elongated with a Gal-GlcNAc unit | H-1 : 4.46 (32, 33, 34, 36) |
| | —elongated with a Gal-GlcNAc and GalNAc-ol- *O*-6 substituted | H-1 : 4.45 (38) |
| | —*O*-2 fucosylated (H determinant) | H-1 : 4.57–4.58 (39, 40, 41, 42) |
| Gal[3]IV | —terminal | H-1 : 4.45 (32) |
| Gal[4]IV | —terminal | H-1 : 4.48 (33) |
| | —with Le[x] determinant | H-1 : 4.46 (34) |
| | —with Le[y] determinant | H-1 : 4.51 (36) |
| Gal[4]II′ | —terminal | H-1 : 4.47 (37, 38, 39, 43, 46, 49) |
| | —with H determinant | H-1 : 4.57 (41, 44) |
| | —with Le[x] determinant | H-1 : 4.45 (40) |
| | —with Le[y] determinant | H-1 : 4.50 (42, 45) |
| | —α-2,3-sialylated | H-1 : 4.548 |
| | —with sialyl Le[x] determinant | H-1 : 4.52 (51) |
| GlcNAc[3]III | —terminal | H-1 : 4.604 (48); 4.597 (49) |
| | —*O*-3 substituted with Gal | H-1 : 4.70 (32); 4.653 (50) |
| | —*O*-4 substituted with Gal | H-1 : 4.62 (33) |
| | —with Le[x] determinant | H-1 : 4.70 (34) |
| | —with Le[y] determinant | H-1 : 4.75 (35) |
| | —with sialyl Le[x] determinant | H-1 : 4.684 (52) |
| GlcNAc[6]II′ | —terminal | H-1 : 4.537 (35) |
| | —*O*-4 substituted with Gal | H-1 : 4.56 (37, 38, 39, 43, 49) |
| | —with H determinant | H-1 : 4.54 (41, 44) |
| | —with Le[x] determinant | H-1 : 4.57 (40) |
| | —with Le[y] determinant | H-1 : 4.56 (42); 4.55 (45) |
| | —with sialyl Le[x] determinant | H-1 : 4.56 (51) |
| GlcNAc[3]II | —terminal | H-1 : 4.600 (48, 49, 51) |
| | —with H determinant | H-1 : 4.602 (44) |
| | —*O*-4 substituted with Gal | H-1 : 4.623 (43, 45) |
| | —with Le[x] determinant | H-1 : 4.64 (47) |
| | —with Le[y] determinant | H-1 : 4.54 (46) |
| Gal[4]III | —termainl | H-1 : 4.454 (43, 45) |
| | —with H determinant | H-1 : 4.52 (44) |
| | —with Le[x] determinant | H-1 : 4.444 (47) |
| | —with Le[y] determinant | H-1 : 4.48 (46) |
| | —α-2,3-sialylated | H-1 : 4.533 |
| | —with sialyl Le[x] determinant | H-1 : 4.528 (52) |

[a] Numbers in brackets refer to the compounds in *Figure 30*.

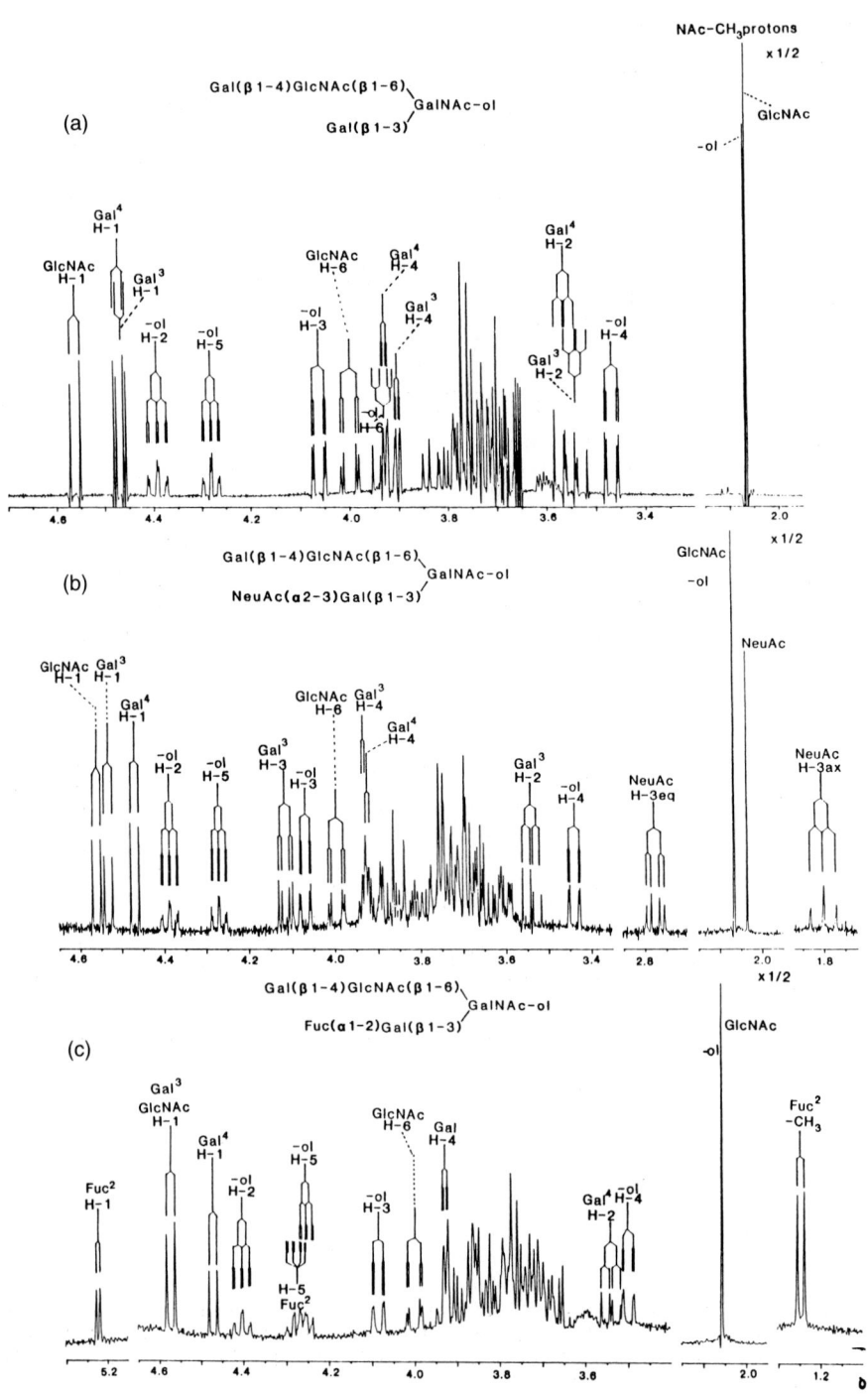

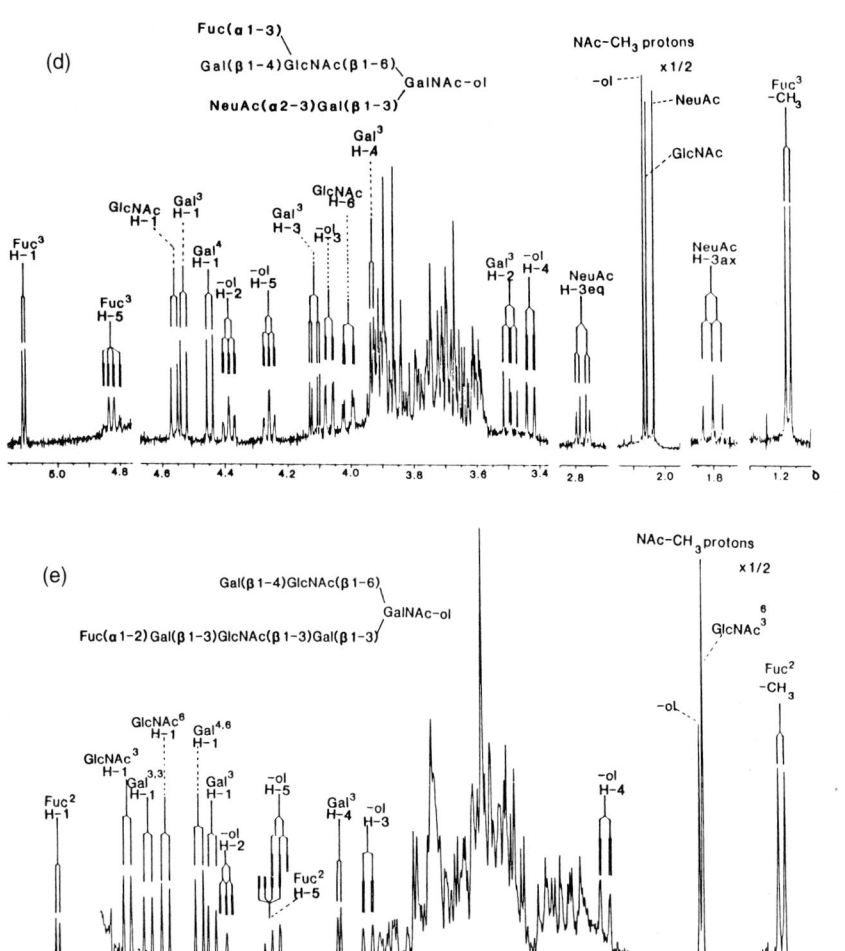

**Figure 31.** Comparison of the ¹H-NMR spectra of oligosaccharide-alditols showing the effect of fucosylation, sialylation, and elongation by a Gal-GlcNAc unit. (a) Basic structure. (b) Downfield shift of Gal H-3. (c) Downfield shift of Gal³ H-1. (d) Upfield shift of GlcNAc NAc signal. (e) Downfield shift of Gal³ H-4.

identification of the core type, according to the specific values of H-2 and H-5 atom resonances of GalNAc-ol given in *Table 26*. For instance, a type 4 core glycan containing two Gal residues should necessarily possess the structure designated as **43** in *Figure 30*. Starting from this reference, the position of the Fuc residue can be rapidly established according to the set of Fuc H-1, H-5, and CH$_3$ resonances (*Table 28*) which is characteristic of the H, Le$^a$, Le$^b$, Le$^x$, and Le$^y$ determinants.

The attachment of α-1,2-Fuc to Gal of Gal-GlcNAc causes a downfield shift of the Gal H-1 resonance (+0.07 to 0.1 p.p.m.) and an upfield shift of GlcNAc H resonance (−0.02 p.p.m.), but has no significant effect on NAc chemical shift. The attachment of α-1,3-Fuc to GlcNAc has no effect on GlcNAc H-1, but induces upfield shifts of the NAc (−0.010 p.p.m.) and of the Gal H-1 (−0.024 p.p.m.) atom resonances of the Gal-GlcNAc unit.

In *Table 27* most of $^1$H-atom resonances observed in fucosylated oligo-saccharide-alditols are given. Although numerous values are superimposable, they can be discriminated according to the type of core of the molecule and to the nature of substituents. For instance, the anomeric proton observed at ~ 4.45 p.p.m. can originate from terminal Gal(β1 → 3) IV (compound **32**), Gal(β1 → 4) II′ (Le$^x$ determinant in compound **40**), or terminal Gal(β1 → 3) III (compounds **43**, **45**).

For Gal IV, this parameter must be associated with the signals relative to GlcNAc III (4.70 p.p.m.) and Gal II (4.46 p.p.m.). For Gal II′, the presence of Fuc(α1 → 3) and GlcNAc(β1 → 6) should be verified. For terminal Gal III, the H-1 atom resonance of GlcNAc III should be observed at 4.363 p.p.m. from GlcNAc II H-1.

The attachment of α-2,3-sialic acid also leads to a characteristic downfield shift effect on the Gal H-1, and to the deshielding of Gal H-3 from 3.6–3.7 p.p.m. to 4.08–4.11 p.p.m. The *O*-3 substitution of Gal by a GlcNAc residue leads to a resonance at 4.12 p.p.m. These observations are illustrated in *Figure 31*, which shows the shift effects due to the attachment of Fuc, NeuAc, or Gal-GlcNAc taking as reference the spectrum of compound **37**.

**Table 28.** Structural reporter groups of fucose residues

|  |  | H | Le$^a$ | Le$^x$ | Le$^b$ | Le$^y$ | Sialyl-Le$^x$ |
|---|---|---|---|---|---|---|---|
| **F²** | H-1 | 5.20–5.30 | — | — | 5.15 | 5.21–5.28 | — |
|  | CH$_3$ | 1.23–1.24 | — | — | 1.27 | 1.27 | — |
|  | H-5 | 4.22–4.28 | — | — | 4.34 | 4.25–4.27 | — |
| **F³** | H-1 | — | — | 5.10–5.14 | — | 5.10–5.11 | 5.11–5.13 |
|  | CH$_3$ | — | — | 1.15 | — | 1.24 | 1.17 |
|  | H-5 | — | — | 4.81–4.83 | — | 4.86 | 4.83 |
| **F⁴** | H-1 | — | 5.03 | — | 5.03 | — | — |
|  | CH$_3$ | — | 1.18 | — | 1.26 | — | — |
|  | H-5 | — | 4.86 | — | 4.86 | — | — |

# Acknowledgements

We would like to express our sincere gratitude to Prof. André Verbert for his critical comments. We wish to thank Mrs Catherine Alonso for excellent technical assistance in the typing of the manuscript.

# References

1. Montreuil, J. (1980). *Adv. Carbohydr. Chem. Biochem.*, **37**, 157; (1982). In *Comprehensive biochemistry* (ed. A. Neuberger and L. L. M. Van Deenen), Vol. 19 B, Part II, pp. 1–188. Elsevier, Amsterdam; (1984). *Pure Appl. Chem.*, **56**, 859; (1984). *Biol. Cell*, **51**, 115.
2. Kobata, A. (1992). *Eur. J. Biochem.*, **209**, 483.
3. Allen, H. J. and Kisailus, E.C. (1992). *Glycoconjugates: composition, structure, and function*, Marcel Dekker, New York.
4. Varki, A. (1993). *Glycobiology*, **3**, 97.
5. Cumming, D. A. (1991). *Glycobiology*, **1**, 115.
6. Ferguson, M. A. J. (1992). *Biochem. Soc. Trans.*, **20**, 243.
7. Vliegenthart, J. F. G., Van Halbeek, H., and Dorland, L. (1988). *Pure Appl. Chem.*, **53**, 45.
8. Vliegenthart, J. F. G., Dorland, L., and Van Halbeek, H. (1983). *Adv. Carbohydr. Chem. Biochem.*, **41**, 209.
9. Dell, A. (1987). *Adv. Carbohydr. Chem. Biochem.*, **45**, 20.
10. Hillenkamp, F., Karas, M., Holtkamp, D., and Klüsener, P. (1986). *Int. J. Mass. Spectr. Ion Proc.*, **69**, 265.
11. Yamashita, K., Mizuochi, T., and Kobata, A. (1982). In *Methods in enzymology* (ed. V. Ginsburg), Vol. 83, pp. 105–26. Academic Press Inc., New York.
12. Iyer, R. N. and Carlson, D. M. (1971). *Arch. Biochem. Biophys.*, **142**, 101.
13. Lee, Y. C. and Scocca, J. R. (1972). *J. Biol. Chem.*, **247**, 5753.
14. Yosizawa, Z., Sato, T., and Schmid, K. (1966). *Biophys. Biochim. Acta*, **121**, 417.
15. Paz-Parente, J., Leroy, Y., Montreuil, J., and Fournet, B. (1984). *J. Chromatogr.*, **948**, 147.
16. Baenziger, J. U. and Natowicz, M. (1981). *Anal. Biochem.*, **112**, 357.
17. Van Pelt, J., Damm, J. B. C., Kamerling, J. P., and Vliegenthart, J. F. G. (1987). *Carbohydr. Res.*, **169**, 43.
18. Paz-Parente, J., Strecker, G., Leroy, Y., Montreuil, J., and Fournet, B. (1982). *J. Chromatogr.*, **249**, 199.
19. Bergh, M. L. E., Koppen, P., and Van den Eijden, D. H. (1981). *Carbohydr. Res.*, **94**, 225.
20. Michalski, J. C., Haeuw, J. F., Wieruszeski, J. M., and Strecker, G. (1990). *Eur. J. Biochem.*, **189**, 369.
21. Blumberg, K., Liniere, F., Pustilnik, L., and Bush, C. A. (1982). *Anal. Biochem.*, **119**, 409.
22. Towsend, R. R., Hardy, M. R., and Lee, Y. C. (1989). In *Methods in*

*enzymology*, (ed. V. Ginsberg), Vol. 179, pp. 65–76. Academic Press, New York.

23. Hardy, M. R. and Towsend, R. R. (1988). *Proc. Natl. Acad. Sci. USA*, **85**, 3289.
24. Reddy, G. P. and Bush, C. A. (1991). *Anal. Biochem.*, **198**, 278.
25. Hase, S., Ikenaka, K., Mikoshiba, K., and Ikenata, T. (1988). *J. Chromatogr.*, **434**, 51.
26. Goldstein, I. J. and Hayes, C. E. (1978). *Adv. Carbohydr. Chem. Biochem.*, **35**, 127.
27. Blake, D. A. and Goldstein, I. J. (1983). In *Methods in enzymology* (ed. V. Ginsburg), Vol. 83, pp. 127–32. Academic Press, New York.
28. Sharon, N. and Lis, H. (1989). *Lectins*. Chapman and Hall, London.
29. Montreuil, J., Debray, H., Debeire, Ph., and Delannoy, Ph. (1983). In *Structural carbohydrates in the liver*, (ed. H. Popper, W. Reutter, F. Gudat, and E. Köttgen), Vol. 34, pp. 239–58. MTP Press, London, Falk Symposium.
30. Kornfeld, R. and Ferri, C. (1975). *J. Biol. Chem.*, **250**, 2614.
31. Debray, H., Decout, D., Strecker, G., Spik, G., and Montreuil, J. (1981). *Eur. J. Biochem.*, **117**, 41.
32. Debray, H., Pierce-Crétel, A., Spik, G., and Montreuil, J. (1983). In *Lectins— biology, biochemistry, clinical biochemistry* (ed. T. C. Bøg-Hansen, and G. A. Spengler), Vol. 3, pp. 335–50. De Gruyter, Berlin.
33. Debray, H. and Montreuil, J. (1991). *Adv. Lectin Res.*, **4**, 51.
34. Debray, H. and Montreuil, J. (1992). In *Affinity electrophoresis: principles and applications* (ed. J. Breborowicz and A. Mackiewicz), pp. 23–57. CRC Press, London.
35. March, S. C., Parikh, I., and Cuatrecasas, P. (1974). *Anal. Biochem.*, **60**, 149.
36. Wilchek, M. and Miron, T. (1974). *Mol. Cell. Biochem.*, **4**, 181.
37. Dulaney, J. T. (1979). *Mol. Cell. Biochem.*, **21**, 43.
38. Lotan, R. and Nicolson, G. L. (1979). *Biochim. Biophys. Acta*, **559**, 329.
39. Bøg-Hansen, T. C. (1973). *Anal. Biochem.*, **56**, 480.
40. Heegaard, P. M. H., Heegaard, N. H. H. and Bøg-Hansen, T. C. (1992). In *Affinity electrophoresis: principles and applications* (ed. J. Breborowicz, and A. Mackiewicz), pp. 3–21. CRC Press, London.
41. Lotan, R., Gussin, A. E. S., Lis, H., and Sharon, N. (1973). *Biochem. Biophys. Res. Commun.*, **52**, 656.
42. Toyoshima, S., Osawa, T., and Tonomura, A. (1970). *Biochim. Biophys. Acta*, **221**, 514.
43. Murphy, L. A. and Goldstein, I. J. (1977). *J. Biol. Chem.*, **252**, 4739.
44. Eckhardt, A. F. and Goldstein, I. J. (1983). *Biochemistry*, **22**, 5290.
45. Shankar, Yyer, P. N., Wilkinson, K. D., and Goldstein, I. J. (1976). *Arch. Biochem. Biophys.*, **177**, 330.
46. Endo, T., Ohbayashi, H., Kanazawa, K., Kochibe, N., and Kobata, A. (1992). *J. Biol. Chem.*, **267**, 707.
47. Torres, B. V., Mc Crumb, D. K., and Smith, D. F. (1988). *Arch. Biochem. Biophys.*, **262**, 1.
48. Animashaun, T. and Hughes, R. C. (1989). *J. Biol. Chem.*, **264**, 4657.
49. Wang, W. C. and Cummings, R. D. (1988). *J. Biol. Chem.*, **263**, 4576.
50. Shibuya, N., Goldstein, I. J., Broekaert, W. F., Nsimba-Lubaki, M., Peeters, B., and Peumans, W. J. (1987). *Arch. Biochem. Biophys.*, **254**, 1.

51. Yamashita, K., Kobata, A., Suzuki, T., and Umetsu, K. (1989). In *Methods in enzymology* (ed. V. Ginsburg), Vol. 179, pp. 331–41. Academic Press, New York.
52. Yamashita, K., Kochibe, N., Ohkura, I., Veda, I., and Kobata, A. (1985). *J. Biol. Chem.*, **260**, 4688.
53. Debray, H. and Montreuil, J. (1989). *Carbohydr. Res.*, **185**, 15.
54. Cummings, R. D. and Kornfeld, S. (1982). *J. Biol. Chem.*, **257**, 11235.
55. Merkle, R. K. and Cummings, S. (1987). In *Methods in enzymology* (ed. V. Ginsburg), Vol.138, pp. 232–59. Academic Press, New York.
56. Osawa, T. and Tsuji, T. (1987). *Annu. Rev. Biochem.*, **56**, 21.
57. Dische, Z. (1955). In *Methods biochem. anal.* (ed. D. Glick), Vol. 2, pp. 313–48. Interscience, New York.
58. Montreuil, J. and Spik, G. (1963). *Méthodes colorimétriques de dosage des glucides totaux.* Monog. Lab. Chim. Biol. Fac. Sci. Lille.
59. White, C. A. and Kennedy, J. F. (1981). In *Techniques in carbohydrate metabolism* (ed. H. L. Kornberg, J. C. Metcalfe, D. H. Northcote, C. I. Pogson, and K. F. Tipton), Vol. B3, B312. Elsevier, Amsterdam.
60. Montreuil, J., Spik, G., Fournet, B., and Tollier, M. T. (1981). In *Techniques d'analyse et de contrôle dans les industries agro-alimentaires* (ed. J. L. Multon), Vol. 4, pp. 99–144. Lavoisier, Paris.
61. Belcher, R., Nutten, A. J., and Sambrook, C. M. (1954). *Analyst*, **79**, 201.
62. Niazi, S. and State, D. (1948). *Cancer Res.*, **8**, 653.
63. Schauer, R. (1978). In *Methods in enzymology* (ed. V. Ginsburg), Vol. 50, pp. 64–89. Academic Press, New York.
64. Gombas, G. and Zanetta, J.-P. (1978). In *Research methods in neurochemistry* (ed. N. Marks and R. Rodnight), Vol. 4, pp. 307–43.
65. Hounsell, E. F. (1993). *Glycoprotein analysis in biomedicine* Humana Press, Ottawa.
66. Lindberg, B. (1972). In *Methods in enzymology* (ed. V. Ginsburg), Vol. 28, pp. 178–95. Academic Press, New York.
67. Lindbergh, B. and Lönngren, J. (1978). In *Methods in enzymology* (ed. V. Ginsburg), Vol 50, pp. 3–33. Academic Press, New York.
68. Hakomori, S. I. (1964). *J. Biochem.* (Tokyo), **55**, 205.
69. Paz-Parente, J., Cardon, P., Leroy, Y., Montreuil, J., Fournet, B., and Ricart, G. (1985). *Carbohydr. Res.*, **141**, 41.
70. Finne, J., Krusius, T., and Rauvala, H. (1980). *Carbohydr. Res.*, **80**, 336.
71. Ciucanu, I. and Kerek, F. (1975). *Carbohydr. Res.*, **41**, 235.
72. Zolotaren, B. M., Ott, A. Ya., and Chizhov, O. S. (1978). *Adv. Mass Spectrom.*, **7B**, 1371.
73. Fournet, B., Strecker, G., Leroy, Y., and Montreuil, J. (1981). *Anal. Biochem.*, **116**, 489.
74. Domon, B. and Costello, C. E. (1988). *Glycoconjugate J.*, **5**, 397.
75. Reason, A. J., Dell, A., Romero, P. A., and Herscovics, A. (1991). *Glycobiology*, **1**, 387.
76. Strecker, G., Wieruszeski, J. M., Michalski, J. C., and Montreuil, J. (1988). *Glycoconjugate J.*, **5**, 385.
77. Hillenkamp, F. and Karas, M. (1990). In *Methods in enzymology* (ed. V. Ginsburg), Vol. 193, pp. 280–95. Academic Press, New York.

78. Sutton, C. W., Poole, A. C., Cottrell, J. S., Crabbe, T., and Docherty, A. J. P. (1993). *Finnigan MAT Poster Compilation*, **3**, 17.
79. Corfield, A. P., Michalski, J. C., and Schauer, R. (1981). *Perspect. Inherited Metab. Dis.*, **4**, 3.
80. Schauer, R. (1982). *Adv. Carbohydr. Chem. Biochem.*, **40**, 131.
81. Li, S. C., Mazotta, M. Y., Chien, S. F., and Li, Y. T. (1975). *J. Biol. Chem.*, **250**, 6786.
82. Arakawa, M., Ogata, S. I., Muramatsu, T., and Kobata, A. (1974). *J. Biochem.* (Tokyo), **75**, 707.
83. Li, S. C. and Li, Y. T. (1970). *J. Biol. Chem.*, **245**, 5153.
84. Li, Y. T. (1967). *J. Biol. Chem.*, **242**, 5474.
85. Bouquelet, S., Spik, G., and Montreuil, J. (1978). *Biochim. Biophys. Acta*, **522**, 521.
86. Dugal, B. (1978). *Biochem. J.*, **171**, 799.
87. Takashi, N. and Nishibe, H. (1978). *J. Biochem.*, **84**, 1467.
88. Plummer, T. H., Elder, J. H., Alexander, S., Phelan, A. W., and Tarentino, A. L. (1984). *J. Biol. Chem.*, **259**, 10700.
89. Scuder, P., Hanfland, P., Uemura, K. I., and Feizi, T. (1984). *J. Biol. Chem.*, **259**, 6586.
90. Tarentino, A. L. and Maley, F. (1974). *J. Biol. Chem.*, **249**, 811.
91. Elder, J. H. and Alexander, S. (1982). *Proc. Natl. Acad. Sci. USA*, **79**, 4540.
92. Kol, O., Brassart, C., Spik, G., Montreuil, J., and Bouquelet, S. (1989). *Glycoconjugate J.*, **6**, 333.
93. Plummer, T. H. and Tarentino, A. L. (1991). *Glycobiology*, **1**, 257.
94. Koide, N. and Muramatsu, T. (1974). *J. Biol. Chem.*, **249**, 4897.
95. Dorland, L., Haverkamp, J., Schut, B., Vliegenthart, J. F. G., Spik, G., Strecker, G., Fournet, B., and Montreuil, J. (1977). *FEBS Lett.*, **77**, 15.
96. Paz-Parente, J., Wieruszeski, J. M., Strecker, G., Montreuil, J., Fournet, B., Van Halbeek, H., Dorland, L., and Vliegenthart, J. F. G. (1982). *J. Biol. Chem.*, **257**, 13173.
97. Van Pelt, J., Van Kurk, J. A., Kamerling, J. P., Vliegenthart, J. F. G., Van Diggelen, O. P., and Galjaard, H. (1988). *Eur. J. Biochem.*, **177**, 327.
98. Green, E. D., Adelt, G., Baenziger, J. U., Wilson, S., and Van Halbeek, H. (1988). *J. Biol. Chem.*, **263**, 18253.
99. Bergwerff, A. A., Thomas-Oates, J. L., Oostrum, J. V., Kamerling, J. P., and Vliegenthart, J. F. G. (1992). *FEBS Lett.*, **314**, 389.
100. Van Halbeek, H., Dorland, L., Vliegenthart, J. F. G., Fiat, A. M., and Jollès, P. (1980). *Biochim. Biophys. Acta*, **823**, 295.
101. Van Halbeek, H., Dorland, L., Vliegenthart, J. F. G., Hull, W. E., Lamblin, G., Lhermitte, M., Boersma, A., and Roussel, P. (1982). *Eur. J. Biochem.*, **127**, 7.
102. Dua, V. K., Dube, V. E., and Bush, A. (1984). *Biochim. Biophys. Acta*, **802**, 29.
103. Mutsaers, J. H. G. M., Van Halbeek, H., Vliegenthart, J. F. G., Wu, A. H., and Kabat, E. A. (1986). *Eur J. Biochem.*, **157**, 139.
104. Breg, J., Van Halbeek, H., Vliegenthart, J. F. G., Klein, A., Lamblin, G., and Roussel, Ph. (1988). *Eur. J. Biochem.*, **171**, 643.

105. Hounsell, E. F., Lawson, A. M., Stoll, M. S., Kane, D. P., Cashmore, G. C., Carruthers, R. A., Feeney, J., and Feizi, T. (1989). *Eur. J. Biochem.*, **186**, 597.
106. Hanish, F. G., Uhlenbruck, G., Peter-Katalinic, J., Egge, H., Dabrowski, J., and Dabrowski, U. (1989). *J. Biol. Chem.*, **264**, 872.
107. Pierce-Crétel, A., Decottignies, J. P., Wieruszeski, J. M., Strecker, G., Montreuil, J., and Spik, G. (1989). *Eur. J. Biochem.*, **182**, 457.

# 6

# Glycolipids

I. M. MORRISON

## 1. Introduction

Glycolipid is the general term applied to a wide range of compounds. The most important and complex are of biological origin, being distributed throughout animal, plant, and microbial cells, but some synthetic examples have industrial applications. Along with glycoproteins, they form the class of glycoconjugates. The carbohydrate, or 'glyco-', components are oligosaccharide chains which are frequently branched. They can also carry covalently-bound, non-carbohydrate substituents such as acetate or sulfate groups. It is the precise carbohydrate structure which confers the biological specificity to a particular glycolipid.

The lipid portion is far more variable since there is no consensus about which compounds should be classed as 'lipids': this text will be confined to derivatives of glycerol, sterols, polyprenols, and sphingenine-type bases. Examples are shown in *Figure 1*. In the glycoglycerolipids, the glycerol moiety, in the *sn* configuration, is substituted at C-1 and C-2 by fatty acids

---

**Gal-β-(1->3)-GlcNAc-β-(1->3)-Gal-β-(1->4)-Glc-β-(1->1)-Ceramide**

**Glycosphingolipid: Le₄Cer, a tetraosylceramide of the lacto series**

**NeuAc-α-(2->8)-NeuAc-α-(2->3)-Gal-β-(1->4)-Glc-β-(1->1)-Ceramide**

**Ganglioside: G$_{lac}$2, a dineuraminosyl ganglioside of the lacto series**

**1,2-di-O-acyl-[Gal-α-(1->6)-Gal-β-(1->3)]-*sn*-Glycerol**

**Plant glycolipid: a digalactosylglyceride**

**Figure 1.** The major families of animal and plant glycolipids.

which are linked by ester bonds: the fatty acids are invariably different. C-3 is always reserved for the carbohydrate chain which is linked by a glycosidic bond. In the sphingenine-type compounds, which are normally unsaturated, the carbohydrate chain is linked by a glycosidic bond to the primary hydroxyl group and the fatty acids are linked by an amide bond to the amino group of the base. Molecules from the one source can have the same oligosaccharide chain and the same sphingenine base but different fatty acids. The linkage between sterols and the carbohydrate chains are also through glycosidic bonds while the linkages involving the polyprenol-types are unique in the glycolipids being linked via a phosphate bridge.

The sterol- and polyprenol-types are not present in tissues and cells at a high concentration but the latter are widespread in nature where they act as transporters for the carbohydrate components of biologically important molecules. Of the other types, the glycoglycerolipids are the dominant types in plant tissue but they are also found in microbial and, more recently, in certain specialized types of animal tissue. The glycosphingolipids are widespread in animal tissues and are usually classified into the neutral glycosphingolipids and the acidic versions (gangliosides). The gangliosides are characterized by the presence of sialic acid residues. The animal glycolipids are known to carry blood group specificity and antigenicity as well as being responsible for many metabolic disorders.

A number of general problems are associated with the analysis of glycolipids. During the isolation and separation procedures, care must be taken to ensure that the removal or migration of labile substituents does not occur. It may be difficult to know if a labile substituent is present on a native glycolipid but, once such a group has been identified, acidic and basic conditions must be reduced to a minimum. The operator should also be aware of the possibility of enzymatic degradation/migration.

The emphasis here will be placed on the analysis of the carbohydrate component: routine analyses of the lipid portion are cited but rigorous details are only given when the lipid component has physiological and biochemical implications.

## 2. Separation methods

### 2.1 Extraction of glycolipids

The carbohydrate components of glycolipids have hydrophilic properties due, mainly, to the high concentration of hydroxyl groups present, while the lipid component is strongly hydrophobic. The ambivalent or biphasic nature of these conjugates causes great extraction and separation problems. No single method is appropriate for all lipids: those with short carbohydrate chains require solvent systems with higher proportions of lipophilic components while those with longer and/or more branched carbohydrate components require higher proportions or an aqueous component. Plant glycolipids, which have short carbohydrate chains, are more easily extracted *in toto* by a single extraction procedure than the more complex animal glycolipids. However, because of the greater complexity in structure and biological significance, most emphasis will be placed on the animal glycolipids.

The extraction method which has gained most acceptance over the last 35 years is that introduced by Folch *et al.* (1) (*Protocol 1*). Optimum results are obtained when using fresh tissue. If only dried tissue is available, extractions must be carried out for periods of up to 24 hours at room temperature and with continuous stirring.

---

**Protocol 1.** Extraction of glycolipids

1. Treat tissue or cells (*c.*1 g wet weight) with chloroform/methanol (2:1 v/v, 19 vol.) in a Waring blender for 3 min at 4 °C.

2. Centrifuge off the insoluble material at 400 *g* or filter through sintered glass, glass fibre, or a layer of celite.[a]

3. Subsequently extract the insoluble material with chloroform/methanol (1:1 v/v) and (1:2 v/v).[b] Add 5% water to facilitate each of these subsequent extractions.

4. Combine the organic phases for further fractionation.[c]

[a] Filtration through filter paper must be avoided if gangliosides are present since the sialic acids are adsorbed onto the cellulose fibres.
[b] The extractions with higher methanol concentrations were proposed by Susuki (2) and are essential when polyglycosyl constituents and gangliosides are present.
[c] The insoluble material is mainly protein and other non-lipid components.

---

The extractions with higher methanol concentrations can be replaced by treatment with 0.4 M sodium acetate in chloroform/methanol/water (30:15:4 by vol.; 1 g, 40 ml) at room temperature for 24 hours as proposed by Slomiany and Slomiany (3). This procedure is particularly efficient when gangliosides are present.

An alternative overall procedure has been proposed by Tettamanti *et al.* (4). The tissue (1 g) is homogenized in a blender with 0.01 M potassium phosphate buffer (pH 6.8, 1 ml) for 1 min, then tetrahydrofuran (8 ml) is added and homogenization continued for a further 1 min. The material is centrifuged at 12 000 *g* for 10 min at 15 °C and the pellet is re-extracted with the phosphate buffer (1 ml) and tetrahydrofuran (4 ml) for 2 min before centrifuging again. The procedure is repeated twice with the gangliosides being located in the supernatant.

Special precautions need to be taken, particularly for photosynthetic plant tissue, to prevent the action of phospholipases and galactolipases, as well as the oxidation of unsaturated fatty acids. The fresh tissue is broken up in dry ice then homogenized in a Waring blender with hot propan-2-ol for a few minutes (5).

## 2.2 Purification of glycolipids

No single method can be used to separate all the glycolipids in an extract from the other extracted contaminants. Each procedure must be tested for the particular fraction required. As a general procedure, the chloroform/methanol extracts obtained from either the Folch or Susuki procedures are adjusted to a ratio of 2:1 and 0.2 vol. of 0.12 M NaCl or KCl are added. The extracts partition into two phases. The lower phase contains the neutral glycolipids, particularly those with short carbohydrate chains, along with neutral lipids and phospholipids, while the upper phase contains not only the gangliosides and the neutral glycolipids with longer carbohydrate chains but also contaminating glycoproteins and glycopeptides. The precise shape and charge of a particular glycolipid will affect its partition into the two phases. For maximum purification, each phase should be backwashed with the solvent of the other phase and similar washings combined.

The gangliosides can be recovered from the aqueous phase of the Folch extract by dialysis at 2–4 °C for three days against distilled water. The gangliosides are retained, lyophilized, and stored at −20 °C. However, the optimum method is the solid phase extraction (SPE) method first proposed by Williams and McCluer (6) (*Protocol 2*). The use of SPE has been reviewed elsewhere (7).

---

**Protocol 2.** Solid phase extraction of gangliosides

1. Wash an SPE column (Sep-Pak$^{TM}$C$_{18}$ cartridge)[a] with 10 ml portions of methanol (3 ×), chloroform/methanol (2:1 v/v), methanol, and chloroform/methanol/water (3:48:47 by vol.) containing 0.1 M KCl.

2. Apply the aqueous (upper) phase of the Folch extraction to the column and remove salts by eluting with water.

3. Elute the gangliosides with methanol or chloroform/methanol (2:1 v/v).

[a] With care, the cartridges are reusable for a number of times after appropriate washings.

---

In the Tettamanti procedure, the phosphate/tetrahydrofuran extract is treated with 0.3 vol. diethylether and, after centrifuging at 600 *g* for 20 min at 15 °C, the gangliosides are recovered from the aqueous phase.

Glycolipids (plus lipids) are also separated from contaminants by column chromatography. The total extract is applied to a Sephadex LH20 (100 × 2 cm) column and eluted with chloroform/methanol/water (40:20:1 by vol.). The lipids are all eluted at the void volume while non-lipids are retained. Separation of total lipids from non-lipids can also be achieved on short columns (15 × 1.0 cm) of Sephadex G-25 in the bead form. A chloroform/ methanol/water (8:4:3 by vol.) mixture is allowed to separate into two phases. The gel is swollen and packed in the upper phase while elution is carried out with the lower phase. The lipids are eluted at the void volume.

## 2.3 Separation of glycolipid mixtures

### 2.3.1 Separation into neutral glycolipids and acidic glycolipids (gangliosides)

Such separations involve the use of diethylaminoethyl (DEAE) supports and one using DEAE–Sephadex was proposed by Yu and Ledeen (8) (*Protocol 3*).

---

**Protocol 3.** Separation of glycolipid classes

1. Generate DEAE–Sephadex A-25 (2.2 g) into the acetate form by washing with 0.8 M sodium acetate in methanol then pack into a 1.4 cm diameter column.

2. Wash the column with chloroform/methanol/water (15:30:4 by vol.) and apply the sample in the same solvent.

3. Elute with the same solvent to remove the neutral glycolipids while eluting with 0.8 M sodium acetate in methanol removes the gangliosides.

---

Since gangliosides can contain a number of sialic acid residues, DEAE– Sephadex can also be used to separate mono-, di-, and trisialoganglioside mixtures (9). A crude ganglioside mixture (2–5 g) is dispersed in chloroform/ methanol (1:1 v/v 60 ml) and mixed with DEAE–Sephadex A-25 (acetate form: 5 g) with shaking. The suspension is diluted with methanol (30 ml) and water (8 ml) then packed in a 27 × 4.6 cm column. Neutral lipids are washed off with 3 vol. chloroform/methanol/water (15:30:4 by vol.) and neutral glycolipids with 1 vol. methanol. The gangliosides are separated with a concave gradient of 0 to 0.45 M ammonium acetate in methanol. The gradient was achieved by connecting, in series, vessels with 0.05, 0.15, and 0.45 M ammonium acetate in methanol. The higher the molarity of the ammonium acetate, the greater the number of sialic acid residues in the molecule.

A procedure using DEAE–cellulose has also been published (10) (*Protocol 4*).

---

**Protocol 4.** Separation of gangliosides

1. Suspend dry DEAE–cellulose, in the hydroxide form, in glacial acetic acid and pack into a 30 × 3 cm column.

2. Remove the acetic acid by washing with methanol and chloroform/ methanol (2:1 v/v), then equilibrate with chloroform/methanol/water (450:50:1 by vol.).

3. Apply the glycolipid mixture (< 500 mg) in the same solvent. The neutral glycolipids are eluted (and partially fractionated) by stepwise use of solvents with decreasing proportions of chloroform.

4. Elute the gangliosides with chloroform/methanol/0.05 M ammonium acetate (4:1:1 by vol.), and finally acetic acid.

---

### 2.3.2 Separation of partially purified glycolipid mixtures

Partially purified glycolipid mixtures, both neutral and acidic, are fully separated by TLC, HPLC, or conventional column chromatography, and the stationary phase is invariably a silica gel.

*i. Column chromatography*

Although not now used as frequently as TLC or HPLC, low pressure column chromatography still offers many advantages especially when larger quantities of product are required. Many grades of silica gel can be used and elution can be either stepwise or with a linear gradient. As an example, silica gel spheres (Iatrobeads) were used to separate mono- to tetraglycosylceramides using a linear gradient of chloroform/methanol/water with increasing proportions of methanol and water. The effluent is monitored either by a colour reaction for hexose or by TLC of the fractions. If tailing occurs due to the glycolipids being too polar or separation is less than optimum due to the components being closely related, the chromatography can be improved by acetylating the mixture. Acetylation of glycolipids is carried out by sonicating with pyridine/ acetic anhydride (2:1 1.0 ml) for ten minutes then leaving overnight at 25 °C. The solvents are removed on a rotary evaporator below 40 °C and final traces are removed by repeated evaporation with toluene before chromatography. Deacetylation, if necessary, is achieved by treating the individual glycolipids with 0.1% sodium methoxide in chloroform/methanol (2:1 v/v). Care must be taken to ensure that the acetylation and deacetylation steps do not cause any hydrolysis of chemical bonds or migration of unstable substituent groups.

*ii. TLC*

Using silica gel plates, neutral solvent systems such as chloroform/methanol/ water (65:25:4 by vol.) are preferred for neutral glycolipids while basic systems, such as chloroform/methanol/water/ammonia (60:35:7:1 by vol.), or systems containing $Ca^{2+}$ ions, such as chloroform/methanol/0.2% aqueous calcium chloride (60:40:9 by vol.) give better separations for gangliosides (11).

Bands or spots on TLC plates are detected by conventional spray reagents for non-reducing sugars (see Chapter 1, *Table 1*). Contrary to column practice, acetylation of glycolipids does not appear to offer any advantages for TLC. When preparative TLC is carried out, the plates should not be allowed to dry out since it becomes impossible to elute the bands in a quantitative manner even by sonicating.

### iii. HPLC

Due to its adaptability in the use of different solvents with different phases and at different temperatures, HPLC has largely superseded all other techniques for the analysis of complex glycolipid mixtures. Its application to native and acetylated glycolipid mixtures has been reviewed (12). Since the technique needs to be tailored to suit each application, the general method will not be considered further. The major problem with glycolipids is sensitivity of detection. Most commercial instruments are equipped with either an RI or a UV detector. RI detectors are universal but insensitive while UV detectors rely on a good chromophore which is not present on glycolipids. Benzoyl, or *p*-nitrobenzoyl, derivatives have been used successfully (13) (*Protocol 5*).

---

**Protocol 5.** HPLC of glycolipids

1. Dissolve the glycolipid (< 1.0 mg) in 20% benzoyl chloride[a] in 0.6 ml dry pyridine, and heat at 60 °C for 1 h.

2. Remove the solvent in a stream of $N_2$. Take up the residue in hexane (5 ml), and wash successively with 95% methanol containing saturated sodium carbonate, 0.6 M HCl, and water.

3. Evaporate the hexane solution to dryness under $N_2$ and inject into the HPLC in a few drops of hexane.

4. Using a Zipax column (E. I. du Pont de Nemours and Co. Inc.), a gradient of 0.20% to 0.75% methanol in hexane, and detection at 254 nm, separation of mono- to tetraglycosylceramides is achieved in 25 min at 10 pmol concentrations.

---

[a] The *p*-nitrobenzoyl derivatives are prepared in the same way. They are more sensitive for detection but the benzoyl derivatives give clearer separations. It is not too difficult to remove *O*-benzoyl groups but *N*-benzoyl groups on amino sugars and the ceramide component are very difficult to remove. This drawback can be overcome by using a catalyst such as 4-dimethylaminopyridine with benzoyl anhydride which produces only *O*-substitution.

In the future, moving wire flame ionization detectors may offer increased sensitivity for native glycolipids while the evaporative light-scattering detector has been used to separate lipid classes, including glycolipids, but not, to our knowledge, to separate complex glycolipid mixtures.

### iv. Supercritical fluid chromatography

Supercritical fluid chromatography (SFC) is a relatively new technique which can be used to analyse thermally labile and non-volatile compounds at lower temperatures than GLC while a range of detectors, such as FID, has advantages over HPLC. The ease of interface to MS is another advantage. Its application to glycolipid analysis owes much to Reinhold and his group (14). They have demonstrated the separation of the mono-, di-, and trisialoganglio-sides $G_{M1}$, $G_{D1a,b}$, and $G_{T1b}$ on an SB-phenyl-5™ column at 120 °C with carbon dioxide and characterization by MS.

# 3. Chemical methods of analysis

## 3.1 Composition of glycolipids

### 3.1.1 Carbohydrate components

To determine the carbohydrate composition of an individual glycolipid, the biphasic nature of a glycolipid means that special hydrolysis conditions are required for depolymerization to occur under homogeneous conditions (*Protocols 6* and *7*).

---

**Protocol 6.** Alditol acetate method for glycolipids

1. Treat the glycolipid (50–200 µg) with glacial acetic acid/0.15 M sulfuric acid (9:1 v/v, 0.3 ml) at 80 °C for 6 h in a Teflon lined screw-capped vial.

2. Cool the hydrolysate and pass through a Dowex 1-X8 (200 mg) column washed with methanol.

2. Evaporate the filtrate to dryness under $N_2$. Reduce the released sugars to alditols with sodium borohydride (10 mg) in 0.1 M ammonia (0.3 ml) for 3 h.

4. Add two drops of acetic acid and evaporate the solution to dryness.

5. Repeat the evaporations five times with 5% acetic acid in methanol to remove the borate ion as the volatile methyl borate.

6. Acetylate the alditols with acetic anhydride (0.3 ml) at 100 °C for 2 h and remove the excess by repeated evaporations with toluene.

7. Partition the residue between chloroform and water (1:1 v/v, 10 ml), discard the aqueous layer, and dry the chloroform layer (15).

---

8. The peracetylated alditols are analysed by GLC or GLC–MS by methods outlined in detail in Chapter 1, Section 7. An alternative acetylation procedure involving 1-methylimidazole as catalyst and carried out at room temperature has been proposed (16).

---

**Protocol 7.** TMS methylglycoside method for glycolipids

1. Dissolve the glycolipid (50–200 µg) in 1.0 M methanolic hydrogen chloride (1.0 ml) (see Chapter 1, *Protocol 14*).

2. Heat at 85 °C for 18 h. Methanolysis is carried out in a sealed tube under $N_2$.

3. Add 50 mg of silver carbonate to neutralize the HCl and 50 µl acetic anhydride to re-*N*-acetylate amino sugars.

4. Leave the mixture for 6 h then centrifuge before removing the methanol layer.

5. Wash with methanol (1.0 ml 3 ×) and evaporate the combined methanol solutions to dryness.

6. Prepare the trimethylsilyl derivatives of the methylglycosides by adding a mixture of dry pyridine/hexamethyldisilazane/trimethyl-chlorosilane (5:1:1 by vol., 0.05 ml) and leave at room temperature for 30 min.

7. The TMS methylglycosides are analysed by GLC using procedures outlined in Chapter 1, Section 7. An alternative trimethylsilylation procedure is also given.

---

The hydrolysis procedure with sulfuric acid is necessary to fully release the amino sugars from the glycolipids: methanolysis is not sufficiently strong. Conversely, the hydrolysis procedure causes partial destruction of any sialic acid residues present. Methanolysis is necessary to quantitatively determine these residues. If internal sialic acids are present, the methanolysis conditions are made even milder to preserve these groups.

Sialic acid residues can be present as *N*-acetyl and *N*-glycolyl derivatives as well as some disubstituted residues, depending on the source. They can be determined by a GLC procedure (17), by HPLC (see Chapter 1, Section 6), by a colorimetric method as outlined in Chapter 1, Section 3, or by TLC (see Chapter 1, *Protocol 11*).

## 3.1.2 Other components

*Protocols 8–11* describe methods for the determination of some of the other components in glycolipids.

---

**Protocol 8.** Determination of the fatty acid components in glycolipids

1. Heat the glycolipid (2–5 mg) with a mixture of methanol/water/conc.HCl (29:4:3 by vol., 2.0 ml) at 80 °C for 18 h in a sealed tube under $N_2$.
2. Extract the released fatty acids and methyl esters from the cooled hydrolysate with petroleum ether (BP 40–60 °C; 3 × 2.0 ml) and evaporate to dryness.
3. Add 1.5 M methanolic HCl (1.0 ml) and heat for 2 h at 40 °C in a sealed tube to esterify the fatty acids.
4. Add water (1.0 ml) and extract the esters into hexane (3 × 2.0 ml).
5. Wash the hexane solution with 2% potassium carbonate and dry over anhydrous sodium sulfate.
6. Analyse the fatty acid methyl esters by capillary GLC on a column such as BP1 or BPX70.
7. If hydroxy fatty acids are thought to be present, the methyl esters are trimethylsilylated as described in *Protocol 7*. This allows faster separations and reduces tailing from these components. If the fatty acid composition is to be further characterized by GLC-MS, a nitrogen-containing ester group, such as the picolinyl group, is preferred for the MS (18).

---

**Protocol 9.** Determination of sphingosine base components

1. The conditions for the release of the sphingosine bases are as described in *Protocol 8* for the release of the fatty acids. After removal of the fatty acids, take the aqueous phase to dryness, and dissolve in 1.0 M NaOH (2.0 ml).
2. Extract the sphingoid bases with diethylether (3 × 4.0 ml) and evaporate to dryness.
3. Prepare the trimethylsilyl derivatives as described in *Protocol 7* and analyse by GLC on 3% SE-30 on Gas-Chrom Q (19).

---

**Protocol 10.** Determination of glycerol by GLC

1. Dissolve glycoglycerolipid (1.0 mg) in dry diethylether (1.0 ml).

2. Add lithium aluminium hydride (3.0 mg) in diethylether (0.5 ml) drop-wise until boiling stops.

3. Add a further aliquot of the hydride and reflux for 1 h.

4. Add acetic anhydride (0.5 ml) drop-wise to destroy excess hydride, and 'Analar' xylene, and reflux for 6 h.

5. Evaporate the solution to dryness and analyse by GLC on an SE-30 column similar to that used for the sphingosine bases above.

---

Glycerol can also be released quantitatively by the action of a lipase on the intact glycoglycerolipid and determined by a spectrophotometric assay (*Protocol 11*).

---

**Protocol 11.** Determination of glycerol using lipase

1. To a solution of glycerol (0.1 ml), obtained either by the action of lipase or alkaline hydrolysis of the glyceride, add 0.1 M triethanolamine buffer (0.4 ml, pH 7.6) and the following reagent mixture (0.5 ml).

   • 71.4 mM triethanolamine (pH 7.6) buffer

   • 8.5 mM $Mg^{2+}$

   • 1.0 mM phosphoenol pyruvate

   • 0.4 mM NADH

   • 2.0 mM ATP

   • 20 µg/ml pyruvate kinase (> 3 U/ml)

   • 10 µg/ml lactate dehydrogenase (> 3.6 U/ml).

2. Measure the change in absorbance at 340 nm for 10 min.

3. Add 5 µl of 5 µg/ml glycerokinase (> 3.6 U/ml) and measure the new absorbance after 10 min. The difference in absorbance is proportional to the glycerol content (20).

---

Sterols are isolated by a similar method to that for sphingosine bases (see *Protocol 9*) and analysed by GLC either as the free sterol or, preferably, as the acetate derivative by capillary GLC on a BP1 or similar column.

## 3.2 Methylation analysis of oligosaccharide chains

### 3.2.1 Methylation procedure

Methylation of oligosaccharide chains in intact glycolipids or those released from glycolipids (see Section 3.3) has only become practical as a method of sequence analysis since the work of Hakomori (21). Prior to that, methylation analysis required either too large a sample or caused too much degradation to give meaningful results. The preparation of the methylsulfinyl anion is essentially that described in Chapter 3, *Protocol 4* and is not repeated. A procedure for the methylation of an intact glycolipid is described (*Protocol 12*).

---

**Protocol 12.** Methylation analysis of glycolipids

1. Dissolve the glycolipid (1.0 mg) in dry dimethyl sulfoxide (DMSO, 1.0 ml) in a screw-capped vial with a Teflon-coated liner and a small magnetic stirrer.

2. Add the methylsulfinyl anion (1.0 ml, 1.6 M) and flush the vial with $N_2$ before capping.

3. Stir for 3 h at room temperature then add redistilled methyl iodide (1.0 ml). Continue stirring for a further 2 h.

4. Remove the excess methyl iodide in a stream of $N_2$ and remove the excess DMSO *in vacuo* over $P_2O_5$. Alternatively, add chloroform/ methanol (1:1 v/v 1.0 ml) and apply the sample to a Sephadex LH-20 column (30 × 1.0 cm) equilibrated in the same solvent. Collect fractions and analyse by a colorimetric method such as the phenol–sulfuric acid procedure or by TLC. Pool similar fractions and evaporate to dryness.

---

Many modifications of this procedure have been published: those involving the potassium methylsulfinyl anion (prepared from potassium hydride or potassium *t*-butoxide) are reported to give cleaner products (22). Extreme caution must be taken in the handling of these carbanions, especially the potassium ones. The hydrides are relatively stable when purchased as dispersions in mineral oil but pure hydrides are extremely reactive and must only be used in small quantities and with great care. The carbanions are also reactive with oxygen, carbon dioxide, and water so must be stored under $N_2$ or argon.

Due to the small sample size which can be derivatized, it is not practical to determine the methoxyl content by a chemical method. The absence of an absorbance in the IR spectrum at 3500 $cm^{-1}$ is the most useful criterion.

## 3.2.2 Separation and characterization of methylated sugars

Due to the biphasic nature of glycolipids, special hydrolysis conditions are necessary. If hexosamines are present, *Protocol 13* is preferred.

---

**Protocol 13.** Hydrolysis of methylated glycolipids containing hexosamines

1. Dissolve the permethylated glycolipid in glacial acetic acid/5 M sulfuric acid (19:1 v/v, 0.3 ml) and heat in a capped vial at 80 °C for 18 h.

2. Add water (0.3 ml) and continue heating for a further 5 h.

3. Pass the cooled hydrolysate down a Dowex 50W-X8 ion-exchange resin column (200 mg), acetate form, in a Pasteur pipette, and wash with methanol (4 ml) to remove the sulfate ion.

4. Collect the washings and dry in a stream of $N_2$ (23).

---

If only neutral sugars are present, permethylated glycolipids can be depolymerized by heating with 95% formic acid (1.0 ml) at 100 °C for two hours. Formic acid is removed on a rotary evaporator and the formyl esters are hydrolysed with 0.1 M sulfuric acid at 80 °C for 18 hours. The clean-up follows *Protocol 13*. Another convenient method for neutral sugar chains is to autoclave with 2 M trifluoroacetic acid at 121 °C for one hour. This acid is volatile and can be removed in a stream of $N_2$. Methanolysis with methanolic HCl is avoided since loss of methoxyl groups can occur.

Derivatization is necessary for satisfactory GLC and GLC–MS separations. Reduction of the free sugars to the corresponding alditols followed by acetylation of the newly generated hydroxyl groups is the most frequently used method and the procedures are fully described in Chapter 1, Section 7.

The only frequent constituent of glycolipids which can be misinterpreted by this procedure is D-galactose. Reduction with borodeuteride solves the problem if MS is used for detection. It can also be solved by converting the methylated sugars into their partially acetylated aldononitriles but the data for these derivatives is still not as complete as for the corresponding alditols. No special problems are encountered in the analysis by GLC or GLC–MS of these derivatives from glycolipids and the procedures are fully considered in Chapter 1, Section 7. The procedure, as applied to glycolipids, has been reviewed (24).

Another analytical variation is to selectively hydrolyse the permethylated sample at a labile bond. The oligosaccharides are reduced and peralkylated (usually perethylated) as described above and separated by HPLC. Further characterization by GLC–MS or by MS alone allows structural features of the oligosaccharides to be determined.

## 3.3 Isolation of oligosaccharide chains

It is frequently desirable to isolate an intact oligosaccharide chain from a glycolipid and, fortunately, most sphingosine bases found in nature have an unsaturated bond which is essential for *Protocols 14* and *15*.

---

**Protocol 14.** Osmium tetroxide—periodic acid oxidation of glycolipids

1. Acetylate the glycolipid (1–3 mg) with acetic anhydride (1.0 ml) and pyridine (0.5 ml) at 37 °C for 18 h.

2. Evaporate to dryness and repeat with toluene (3 ×).

3. Dissolve the product in methanol (0.5 ml), add 0.2 M periodic acid in methanol (0.2 ml), then add 5% osmium tetroxide in diethylether (25 µl), and leave at 5 °C for 18 h.

4. Add a drop of glycerol to destroy any excess reagent and partition the mixture by adding chloroform/methanol (2:1 v/v, 6 ml) and water (1.5 ml).

5. Discard the upper (aqueous) phase, wash the organic phase with chloroform/methanol/water (1:10:10 by vol.), and evaporate to dryness.

6. Release the oligosaccharides from the aldehyde by dissolving in methanol (0.3 ml) and adding 0.2 M sodium methoxide in methanol (50 µl) before leaving at room temperature for 1 h.

7. Add a drop of acetic acid, evaporate the mixture to dryness, and take up in chloroform/methanol/water (1:10:10 by vol.) before washing with chloroform/methanol/water (60:35:8 by vol.).

8. The oligosaccharide is recovered from the aqueous layer (25).

---

The oligosaccharide yield from ceramide dihexoside was 44% while that from ceramide trihexoside was 70%.

---

**Protocol 15.** Ozone oxidation of glycolipids

1. Dissolve the glycolipid (5 mg) in methanol (2.0 ml) and slowly saturate the solution with ozone (from an ozone generator) at 20 °C.

2. Evaporate off the solvent and dissolve the residue in 0.1 M sodium hydroxide or carbonate (2 ml).

3. Leave at room temperature for 18 h.

---

4. Treat the product with Dowex 50W-X8 cation-exchange resin (H⁺ form) to remove Na⁺ ions and extract with chloroform to remove fatty aldehydes.

5. Fractionate the products on a Sephadex G-25 column using 0.1 M acetic acid as eluent (26).

### 3.3.1 Trifluoroacetolysis

When glycolipids react with a mixture of trifluoroacetic acid and trifluoroacetic anhydride, a number of reactions take place (27). *O*-Trifluoroacetylation is complete and amino functions are quantitatively *N*-trifluoroacetylated with *N*-acetyl groups undergo transamidation to *N*-trifluoroacetyl groups. Sialic acid residues are hydrolysed from gangliosides while ceramides, provided they contain the more usual unsaturated sphingosine base, are degraded and yield the asialo-oligosaccharide. The yield of oligosaccharide is not quantitative and depends on the ratio of acid to anhydride. Monoglycosylceramides give a better yield of glucose or galactose with a 1:1 mixture while a triglycosylceramide gave a superior yield of trisaccharide with a 1:50 mixture. The oligosaccharides can be characterized by methylation analysis as previously described: the *N*-trifluoroacetyl groups are retained during the procedure and enhance the separation of amino sugar derivatives by GLC.

## 3.4 Determination of anomeric configuration

One of the few chemical methods of determining the anomeric configuration is by oxidation with chromium trioxide (28). In neutral polysaccharides, *β*-glycosidic bonds are oxidized while α-glycosidic bonds are resistant. For reasons yet unknown, the method has not been generally useful in glycolipid methodology since some *β*-glycosidic bonds are resistant and some α-glycosidic bonds are oxidized.

# 4. Physical methods of analysis

## 4.1 Nuclear magnetic resonance spectroscopy

Some aspects of the use of NMR in carbohydrate analysis are given in detail in Chapter 5 but certain aspects which are specific to glycolipids are cited below.

### 4.1.1 ¹H-NMR

Native glycolipids, free oligosaccharides released from glycolipids, or derivatives of glycolipids, such as permethyl or peracetyl, can all be analysed by ¹H-NMR spectroscopy. There are advantages in each procedure but the method of choice, especially when information is required about secondary structure, is to use the native glycolipid. All the ${}^{1}$H signal from the ceramide component can be assigned while the proton signals from the oligosaccharide component, particularly those from the glycosidic groups, neither have any effect on nor overlap with the ceramide signals.

As with other procedures, the biphasic nature of glycolipids means that special solvent systems are necessary for dissolution. Neither water nor chloroform are suitable and most workers use 2% D$_2$O in deuterated dimethyl sulfoxide. Some worker use pyridine-$d_5$. High resolution 500 or 600 MHz instruments can give unambiguous results with as little as 100 μg of glycolipid but future generations of instruments are unlikely to improve significantly on this lower limit. The literature on ${}^{1}$H-NMR spectroscopy of glycolipids is extensive and has been reviewed (29).

H-1 signals in the ${}^{1}$H-NMr spectra of glycolipids are well separated from the other ring proton signals and their chemical shifts are characteristic of the sugar component. While it is not possible to determine unambiguously the carbohydrate structure in an unknown glycolipid by one-dimensional ${}^{1}$H-NMR spectroscopy, a comparison of the spectrum of the unknown glycolipid with those of other simpler members of the same family is usually very accurate. This is illustrated in *Table 1* for the globo-series up to the pentasaccharide (Forssman antigen). The data clearly show the differences in chemical shifts of the two galactose and two galactosamine residues, which are substituted in different ways, and also the different coupling constants for the same residues which allow the anomeric configuration to be assigned (30). In general, α-glycosidic protons have $J_{1,2}$ coupling constants of 3–4 Hz while β-anomers show values of 7–9 Hz. Some ambiguities do occur, especially between residues with the α-galacto configuration. These can be clarified by examining other resonances for each particular sugar. α-*N*-Acetylgalactosamine is confirmed from the signals of the acetamido methyl group and α-fucose from the C-6 methyl group. Other ambiguities, such as β-glucose and β-galactose, are resolved from other resonances, in this example the H-2 resonances.

Recent advances have made extensive use of various forms of two-dimensional spectroscopy. Not only is the structure of the glycolipid determined by these methods but information is also available on the two- and three-dimensional conformation and interaction within and between glycolipids. Although some principles are detailed in Chapter 5, the specific application to glycolipids has been reviewed (31).

## 4.1.2 ${}^{13}$C-NMR

Far more structural information can be obtained from the ${}^{13}$C-NMR spectrum than from ${}^{1}$H-NMR but the technique suffers from the low natural abundance of the ${}^{13}$C isotope. This aspect of NMR is also covered in Chapter 5 but it should be noted that the absolute configuration of the glucose residue in glucosylceramide has been established by the technique.

**Table 1.** H-1 chemical shifts (p.p.m. from $Me_4Si$) and $J_{1,2}$ coupling constants (Hz) of glycosphingolipids in dimethyl sulfoxide-$d_6$[a]

| Compound | GalNAc ($\alpha$-1 → 3) | | GalNAc ($\beta$-1 → 3) | | Gal ($\alpha$-1 → 4) | | Gal ($\beta$-1 → 4) | | Glc ($\beta$-1 → 1) Cer | |
|---|---|---|---|---|---|---|---|---|---|---|
| | $\delta$ | J | $\delta$ | J | $\delta$ | J | $\delta$ | J | $\delta$ | J |
| 1 | — | — | — | — | — | — | — | — | 4.10 | 7.7 |
| 2 | — | — | — | — | — | — | 4.22 | 7.3 | 4.17 | 7.7 |
| 3 | — | — | — | — | 4.81 | 4.0 | 4.27 | 7.7 | 4.16 | 8.1 |
| 4 | — | — | 4.52 | 8.1 | 4.81 | 3.6 | 4.26 | 7.7 | 4.16 | 7.7 |
| 5 | 4.74 | 3.6 | 4.56 | 8.5 | 4.82 | 3.6 | 4.27 | 7.7 | 4.17 | 7.7 |

[a] Ref. 30, reprinted with permission of the publishers.

## 4.2 Mass spectrometry

One of the major applications of MS to glycolipid analysis is in combination with GLC (and nowadays HPLC) for the structural determination of methylated glycolipids. The peracetyl partially methylated alditols obtained after hydrolysis are unambiguously characterized. This application is discussed fully in Chapter 1, Section 8. Applications of MS to native glycolipids and intact oligosaccharides derived from them will be discussed here.

### 4.2.1 Electron impact mass spectrometry

This technique was used to determine the structure of a permethylated Le[b]-active glycosphingolipid based on the prior determination by the same method of the structure of the related neolactotetraosylceramide (32). The spectrum is shown in *Figure 2a*. The largest ion that could be positively identified had an m/z of 1696 but its intensity is very weak. In theory, the procedure can be used for peracetyl, pertrimethylsilyl, or pertrifluoroacetyl

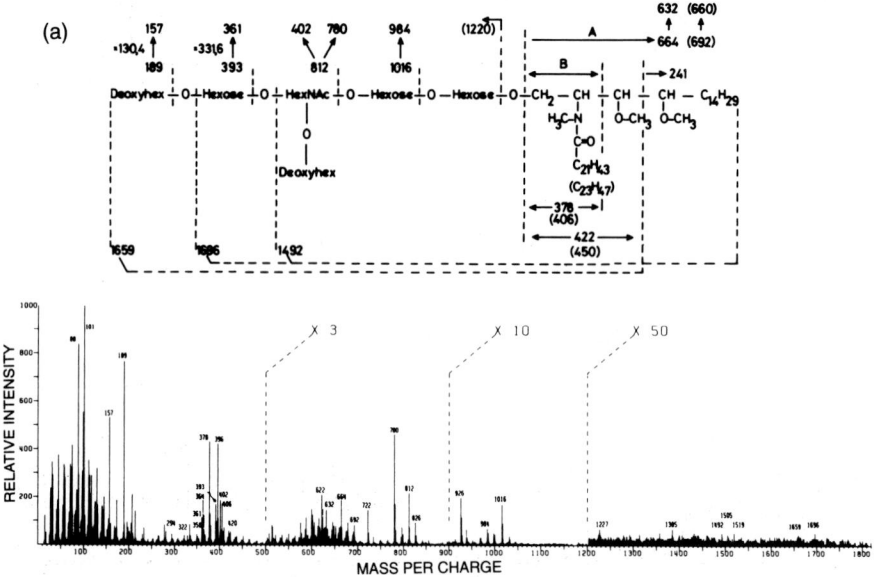

**Figure 2.** Comparisons of different types of mass spectra. (a) Electron impact mass spectrum of permethylated Le$_b$-active glycosphingolipid. Reprinted from ref. 32 with permission. (b) Ammonia direct chemical ionization mass spectrum of permethylated and reduced ganglioside G$_{M1}$. Reprinted from ref. 33 with permission. (c) Negative ion fast atom bombardment mass spectrum of ganglioside G$_{M1}$. Reprinted from ref. 35 with permission.

(b)

(NHCH3–C2H2–(CH2)x–CH3 + H)+
x=16:0
x=18:0
((RO–NANA–OH–H2O) + H)+
452 –H2O
(RO–Gal–O–GalNAc–OH + H)+
HO–Glc–O–Ceramide
x 30
Relative intensity (%)
(RO–Gal–O–GalNAc–O–Gal–O–Glc–OCH3 + H)+
NANA
(–OCH3, +H)
[(MH–NANA) +H]+
[(M+H)–253]+
[M+H]+
m/z


(c)

Gal-o-GalNAc-o-Gal-o-Glc-o-Cer
NeuAc
*3.0
M/Z

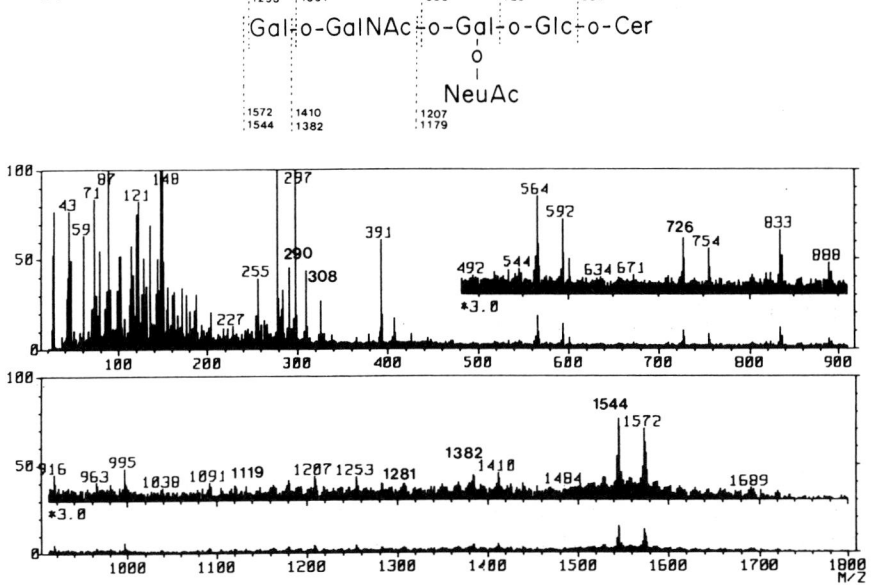

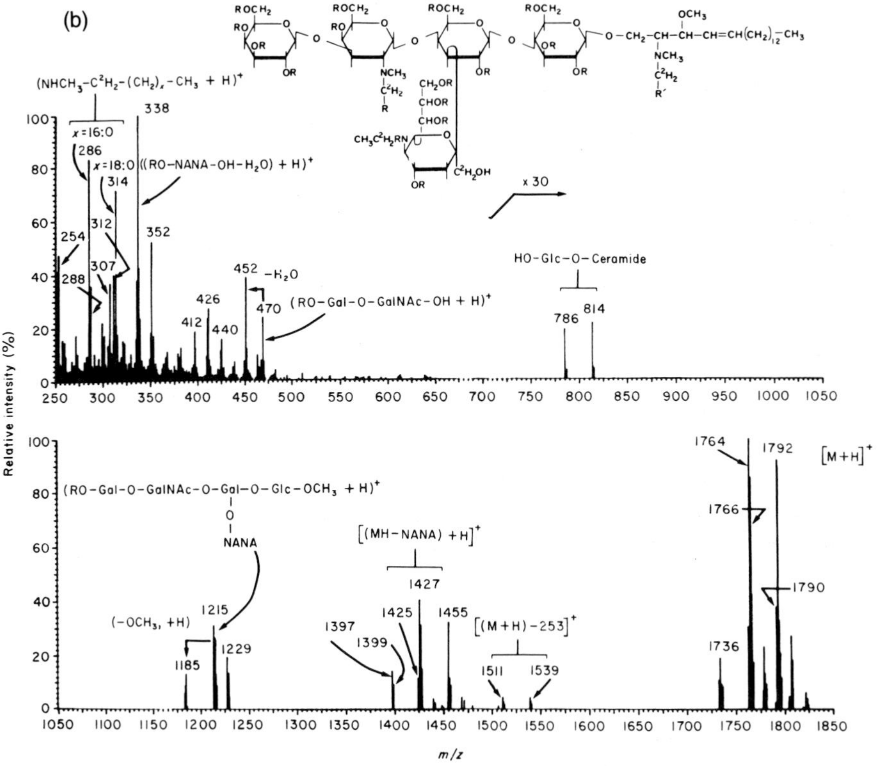

derivatives but, since each methyl substituent only adds 14 mass units to the molecule, permethylation allows the analysis of glycolipids with longer chain lengths. However, this procedure can only be carried out in the gas phase and the high energy ionization required in this phase causes excess fragmentation. The spectra are dominated by small ions so other procedures are now preferred.

### 4.2.2 Chemical ionization mass spectrophotometry

This procedure involves the ionization of a simple compound such as methane or ammonia by a much lower source of energy. The ions, $CH_5^+$ or $NH_4^+$, then react with the derivatized glycolipid to form adducts which are relatively stable and produce a high concentration of ions with large m/z ratios. If carried out in the ionization chamber, it is termed direct chemical ionization (DCI). The first application to a glycolipid allowed not only the size and oligosaccharide sequence to be determined but also details of the lipid component (33). The ammonia DCI-MS of the permethylated and reduced ganglioside $G_{M1}$ is shown in *Figure 2b* (33) and the subject has been reviewed by Reinhold (34).

### 4.2.3 Fast atom bombardment mass spectrometry

This technique is the method of choice for the MS of underivatized glycolipids. The topic has recently been reviewed for lipids in general and sections on specific types of glycolipids are included (35). A small sample (c.5 μg) can be used and the sample is bombarded with fast $Ar^O$ and $Xe^O$ atoms at 3–8 keV. The earlier reports used glycerol as the matrix but triethanolamine is now the preferred matrix. The method can be used in both the positive and negative ion modes, and the molecular weights of both glycosphingolipids and gangliosides are determined from the $[M + H]^+$, $[M + Na]^+$, $[M + H - H_2O]^+$, and $[M - H]^-$ ions respectively. The negative ion FAB-MS of the ganglioside $G_{M1}$ is shown in *Figure 2c*. The structure of an unknown glycolipid can be deduced from the FAB-MS and structurally similar gangliosides such as $G_{M1a}$ and $G_{M1b}$ are distinguishable from their fragmentation pattern.

Other MS techniques, such as high performance tandem MS, are currently being investigated while special strategies involving MS have been proposed to elucidate problems such as the presence of inner esters in gangliosides.

## 4.3 Infra-red spectroscopy

IR spectroscopy has not been applied to the analysis of glycolipids to any extent since the dominance of OH absorbances and similarities in structure do not favour the method. It has been used for the identification of molecular species with unsaturated fatty acids and for the confirmation of ester substituents.

# 5. Biochemical methods of analysis

## 5.1 Enzymatic methods

The sequential treatment of a glycolipid with specific exoglycosidases has been used for many years to determine the sequence and anomeric configuration of the various sugar residues. It will not give data on the actual position of linkage. Since each enzyme requires different optimum buffer conditions, a standard protocol is not appropriate but the following points are useful. The method requires only about 100 μg of glycolipid, depending on size and complexity, while the buffers need to be supplemented with, for example, sodium taurocholate, to maintain a homogeneous system. The sugar residue released is extracted by partitioning into an aqueous and an organic phase and determined—the speed and sensitivity of TLC makes it the method of choice. The sequence of the A-active glycolipid from hog stomach mucosa (36) was confirmed by the sequential use of α-N-acetylgalactos-aminidase, β-N-acetylhexosaminidase, α-L-fucosidase, β-D-galactosidase, and β-N-acetylglucosaminidase as shown in *Figure 3*.

**GalNAc-α-(1->3)-Gal-β-(1->3)-GlcNAc-β-(1->3)-Gal-β-(1->4)-Gal-β-(1->4)-Glc-β-(1->1)-Ceramide**
        2
        ^
        |
        1
     α-**Fuc**

**Figure 3.** Structure of the A-active glycolipid from hog stomach mucosa as determined by sequential enzymatic hydrolysis (ref. 36).

Many exoglycosidases are now commercially available. The failure of a specific enzyme must not be taken, however, as confirmation that the particular residue is not present at the non-reducing terminus: the β-galactosidase from Jack bean hydrolyses a $(1 \rightarrow 4)$ bond far more readily than a $(1 \rightarrow 3)$ bond. In addition, some enzymes will show multiple activities.

The analysis of gangliosides has proved even more complex. Many sialidases have been isolated but their specificity seems complex and some require the presence of cholate for activity.

## 5.2 Immunological methods

The review of Hakomori and Young (37) is still the base for a method of glycolipid analysis which continues to expand rapidly. As an example, antibodies have been raised by conventional methods and are specific to the structures Gal-β-$(1 \rightarrow 4)$-GlcNAc-β-$(1 \rightarrow R)$ and Fuc-α-$(1 \rightarrow 2)$-Gal-β-$(1 \rightarrow 4)$-GlcNAc-β-$(1 \rightarrow R)$ (38). Antibodies have now been raised against many oligosaccharide sequences found in glycolipids and are used to investigate such aspects as the distribution, quantity, and function of glycolipids in specific tissues. The combination of immunological techniques coupled to sensitive methods of analysis, such as TLC, offer exciting prospects for future developments.

# References

1. Folch, J., Lees, M., and Sloane-Stanley, G. H. (1957). *J. Biol. Chem.*, **226**, 497.
2. Susuki, K. (1965). *J. Neurochem.*, **12**, 629.
3. Slomiany, B. L. and Slomiany, A. (1977). *Biochim. Biophys. Acta*, **486**, 531.
4. Tettamanti, G., Bonaldi, F., Marchesini, S., and Zambotti, V. (1973). *Biochim. Biophys. Acta*, **296**, 160.
5. Kates, M. (ed.) (1986). *Techniques of lipidology* 2nd edn. Elsevier, Amsterdam.
6. Williams, M. A. and McCluer, R. H. (1980). *J. Neurochem.*, **35**, 266.
7. Christie, W. W. (1992). In *Advances in lipid methodology—One* (ed. W. W. Christie), pp. 1–17. The Oily Press Ltd., Ayr.
8. Yu, R. K. and Ledeen, R. W. (1972). *J. Lipid Res.*, **13**, 680.
9. Momoi, T., Ando, S., and Nagai, Y. (1976). *Biochim. Biophys. Acta*, **441**, 488.
10. Rouser, G., Kritchevsky, G., Yamamoto, A., Simon, G., Galli, C., and Bauman, A. J. (1969). In *Methods in enzymology*, Vol. 14, (ed. J. M. Lowenstein), pp. 272–317. Academic Press, New York.
11. Hakomori, S. (1983). In *Handbook of lipid research 3: sphingolipid biochemistry* (ed. J. N. Kanfer and S. Hakomori), p. 34. Plenum Press, New York and London.
12. Kannagi, R., Watanabe, K., and Hakomori, S. (1987). In *Methods in enzymology*, Vol. 138, (ed. V. Ginsburg), pp. 3–12. Academic Press, New York.
13. Gross, S. K. and McCluer, R. H. (1980). *Anal. Biochem.*, **102**, 429.
14. Merritt, M. V., Sheeley, D. M., and Reinhold, V. N. (1991). *Anal. Biochem.*, **193**, 24.
15. Watanabe, K., Hakomori, S., Childs, R. A., and Feizi, T. (1979). *J. Biol. Chem.*, **254**, 3221.
16. Iwata, S., Narui, T., Takahashi, K., and Shibata, S. (1984). *Carbohydr. Res.*, **133**, 157.
17. Schauer, R. (1978). In *Methods in enzymology*, Vol. 50, (ed. V. Ginsburg). pp. 64–89. Academic Press, New York.
18. Harvey, D. J. (1992). In *Advances in lipid methodology—One* (ed. W. W. Christie), pp. 19–80. The Oily Press Ltd., Ayr.
19. Fujino, Y. and Ohnishi, M. (1976). *Chem. Phys. Lipids*, **17**, 275.
20. Eggstein, M. and Kuhlmann, E. (1974). In *Methods of enzymatic analysis* (ed. H. U. Bergmeyer), Vol. 4, p. 1825. Academic Press, New York.
21. Hakomori, S. (1964). *J. Biochem.*, **55**, 205.
22. Harris, P. J., Henry, R. J., Blakeney, A. B., and Stone, B. A. (1984). *Carbohydr. Res.*, **127**, 59.
23. Stellner, K., Saito, H., and Hakomori, S. (1973). *Arch. Biochem. Biophys.*, **155**, 464.
24. Levery, S. B. and Hakomori, S. (1987). In *Methods in enzymology*, Vol. 138, (ed. V. Ginsburg), pp. 13–25. Academic Press, New York.
25. MacDonald, D. L., Patt, L. M., and Hakomori, S. (1980). *J. Lipid Res.*, **21**, 642.

# 6: Glycolipids

26. Wiegandt, H. and Bucking, H. W. (1970). *Eur. J. Biochem.*, **15**, 287.
27. Lundblad, A., Svensson, S., Löw, B., Messeter, L., and Cedergren, B. (1980). *Eur. J. Biochem.*, **104**, 323.
28. Lindberg, B., Lonngren, J., and Svensson, S. (1975). *Adv. Carbohydr. Chem. Biochem.*, **31**, 185.
29. Dabrowski, J., Hanfland, P., and Egge, H. (1980). In *Methods in enzymology*, Vol. 83, (ed. V. Ginsburg), p. 69. Academic Press, New York.
30. Dabrowski, J., Hanfland, P., and Egge, H. (1980). *Biochemistry*, **19**, 5652.
31. Koerner, T. A. W., Prestegard, J. H., and Yu, R. K. (1987). In *Methods in enzymology*, Vol. 138, (ed. V. Ginsburg), pp. 38–59. Academic Press, New York.
32. Egge, H. and Hanfland, P. (1981). *Arch. Biochem. Biophys.*, **210**, 396.
33. Carr, S. A. and Reinhold, V. N. (1984). *Biomed. Mass Spectrom.*, **11**, 633.
34. Reinhold, V. N. (1987). In *Methods in enzymology*, Vol. 138, (ed. V. Ginsburg), pp. 59–84. Academic Press, New York.
35. Matsubara, T. and Hayashi, A. (1991). *Prog. Lipid Res.*, **30**, 301.
36. Slomiany, B. L., Slomiany, A., and Horowitz, M. I. (1974). *Eur. J. Biochem.*, **43**, 161.
37. Hakomori, S. and Young, W. W. (1983). In *Handbook of lipid research 3: sphingolipid biochemistry* (ed. J. N. Kanfer and S. Hakomori), pp. 381–436. Plenum Press, New York.
38. Young, W. W., Portoukalian, J., and Hakomori, S. (1981). *J. Biol. Chem.*, **19**, 5652.

# Index

Page numbers in **bold** refer to illustrations

2-acetamido-2-deoxy-D-galactose, *see* N-acetyl
    galactosamine
2-acetamido-2-deoxy-D-glucose, *see* N-acetyl
    glucosamine
acetolysis, of polysaccharides 90–2
acetylation
    of glycolipids 300
    *see also* alditol acetates
N-acetylation 31, 85
N-acetyl galactosamine
    O-acetyl-O-methyl derivatives **35**
    GLC of trimethyl silyl ethers **28**, 29
    ¹H-NMR 311
    HPLC **19**
    in protein glycan link 182–3
    *see also* N-acetyl hexosamine
N-acetyl glucosamine
    O-acetyl-O-methyl derivatives **35**, 235
    GLC of trimethyl silyl ethers **28**, 29
    HPLC **19**
    in glycoproteins 183–7
    *see also* N-acetyl hexosamine
N-acetyl hexosamine
    deacetylation and deamination 99–100
    lectin specificity 206–11
    *see also* hexosamine
N-acetyl hexosaminidase 252–3, 262–5
N-acetyllactosamine glycans 186, 266–76,
    **268–9, 270, 272–3**
N-acetyl neuraminic acid, *see* sialic acid
α-acid glycoprotein **186, 230**, 151
adsorption chromatography 50, 300
affinity chromatography 55, 105–8, 195,
    205–22
    *see also* HPLC
alditol acetates
    GLC 32–3, **35**, 36–7, **87, 97**
    glycolipids 302–3
    mass spectrometry 34–9, 85–7, 234–9
alditols
    GLC **22**
    HPLC 20–2
aldinonitriles, GLC–MS 113–14
*Aleuria aurantia* agglutinin (AAA) 206, 220
alkaline degradation
    glycoproteins 188–90, 192
    polysaccharides 95–7
almond emulsin 250
anomeric configuration, determination
    of 101–2, 309
antenna, glycoprotein 182–6

antigenicity 103
arabinogalactan 116–18
L-arabinose
    in extensin 182
    GLC of trimethylsilyl ethers **28**, 29
    HPLC **19**
    paper chromatography 12
    TLC 12
    *see also* pentose

Barry degradation 88
biantennary glycans **187**
blood group glycans **187**
*Bowringia millbreadii* lectin 221

capillary electrophoresis 65–6, 161–5
carbon-13 nuclear magnetic resonance, *see*
    ¹³C-NMR
Carrez reagent 1–2
cellobiose
    ion-exchange chromatography **51**
    paper chromatography **89**
chondroitin sulfate
    capillary electrophoresis 161–5
    linkage to protein **184**
    oligosaccharides 158–65
    polyacrylamide gel electrophoresis **157**,
    158–9, **160**
    structure **126**, 127
chorionic gonadotrophin **186**
chromium trioxide oxidation 102–3, 309
¹³C-NMR
    determination of anomeric linkage 101–2
    glycolipids 310
    oligosaccharides 66
colorimetric assays
    glycoproteins and glycopeptides 223–7
    for glycosidases 248
    monosaccharides 2–10
    oligosaccharides 44–8
    proteoglycans 174–6
    spot tests 195
    total carbohydrate 2–3, 47–8
concanavalin A (Con A) 206–7, **208**, 215,
    217–19
crossed immuno-affinity electrophoresis 213

D- and L-configuration determination 30, 76–9

*Datura stramonium* agglutinin (DSA) **209**, 220
deamination, degradation by 99–100
degree of polymerization 79–81
deoxy hexoses, in polysaccharides 75
  *see also* L-fucose; 6-deoxy-talose; L-rhamnose
6-deoxy-L-talose 97–8
dermatan sulfate 128
dextrose equivalent (DE) 44, 55
disaccharides, *see Chapter 2*

electrophoresis
  crossed immuno-affinity 213
  glycopeptides 195
  *see also* gel electrophoresis
β-elimination 96–7, 283
  *see also* alkaline degradation
endo-*N*-acetyl glucosaminidase **260–1**, 262–5
endogalactosidase 259–61
endoglycosidases 256–65
  *see also* glycosidases
exoglycosidases, *see* glycosidases
extraction
  of glycolipids 296–8
  of monosaccharides 1–2
  of polysaccharides 74–6
  of proteoglycans 129–34

FAB mass spectrometry 239–45, 314
fatty acid determination 304
fluorimetry 9, 248
fluorography 154
Folch extraction 297
D-fructose
  enzymatic assay 9
  GLC 29
  HPLC **22**
  paper chromatography 12
  TLC 12
  *see also* ketose
L-fucose
  *O*-acetyl-*O*-methyl derivatives **35**
  enzymatic assay 10
  GLC of alditol acetates **35**
  GLC of trimethylsilyl ethers **28**, 29
  in glycoproteins **186**, 273–6, 220–1
  HPLC **19**
  hydrolysis 227
  interference in the Warran assay 8
  ion-exchange chromatography **51**
  lectin specificity 206
  paper chromatography 12
  reporter groups 288
  TLC 12
fucosidase 255–6

gas–liquid chromatography, *see* GLC
gas–liquid chromatography mass spectrometry, *see* GLC–MS
galactosamine
  TLC 12
  *see also* hexosamine
D-galactose
  *O*-acetyl-*O*-methyl derivatives **35**, **87**, 236
  in collagen 182
  enzymatic assay 10
  GLC of alditol acetates **35**
  GLC of trimethylsilyl ethers **28**, 29
  in glycoproteins 183
  ¹H-NMR 311
  HPLC **19**, **22**
  ion-exchange chromatography **51**
  lectin specificity 206
  paper chromatography 12, **87**
  TLC 12
  *see also* Hexose
galactosidase 250–3
galacturonic acid
  GLC of trimethylsilyl ethers **28**, 29
  *see also* uronic acids
gangliosides 295, 298
gel electrophoresis
  fluorography 154
  gel preparation 151–2
  proteoglycans 150–61, **153**, **157**
  Western blotting 154–5
gel permeation chromatography
  columns 53
  column packing 148
  glycolipids 299
  glycopeptides 194
  HPLC 61
  oligosaccharides 53–5, **54**, 61
  polysaccharides 80
  proteoglycans 147–50
GLC
  columns 27
  fatty acids 304
  glycerol 305
  glycolipids 302–3
  inositols 34
  methyl alditol and alditol acetates **35**
  methyl glycoside trifluoroacetates 228–9
  monosaccharides 27–37, **28**
  oligosaccharides 63–4
  sphingosine bases 304
  sterols 305
  trimethylsilyl ethers 28–32, **28**, 229–30, **231**, 303
GLC–MS
  aldononitriles 114
  methyl alditol and alditol acetates 34–39, **35**
  permethylated oligosaccharides 66, 234–46, 312–3

# Index

glucosamine
  TLC 12
  *see also* hexosamine
D-glucose
  *O*-acetyl-*O*-methyl derivatives **35**, **87**
  enzymatic assay 8–10
  GLC of alditol acetates **35**
  GLC of trimethylsilyl ethers **28**, 29
  $^1$H-NMR 311
  HPLC **19**, **22**
  lectin specificity 206
  infra-red spectroscopy 67
  ion-exchange chromatography **51**, **52**
  paper chromatography 12, **89**
  TLC 12, **49**
  *see also* hexose
D-glucuronic acid
  cleavage at 116–17
  GLC of trimethylsilyl ethers **28**, 29
  *see also* uronic acids
D-glucurono-6,3-lactone, *see* uronic acids
glycan–protein linkages 182
glycerol
  enzyme assay 305
  GLC 305
  HPLC **22**
  paper chromatography **89**
*Glycine max*, *see* soybean
glycolipids, *see Chapter 6*
glycopeptide fractionation 194–5, 216–22
*N*-glycopeptide hydrolase 257–9
glycoproteins
  acid hydrolysis 23, 227–8
  enzymatic hydrolysis 247–65
  fractionation 213–16
  *see also Chapter 5*
glycosaminoglycans
  dye binding assay 175–6
  *see also Chapter 4*
glycosidic linkage position determination, *see*
  methylation analysis
glycosidases
  activity 248
  endo- 256–65
  origin 247
  specificities 248, 315
  *see also* hydrolysis, enzymatic
*N*-glycosylpeptide structures **210–11**
glycosylphosphatidyl inositol anchor **187**
*Griffonia simplicifolia* agglutinin (GSA) 206,
  221
gum arabic 108–11


Hakomori 232
heparin structure 128–9
heparan sulfate structure 129

hexosamine
  assay in glycoproteins 224–5
  deamination 99–100
  GLC of alditol acetates **35**
  GLC of trimethylsilyl ethers 28
  HPLC 19
  hydrolysis of polysaccharides 93
  Morgan–Elson assay 6
hexose
  assay using L-cysteine in sulfuric acid 47–8
  assay using orcinol in sulfuric acid 44–5
  assay using phenol in sulfuric acid 2–3
  *see also* D-glucose and reducing sugar
high mannose glycans **185**, 276–87
high performance liquid chromatography *see*
  HPLC
$^1$H-NMR
  anomeric linkage determination 101
  glycans 265–89
  glycolipids 309–11
  oligosaccharides 66–7, 283–5
  polysaccharides 113
  reporter groups 266, 271, 287
HPAE
  chondroitin sulfate 166–8
  glycans 201–4
  monosaccharides 20–1, **22**
HPLC
  affinity chromatography 61–2
  columns 18–22
  glycans 196–205
  glycolipids 301
  glycopeptides 194–5
  glycoproteins 194–5
  monosaccharides 17–26
  oligosaccharides 55–63, 196–205
  post-column detection 24–6, 62–3
  pre-column derivatization 23, 25, 56, 63,
    204–5
  preparative 62
hyaluronic acid
  binding region 168
  capillary electrophoresis **165**
  competitive inhibition assay 176–9
  structure 129
hydrazinolysis 190–2
hydrolysis, acid
  glycolipids 303
  glycoproteins 227–8
  hexosaminyl links 224–5
  methylated glycolipids 307
  methylated polysaccharides 84, 98
  polysaccharides 76, 93
  sialic acids 227
  *see also* methanolysis
hydrolysis, alkaline, *see* alkaline degradation
hydrolysis, enzymatic
  artificial substrates 248

# Index

hydrolysis, enzymatic (*cont.*)
glycolipids 305, 315
glycoproteins 247–65
glycosaminoglycan purification 133
hyaluronic acid 171
sialic acids 227, 249–50
oligosaccharides 94–5
polysaccharides 94–5

immobilization of lectins 207–13
immunological methods
crossed immuno-affinity electrophoresis 213
glycolipids 315–16
polysaccharides 103–10, 120–1
technique 104–5
infra-red spectroscopy
FTIR 40
glycolipids 314
monosaccharides 39–40
oligosaccharides 67
inositols
GLC **28**, 29, 34
HPLC 26
ion-exchange chromatography
columns 20–1, 50–1, 143–4
glycolipids 299
glycopeptides 194
monosaccharides 51
oligosaccharides 50–3, 59–60
proteoglycans 142–6
*see also* HPLC
ion moderated partition 20–1
isomaltose, ion-exchange chromatography **52**
isopanose **52**

Jack bean enzymes 250–4

keratan sulfate 128
ketose, assay using phenol–boric acid–sulfuric acid reagent 5

lactose
ion-exchange chromatography **51**
infra-red spectroscopy 67
enzymatic hydrolysis 250, 252
paper chromatography **89**
lectins
affinity chromatography 205–22
immobilization 207–13
specificity **208–11**
*Lens culinaris* agglutinin (LCA) 206, **208**, 217–20
link protein 168–70

link stability of proteoglycan aggregates 169–73
*Lotus tetragonologus* agglutinin (LTA) 206, **209**, 221

MALDI–MS 245–6
malto-oligosaccharides
capillary electrophoresis **65**
gel permeation chromatography **54**
HPLC **58–9**, 62
ion-exchange chromatography **52**
paper chromatography **89**
TLC **49**
maltose
ion-exchange chromatography **51**
TLC **49**
mannans 91
D-mannose
enzymatic assay 9–10
GLC of alditol acetates **35**
GLC of trimethylsilyl ethers **28**, 29
HPLC **19**, **22**
ion-exchange chromatogrpahy **51**
*O*-aeetyl-*O*-methyl derivatives **35**, 236
paper chromatography 12
TLC 12
α-mannosidase 253–4
β-mannosidase 254–5
mass spectrometry
chemical ionization-MS 314
electron impact-MS 34–9, 85–6, **86**, 312–14
fast atom bombardment-MS 66, 239–45
MALDI–MS 245–6
methyl alditol and alditol acetates 34–9, 234–8
methylated glycolipid **312–13**
monosaccharides 34–9
oligosaccharides 66, 239–46
trimethylsilyl derivatives 39
*see also* GLC–MS
methanolysis
glycoproteins 229
methylglycoside trifluoroacetates 228–9
permethylglycans 235
polysaccharides 92–3
reagent 30–1
methylation analysis
glycolipid 306–7
glycoproteins 231–46
lithium methylsulfinyl carbanion reagent 233–45
oligosaccharides 240–5
polysaccharides 81–7
sodium hydroxide reagent 234
sodium methylsulfinyl carbanion reagent 82
*tert*-butoxide reagent 232
microheterogeneity, of glycoproteins 182

**322**

# Index

molecular weight determination
  oligosaccharides 54
  polysaccharides 79–81
monoclonal antibodies 120–1
monosaccharides
  determination of ring structure 100–1
  determination of D- and L-configuration 30, 76–9
  see also Chapter 1
mucopolysaccharide assay 224

neuraminic acid, see sialic acid
neuraminidases 248–51
nomenclature 73
nuclear magnetic resonance, see ¹H-NMR and ¹³C-NMR

oligosaccharides
  enzymatic hydrolysis 94–5
  fractionation 216–22
  in glycoproteins 183
  HPLC 55–63, 196–205
  mass spectrometry 66, 239–46
  methylation analysis 50–3, 59–60
  ion-exchange chromatography 50–3, 59–60
  see also Chapter 2
optical isomer determination 30, 76–9
orosomucoid, see α-acid glycoprotein
osmium tetroxide–periodate oxidation of glycolipids 308
ovomucoid **186**, **199**, **202**, **239**
oxime derivatization 33–4
ozone oxidation 308–9

panose **52**
paper chromatography
  detection 14–15, 195
  monosaccharides 17
  periodate-oxidized polysaccharide **89**
  solvent systems 13–14
pentose
  in polysaccharides 75
  using ferric–orcinol reagent 4–5
  see also reducing sugar
peptide-N-glycosidase, see N-glycopeptide hydrolase
periodate oxidation
  polysaccharides 87–90, **88**
  with osmium tetroxide 308
Phaseolus vulgaris agglutinin (PHA) **208–9**, 220, 221
phosphates, detection 197
Pisum sativum lectin 220
Pneumococcal capsular glucan 112–13
poly (N-acetyllactosamine)-peptides **187**

polysaccharides, see Chapter 3
protein linkage
  glycoprotein 182–5
  proteoglycan 129
proteoglycans, see Chapter 4
proteolysis of glycoproteins 193
proton magnetic resonance, see ¹H-NMR
Psathyrella velutina agglutinin (PVA) 221
pulsed amperometric detection 24–5, 62, 201
pyruvic acid containing glycans 114

Ricinus communis agglutinin (RCA) 206, **208**, 220–1
reducing sugar
  assay using dinitrosalicylic acid 3, 47
  assay using tetrazolium blue 45–6
  Lane and Eynon assay 2, 45
  Nelson–Somogyi assay 4
reductive cleavage 112, 189
L-rhamnose
  GLC of trimethylsilyl ethers **28**, 29
  HPLC **19**
  ion-exchange chromatography **51**
D-ribose
  GLC of trimethylsilyl ethers **28**, 29
  HPLC **22**
  ion-exchange chromatography **51**
  paper chromatography 12
  TLC 12
  see also pentose
ring structure of monosaccharides 100–1
ruthenium tetraoxide, as oxidizing agent 97–9

Sambucus nigra agglutinin 211
sialic acid
  assay in glycoproteins 226–7
  assay using resorcinol–HCl 8
  derivatives 7, 227, 249
  GLC of trimethylsilyl ethers **28**, 29
  HPLC 26
  hydrolysis 227, 248–50
  in glycolipids 303
  in glycoproteins 267, **271**, 276
  lectin specificity 206
  reporter groups 275–6
  Warren assay 7
sialidase, see neuraminidase
solid phase extraction, of gangliosides 298
soybean agglutinin 206
specrophotometric assays
  glycerol 305
  see also colorimetric assays
sphingosine base components 304
sphingenine-type bases 295–6

# Index

starch
  assay 10
  electron microscopy 119–20, **120**
structural reporter groups 266, 271, 288
sterols 305
sucrose
  enzymatic assay 10
  ion-exchange chromatography **51**
sugar
  colorimetric assay 2–3, 47–8
supercritical fluid chromatography 302

Tettamanti procedure 298
TLC
  detection 14–16, 49, 197
  glycolipids 300–1
  monosaccharides 11–17
  oligosaccharide alditols 189
  oligosaccharides 48–50
  solvent systems 13, 49–50
thin-layer chromatography, *see* TLC
threitol
  paper chromatography, from periodate
    oxidation **89**
trifluoroacetolysis of glycolipids 309
trimethylsilylation 30–1, 303
transferrin **186**, 251
*Trypanosoma brucei* glycoprotein **187**

*Ulex europeus* agglutinin (UEA) 206, **209**, 221

ultracentrifugation
  density gradient fractionation 79–80, **132**,
    134–42
  polysaccharides 79
  proteoglycans **132**, 134–42
uronic acids
  carbazole assay 6–7, 174–5
  β-elimination 96
  GLC of trimethylsilyl ethers **28**, 29
  hydrolysis of oligosaccharides containing 93
  in polysaccharides 75
  meta-hydroxydiphenyl-sulfuric acid
    assay 224
*Vicia faba* lectin 220
Vliegenthardt 265

Western blotting 154–5
wheat germ agglutinin (WGA) 206, **209**, 216,
  220–1
*Wisteria floribunda* lectin 221

xanthan 118–119
D-xylose
  GLC of trimethylsilyl ethers **28**, 29
  HPLC **19**
  in protein glycan link 182
  ion-exchange chromatography **51**
  paper chromatography 12
  TLC 12
  *see also* pentose